STRAITS

Felipe Fernández-Armesto

STRAITS

BEYOND THE MYTH OF MAGELLAN

BLOOMSBURY PUBLISHING
LONDON • OXFORD • NEW YORK • NEW DELHI • SYDNEY

BLOOMSBURY PUBLISHING
Bloomsbury Publishing Plc
50 Bedford Square, London, WC1B 3DP, UK
29 Earlsfort Terrace, Dublin 2, Ireland

BLOOMSBURY, BLOOMSBURY PUBLISHING and the Diana logo
are trademarks of Bloomsbury Publishing Plc

First published in 2022 in the United States by University of California Press
First published in Great Britain 2022

A catalogue record for this book is available from the British Library

ISBN: HB: 978-1-5266-3208-1; TPB: 978-1-5266-3207-4; EBOOK: 978-1-5266-3209-8
EPDF: 978-1-5266-5042-9

2 4 6 8 10 9 7 5 3 1

Printed and bound in Great Britain by CPI Group (UK) Ltd, Croydon CR0 4YY

MIX
Paper from
responsible sources
FSC   FSC® C171272

To find out more about our authors and books visit
www.bloomsbury.com and sign up for our newsletters

To the Hakluyt Society,
in tribute and gratitude for 175 years of work in
the service of scholarship on the history of travel
and exploration.

Gradiamur simul, eroque socius itineris tui.

Contents

List of Illustrations

MAPS

Preface

Failure is fatal to happiness but can be fruitful for fame. Metaphorically, resurrections often follow crucifixions. Sometimes partial but spectacular success adds glamor to a downfall, like Alexander's or Napoleon's. Magellan is exceptional because his failure was total. Yet his renown seems impregnable.

Portuguese, among whom he was born, and Spaniards, to whom he belonged by naturalization, compete to ascribe him as kindred. His claims to a literally global role transcend nationalisms. Various countries vie to commemorate the quincentenary of his death.[1] His name—to judge from its popularity with PR professionals—confers instant approval, triggering connotations with science, enterprise, and achievement. It dignifies, at least in aspiration, an expensive cruise ship, a costly private health care business, and firms dedicated to financial management and aerospace engineering. NASA's mission to Venus was named after Magellan. Other, even odder uses or abuses of his name for profit—in varying degrees of ignorance and irrelevance—are mentioned in this book's conclusion (p. 278). To judge from the numbers of businesses and projects called "Magellan," he makes an appealing figurehead.

The tally of his failures is almost as long as the list of his honorific homonyms. On the voyage for which he is celebrated, most of the ships were lost and all but a handful of his men died or deserted. The cash profit usually ascribed to the outcome is a myth. Magellan did not even reach his nominal destination. In his mission to find a short route from

Spain to the Spice Islands, he did more than just fail: he drove on to disaster when failure was already obvious. His ambitions for himself—to conquer a profitable fief—foundered because he made lethal mistakes. He never considered—let alone accomplished—the circumnavigation of the world; but common opinion continues to credit him for it. It strikes his admirers as universally significant, like the first wireless broadcast or the first moon landing: a contribution to knowledge and science, rather than an exercise of crude might or material exploitation, like the work of most other dead white explorers.

His intentions were imperial; yet he has escaped, so far, the scatter-shot of postcolonial revulsion and vengeance. His conduct, though bold and resolute, was as bloody and destructive as that of most would-be conquistadores. But he died without establishing a colony: that was another of his failures. In consequence, unlike Columbus, say, or Cortés, he can more or less dodge arraignment for imperialism. Of all the celebrated or formerly celebrated European explorers of the "Age of Expansion," Magellan seems the most suitable hero, or least obvious villain, for postcolonial times and politically correct scrutineers.

In many ways, he deserved the esteem he has attracted. Qualities he exhibited in abundance include intrepidity, single-mindedness, resilience in misfortune, and devotion to the noble, chivalric ethos in which he was well educated and well read. He also had, for good and ill, an elusive but unmistakable charisma that inspired loyalty in friends and followers in moments of peril and hardship—of which there were plenty in his career. His life, as we shall see, exemplifies reasons why explorers are celebrated, even when they offend officious and anachronistic demands.

But the people who praise him or try to appropriate his renown do not know who Magellan really was. They ought to think again. The best reasons for commemorating him—which appear in the pages that follow—have been overlooked in favor of falsehoods. And anyone prompted by the worldwide celebrations already under way and due to climax in 2022 may care to discard the myths, penetrate the truth, and learn what the explorer's life and times were like.

The sources do not permit the kind of intimate biographies possible for Columbus and Vespucci because Magellan left so little copy in his own words. In the pages that follow, however, I undertake the closest reading ever of the texts that are available. As a result, I think, I can show more of what Magellan was like than any of my predecessors. In particular, I disclose some of the dynamism of his character, as experience changed it; and I discern previously unnoticed influences, which shaped his self-perception and provided models for his life or values to guide his sometimes inscrutable or perplexing behavior. His actions, as far as they can be distinguished among the contradictory narratives, descriptions, and assessments that survive from his day, are evidence of his interactions with the world that surrounded him; so I spend a lot of time on the context—starting with that of the world in his lifetime, or at least the parts of it in which he lived and fought and explored.

Because so little can be said for certain, and because unwarranted assumptions, myths, assertions, and falsehoods dominate the historical tradition, I try to keep the reader in constant touch with the evidence and to provide matter for dissent when I make a judgment that involves imagination or intuition of my own. I make no apology, however, for wielding imagination disciplined by the evidence and informed by long study of the subject. I first worked on the sources when editing *The Times Atlas of World Exploration* some thirty years ago. I did not write the sections of that book in which Magellan appears and would not now endorse everything said there. But the experience sparked an interest sustained ever since. The history of exploration has remained a recurrent theme in my work and provides the broad context in which I try to understand what Magellan did and why and to what extent it matters for the world.

In the tradition of prefaces I have to offer some self-exculpation.

In some places, indicated in the text and the critical apparatus, I have extended the normal limits of scholarly convention by turning reported speech into dialogue. I hope fastidious readers will see why, in each case, this break with convention is justified and how it helps to make the evidence vivid and the events more intelligible than they would otherwise be. In chapter 1, in introducing parts of the background, especially to Magellan's irruptions into the Indian Ocean and Swahili Coast, I have drawn on pages from a previous book of mine, *1492*. As the present book was largely written during the pandemic of 2020–21, I had to juggle with spells of isolation and was not always able to get access to the best editions of all texts. In such cases, I cite in the critical apparatus the text I used at the time and the superior version, where available, against which I may subsequently have been able to check it. Except where indicated, translations are my own; but in quoting translations from Antonio Pigafetta's narrative of the great voyage, I have relied (except where stated in the notes) on the excellent publication by my colleague Theodore Cachey. He worked from a 1987 edition, by the well-known poet and critic Mariarosa Masoero, of the most authoritative manuscript, which I have not been able to consult; I have checked the translation against the famous edition by Mosto (see below, p. 300) and include cross-references in the notes.

In work about Magellan or Magallanes or Magalhães, consistency in rendering names is impossible. At least "Ferdinand Magellan" has the advantage of familiarity to readers of many languages and eludes commitment in the custody battle between Spanish and Portuguese aspirants. Except for the names of Spanish monarchs (for whom I use Castilian for reasons explained below) I use versions naturalized in English, where available. For place-names that postcolonial toponymy has changed in recent years, I give the new equivalents at an early mention.

It would not have been possible to write this book in such adverse circumstances without the prior work of the editors of the sources, who are all acknowledged in the critical apparatus. I pay special tribute to those who have worked hard via the Internet to make available fresh, accurate transcripts or facsimiles of well-known documents, especially Cristóbal Bernal, Tomás Mazón, and Braulio Vázquez and their dedicated collaborators among the professional staffs of the archives concerned. Research, especially by F. Borja Aguinalde, Salvador Bernabeu, José Manuel Garcia, and Juan Gil, has brought important new evidence or reflection to the table in recent years. I sometimes dissent from their interpretations but could not have managed without their work.

I have also to thank Guido van Meersbergen, with whom the idea that I should write this book started; Tessa David, who persuaded me to do it and provided helpful suggestions for how I might go about it; Andrew Gordon, who generously allowed me to work with a rival agent; Michael Fishwick, Niels Hooper, Francisco Reinking, and Elisabeth Magnus—selfless and sagacious editors; and Sebastian Armesto, Theodore Cachey, Joyce Chaplin, Kris Lane, Manuel Lucena Giraldo, Lincoln Paine, Carla Rahn Phillips, and Sanjay Subrahmanyam, who kindly and helpfully read parts or all of this book in draft, saving me from follies, leaving such errors and infelicities as are my fault alone. I am grateful to colleagues, students, and staff at the University of Notre Dame for providing the best possible environment for teaching and learning, and especially to the personnel of the Hesburgh Library for supplying materials in defiance of the virus that has afflicted the world while I have been writing. This book is unsympathetic to "great men"; but it has its heroes.

CULWORTH, NORTHAMPTONSHIRE—NOTRE
DAME, INDIANA
January 2020–August 2021

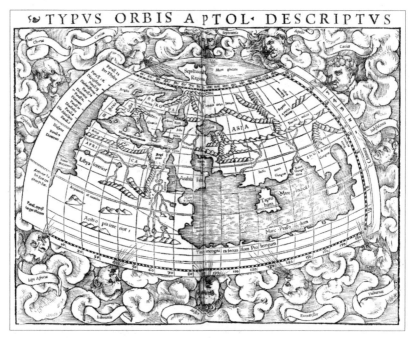

FIGURE 1. The Indian Ocean, where Magellan's overseas adventures began in 1505, was the world's richest arena of commerce at the time. Until the 1490s most fifteenth-century European maps of the world showed the Indian Ocean as enclosed and inaccessible from Europe by sea, thanks to the influence of the *Geography* of the revered second-century Alexandrian cosmographer Claudius Ptolemy. Interpretations of his text informed most educated Europeans in the late fifteenth century, as in this example, reproduced, from a model of the 1480s, in Sebastian Münster's *Cosmographia,* which appeared in successive editions in the 1550s, showing the ocean surrounded by enveloping lands. Putti blow winds onto the world from a cloudy firmament. Magellan retained elements of Ptolemy's image, including the peninsula, here marked "India extra Gangem," to the east of India, and the island of Taprobana, corresponding roughly to modern Sri Lanka. Spanish cosmographers claimed that the meridian of the great bay marked "Sinus Gangeticus" marked the dividing line between zones of navigation agreed upon by Spain and Portugal. Courtesy of the James Ford Bell Library, University of Minnesota.

CHAPTER I

The Globe Around Magellan

The World, 1492 to 1521

And the channels of the sea appeared, the foundations of the
world were discovered, at the rebuking of the Lord, at the
blast of the breath of his nostrils.

—2 Samuel 22:16

"Write what you know," said Robert Graves, who rarely observed the
maxim. It has become a shibboleth for writing teachers but seems con-
temptible: facile, challengeless, narrow-minded, anxious for success.
The unknown is magnetic: an invitation to endlessly unwinding problems, a lure to the receding horizon that seduced Magellan, or a way
into the improbable stories he composed for himself in his head and
tried to act out in his life. In sticky, smelly classrooms, amid Gradgrind-
mark schemes and marginal ticks, history is "about facts." But, for
me, facts are there only to feed problems: insoluble problems—for
preference—that flirt and flit as you grasp at them.

I think I know as much about Magellan as you can know in your head.
I can unpick the contradictions of the evidence. I can reproach predecessors with errors and straighten tangles in the chronology. I can get pat-
ternless details into focus: I know, for instance, as previous inquirers have
known, how many arrows (21,600) and compass needles (thirty-five)
appear in Magellan's ships' manifests, how many hourglasses (eighteen),
how many barrels of anchovies and tons of biscuit (below, p. 110).[1]

I also think I know—or sense convincingly—a lot about his heart: the
tragically flawed social ambition, the heroic self-delusion, the vexing
self-righteousness, the cruelly streaked sense of humor. They all appear
in the pages that follow. I can trace the way his journey through life
changed him, and reconstruct the strange mood of religious exaltation

in which he died. But there's gut knowledge too, which eludes me. Magellan was one of at least 150 men who died on the voyage he led. If you leave out those who survived by deserting or in captivity, the death rate was about 90 percent. Even by the standards of the day, when failure was routine on alarmingly overoptimistic journeys, Magellan's project beggared belief. Objectively considered, the chances of survival, let alone success, were always minimal. As we shall see, the cost in lives bought no quantifiable return. Despite previous historians' assertions, the balance sheet of crude profit and loss ended in the red. The voyage Magellan captained failed in every declared objective.

What made such an egregious adventure attractive, not just to the men who risked it, but also the backers who put money into it? I am not sure I know or can know that. Life was cheap, for reasons, partly intelligible, to which we shall come in a moment. "To set sail," said Luis de Camões, the well-traveled poet who turned Portuguese maritime history into verse in 1572, "is essential. To survive? That's supererogatory."[2] What made such a shocking inversion of common sense seem reasonable? Why were seamen's lives so dispensable—so much cheaper than everyone else's? What made Magellan and some of his men persist as their prospects worsened? What induced the king of Spain and hard-headed merchants in Seville and Burgos to believe in Magellan? Why would they put up money for a proposal from a man who came to them with a reputation for treachery, a dearth of relevant experience, and a scientific sidekick, Rui Faleiro, who, to the psychiatry of the day, was literally, certifiably insane?

. . .

To approach the problems we have to start by trying to understand the constraints and opportunities of the world around Magellan.

That world was riven with paradox. Every textbook will tell you that the centuries in which Magellan lived were an "age of expansion," when stunning new departures happened. In Europe the retrieval of classical tradition intensified in the Renaissance, which equipped minds for new art and thought and endeavor and which spread to much of the rest of the world, including parts of the Americas and sub-Saharan Africa: the first genuinely global intellectual event.[3] The so-called Scientific Revolution was exhibiting early signs when Magellan took to sea, enabling the formerly backward West to catch up with and (in some respects, over the next hundred years or so) to overtake Chinese science and technology. Meanwhile, a global ecological exchange swapped life-forms across a formerly divergent world, spreading creatures, plants, and pathogens, for good and ill, across the globe. A persistent historical tradition even claims that Magellan's lifetime roughly coincided with "the origins of modernity": the distribution and divisions of world religions were taking on something like their present configurations; some of the world's most widely and creatively deployed languages and literatures were taking shape in forms intelligible to today's readers. The world's major civilizations—Christendom, Islam, and the Buddhist world—literally expanded, engrossing territory and people, stretching out to each other across chasms of culture, spreading contacts, conflicts, commerce, and contagion.

How did all that happen in what was also, worldwide, an age of plague and cold?[4]

On the face of it, we should not expect large-scale migrations, long-range conquests, or spells of invention in a period of severe environmental dislocation or harsh conditions.[5] Climate and disease set the context for everything else that happens; and climate is supreme, because diseases depend on it. Magellan's lifetime came roughly in the middle of one of the most severe interruptions in global warming since the end of the Younger Dryas, nearly twelve thousand years ago. The "Little Ice Age," from the mid-fourteenth to the early eighteenth century, inflicted conspicuous and measurable harm on the societies it affected, spreading starvation, worsening wars, provoking rebellions, incubating plagues. The coldest spell was toward the end, generating stories of freak freezings, from the king of France's beard to entire salt seas.[6]

During Magellan's career no conditions or events remotely so extreme occurred, and in the first half of the sixteenth century temperatures seem to have been milder than immediately before or after. Nevertheless, in parts of the world from which quantifiable data are available, spring and winter temperatures were typically between one and two degrees lower, on average, than what we now consider normal (represented by the average in the first half of the twentieth century).[7] As we shall see, Magellan's men complained of the cold they experienced and threatened to mutiny because of it.

Lethal, recurrent bouts of disease accompanied the cold.[8] At the time, people called them plagues: maladies associated with a complex ecology, involving rodent hosts and fleas as vectors.[9] The afflictions in question were persistent and impressive. From the mid-fourteenth century to the early eighteenth—the period exactly matching the cold era—no living subject was portrayed more often in Europe than the Grim Reaper, reveling in his deadly duties, selecting dancing partners without regard to age, sex, or appearance.[10] In 1493, a little before Magellan became a page boy in the Portuguese court, the *Nuremberg Chronicle* included a remarkable text under an image of dancing dead, in various stages of putrescence—gleefully displaying rotting flesh, jangling their bones, with worms wriggling and entrails dangling.

They are irrepressibly jolly. It is tempting to say that the dead exhibit joie de vivre. The words of the song to which the cadavers rise from their graves appear in the text. "There is nothing better than Death." He is just, because all deserve him. He is equitable, because he treats poor and rich alike. He is benign, because he frees the old from sorrow. Visiting prisoners and the sick, he is kind in obedience to commands of Christ. Mercifully, he liberates victims from suffering. He is wise, despising worldly pleasure and deploring the vanity of power and wealth. When the Age of Plague receded, Death became less familiar and therefore, perhaps, more feared. Nowadays, people treat him with evasion. They speak of him in euphemisms, in which "the loved one" is said to "pass away." Magellan and his contemporaries had an uneasy familiarity with Death that it is hard for us to retrieve.

Local recurrences of plague defied, to some extent, expectations that visitations would leave survivors protected by immunity. The pace of microbial mutations, perhaps, explains the persistence. While the bubonic-plague bacillus, *Yersinia pestis*, was responsible for many, if not most or all, of the outbreaks, evidence that survives in victims' DNA includes variations, small and significant, over time.[11] One feature of the environment was constant: the reasons for the way then-prevalent strains of the bacillus responded to fluctuations in temperature have attracted a great deal of inconclusive research; the link with cold, however, is conspicuous. The Black Death of 1348–50 followed a fall in Northern Hemisphere temperatures. The retreat of "plague" coincided with the resumption of global warming.[12]

During Magellan's adulthood, though population was edging upwards in Europe, plague hardly abated. England experienced three outbreaks of "great plague" or "great pestilence" in the first two decades of the sixteenth century. From the frequency with which the city council issued measures of quarantine and of what we should now call "lockdown," Edinburgh seems never to have been free of plague from 1498 to 1514.[13] In Leipzig in 1521 quacks hawked Germany's "first brand-name medicines" as remedies for plague.[14] In 1520 what can fairly be called an international medical conference met in Basel to propose universal measures of prevention.

In the Iberian peninsula, in the harshest winter of Magellan's lifetime, in 1505–6, plague struck Évora, Lisbon, Porto, and Seville. Plagues appeared in Barcelona in 1501, 1507, 1510, and 1515. Those of 1507 and 1510 reached across Spain's eastern seaboard and beyond, into western Andalusia. The years 1507–9 were by some measures the most deadly since the Black Death. The outbreak in 1507 killed a tenth of the prebendaries of Cadiz Cathedral, while the Andalusian priest and sedulous chronicler Andrés de Bernáldez claimed to witness thirty thousand deaths in a single month in May. The previous year, claims that Jews were responsible for spreading plagues helped to provoke a massacre in Lisbon.[15] While Magellan's expedition was at sea, plague broke out in Valladolid and spread to Valencia, Córdoba, and Seville, where his fleet was launched. An earthquake in Játiva in 1519, an exceptionally rainy year, was taken as presaging the plague that followed in Lisbon, Valencia, Zaragoza, and Barcelona, where it persisted until 1521.[16] The plague and cold did not exclude other routine sources of suffering, such as drought in 1513 in the Gudiana valley and more generally in 1515 and 1516. Drought and famine menaced much of Portugal in 1521–22.[17]

In one respect, Magellan's career coincided with an unprecedently aggressive development in the global disease environment: the transmission of Old World pathogens to other continents, along with the other biota of global ecological exchange, which Columbus inaugurated when he swapped products of the Old World and the New on his first two transatlantic voyages.[18] The arrival of Spanish settlers in Hispaniola in 1493, or perhaps of successors in subsequent voyages, introduced unfamiliar diseases with which native immune systems could not cope. In a terrible acceleration of mortality in 1519, the friars to whom the Spanish monarchy had confided control of the colony abandoned hope of keeping the natives alive.[19] Near extinction followed. By the time of Magellan's voyage, Spanish intrusions had generalized the effect around the Caribbean and projected it onto the Central American mainland. Mortality rates of up to 90 percent became normal wherever "the breath of a Spaniard" broadcast disease.[20] The cheapness of life was not the result of glut anywhere in the world of the time. Magellan's era differed from ours not so much, perhaps, because of the low value people put on life as because of the high value they put on death. Death in those days was the real lord of the dance.

. . .

Meanwhile, despite the unpropitious circumstances, in widely separated parts of the world, economic activity and territorial conquests speeded up like springs uncoiling. An "age of expansion" really did begin, but the phenomenon was of an expanding world, not simply, as some historians say, of European expansion. The world did not wait passively for European outreach to transform it, as if touched by a magic wand. Other societies were already working magic of their own, turning states into empires and cultures into civilizations. Beyond the reach of the recurring plagues that stopped demographic growth in much of Eurasia, polities dwarfed those of Latin Christendom. Some of the most dynamic and rapidly expanding societies of the fifteenth century were in the Americas and sub-Saharan Africa. Indeed, in terms of territorial expansion and military effectiveness against opponents, native African and American empires outclassed any state in western Europe until the establishment of the global Spanish monarchy in the sixteenth century.

Even Spain and Portugal, in the three decades or so preceding Magellan's project, seemed feeble by comparison with the rates of expansion of other empires. Aztecs and Incas outperformed Spaniards as conquerors until the newcomers improbably gobbled their empires almost at a gulp in a couple of hectic decades from 1520. The Aztec paramount who greeted Cortés was the most spectacular conqueror in his people's history, shunting armies back and forth between the mouth of the Pánuco River and Soconusco Bay, acquiring tribute—in maize and beans, cacao and gold and jade and the feathers of exotic birds—from forty-four new subject communities. Huayna Capac, who died shortly before Spanish explorers arrived in Peru, made the Inca polity one of the world's fastest-expanding states in Magellan's day, extending the frontier in every direction, conquering the Caranqui in what is now Ecuador and reputedly drowning twenty thousand of their defeated warriors in Lake Yawar-Cocha.

Russia, meanwhile, extended empire to the Eurasian Far North, sending expeditions beyond the Arctic Circle and across the Rivers Perm and Ob in the 1490s. In the decade of Magellan's departure, the tsar conquered Smolensk and extended rule to the Dnieper. While Magellan was readying his fleet, Sigmund von Herberstein visited Moscow as the ambassador of the Holy Roman Empire and heard prophecies from an unconvincing shamaness known as the Old Woman of the Ob, who foretold Russia's future in an icy El Dorado, amid "men of monstrous shape and fishes having the appearance of men."[21]

Just before Magellan's departure on his great voyage the Ottomans conquered Egypt and began to stretch control or influence along the southern shore of the Mediterranean. In West Africa, meanwhile, Muhammad Touré consolidated the preponderance of the empire of Songhay over the vast Niger valley. In East Africa, where Magellan took part in a campaign in 1510 to secure Portuguese coastal toeholds and gain access to trade in gold, salt, and ivory, the empire of Mwene Mutapa grew to occupy the land from the Limpopo to the Zambezi. A Portuguese ambassador to Abyssinia in 1520 thought he had found the fabled realm of Prester John, so impressed was he with the thousands of red tents that housed the Negus's army on the march. The imperial habit was spreading, and new empires were forming in environments that had never experienced imperialism before.

Magellan was born in a small state—Portugal—and died ostensibly trying to build a great empire for Spain. In his lifetime, the two monarchies he served became the foci of allegiance of the world's newest and most dynamic empires, unprecedented for their reach across the globe and the diversity of the environments and cultures they encompassed. His career took him to more of the world than had been accessible to anyone in any previous generation. In early adulthood he spent eight years, from 1505 to 1513, in and around the Indian Ocean, campaigning up and down the coasts of Africa and India to help found the Portuguese Empire, taking part in the conquest of Malacca in 1511, and learning about the islands beyond the ocean that would become the foci of his later career. At intervals from 1513 to 1517 he fought for Portugal in Morocco. His voyage from Spain in 1519 took him further into the South Atlantic than any previous European expedition, and across the Pacific—an ocean never previously traversed, as far as we know, in a single voyage—to archipelagoes at its rim, in what became the Marianas and the Philippines. We shall take a *tour d'horizon* of the places beyond Iberia that mattered in his career before returning to trace the beginnings of his life and education in the next chapter.

. . .

In North Africa, the empire most likely to succeed at the time—the Ottomans'—was unable to expand beyond the deserts that fringed its dependencies. Westward from Egypt, along Africa's Mediterranean coast, numerous small states, founded on the profits of trade or piracy, flourished where Mediterranean and Saharan trade routes met. The first of the straits that channeled Magellan's life was at the western end of North Africa, separating the continent from the Iberian peninsula. Morocco, where he would spend the best part of three years' campaigning, had emerged as a kingdom on the edge of the Islamic world, holding Christendom at bay. The rulers' hereditary viziers, the Banu Marin, seized control in a coup in 1465 but had to fight off pretenders sprung from desert sects and, allegedly, from the stock of the prophet Muhammad. The followers of al-Jazuli, a murdered Sufi, spread rebellion by touring the realm with his embalmed corpse. They weakened the state in the face of invasions from Spain and Portugal. For Iberian rulers, North Africa was an overseas extension of the peninsula, and a legitimate war zone, conquerable on the alleged grounds that it had formed part of the same political unit as Hispania in antiquity and that it was sometime Christian land, recoverable from Muslim usurpers. Portuguese efforts focused on dominating harbors west of the Strait of Gibraltar, where, as we shall see, Magellan served, defending Portuguese acquisitions against revanche by new heirs of al-Jazuli's mantle: a family from the Draa valley, deep in what is now Western Sahara, claimed descent from the Prophet and, from 1509, launched a jihad and organized a tribal confederacy that first manipulated and ultimately displaced the Banu Marin.

Magellan's adventure beyond the Strait of Gibraltar was only a brief episode in an overseas career spent hovering between remoter straits: they bound coastal routes of maritime Asia on the shores of the Indian Ocean, from Ormuz in the west to Malacca (as traditionally known, or Melaka in currently fashionable orthography) in the east. Here the monsoon dominates the environment. Above the equator, northeasterlies prevail in winter. When winter ends, the direction of the winds is reversed. For most of the rest of the year they blow steadily from the south and west, sucked in toward the Asian landmass as air warms and rises over the continent.

The regularity of the wind system made this ocean the most ancient and richest zone of long-range commerce in the world—envied in the relatively poor and confined economies of Christendom. By timing voyages to take advantage of the predictable changes in the direction of the wind, navigators could set sail, confident of a fair wind out and a fair wind home. In the Indian Ocean, moreover, compared with other navigable seas, the reliability of the monsoon season offered the advantage of a speedy passage in both directions. To judge from such ancient and medieval records as survive, a trans-Mediterranean journey from east to west, against the wind, would take fifty to seventy days. With the monsoon, a ship could cross the entire Indian Ocean, from Palembang in Sumatra to the Persian Gulf, in less time. Three to four weeks in either direction sufficed to get between India and a Persian Gulf port.

The monsoon made travel speedy, but the ocean was stormy, unsafe, and hard to get into and out of. Access from the east was barely possible in summer, when typhoons tore into the shores. Fierce storms guarded the southern approaches. No one who knew the reputation of these waters cared to venture between about ten and thirty degrees south and sixty or ninety degrees east during the hurricane season. Arab legends claimed the region was impassable. For most of history, therefore, it was the preserve of peoples whose homes bordered it or who traveled overland—like some European and Armenian traders—to become part of its world. Moreover, all the trade was internal. Indian Ocean merchants took no interest in venturing far beyond the monsoon system to reach other markets or supplies.

As a fictional Spanish mother advised, "Stick to the rich and something may rub off" (Arrimarse a los buenos por ser uno de ellos). Europeans would have wanted to find ways of tapping the wealth of the region, even without the lure of products available nowhere else in the world. In the fifteenth century, however, a conspicuous change in the region was the growing global demand for, and therefore supply of, spices and aromatics—especially pepper. These products, sold to rich buyers in China and Southwest Asia, and, to a lesser extent, in Europe, were the most profitable in the world, in terms of price per unit of weight. No one has ever satisfactorily explained the reasons for the increase in demand. China dominated the market and accounted for well over half the global consumption, but Europe, Persia, and the Ottoman world absorbed ever greater amounts. Population growth contributed—but the increase in demand for spices seems greatly to have exceeded it. The commonplace explanation—that cooks used spices to mask the flavor of bad meat—is nonsense.[22] Produce was far fresher in the medieval world, on average, than in modern urbanized and industrialized societies, and reliable preserving methods were available for what was not consumed fresh. Changing taste has been alleged, but there is no evidence of that: it was the abiding taste for powerful flavors—a taste being revived today as Mexican, Indian, and Sichuanese cuisines go global—that made spices desirable; the same circumstances affected demand for sugar, which boomed at the same time, not because it supplied a new taste, but because it made a long-standing preference attainable in a widening market.[23] The spice boom was part of an ill-understood upturn in economic conditions across Eurasia. In China, especially, increased prosperity made expensive condiments more widely accessible, as the turbulence that brought the Ming to power subsided and the empire settled down to a long period of relative peace and internal stability.

In partial consequence, spice production expanded into new areas. Pepper, traditionally produced on India's Malabar coast, and cinnamon, once largely confined to what is now Sri Lanka, spread around Southeast Asia. Pepper became a major product of Malaya and Sumatra in the fifteenth century. Camphor, sappanwood, and sandalwood, aromatic benzoin and cloves, credited with medical and culinary magic, all overspilled their traditional places of supply. Nonetheless, enough local specialization remained within the region to ensure huge profits for traders and shippers. Commercial opportunities in their home ocean kept seafaring Arabs, Swahili merchant communities, Persians, Indians, and Javanese and other island peoples of the region fully occupied. Indeed, their problem was, if anything, shortage of shipping in relation to the scale of demand for interregional trade. That was why, in the long run, they generally welcomed interlopers from Europe in the sixteenth century, who were truculent, demanding, seemingly barbaric, and often violent, but who added to the shipping stock of the ocean and, therefore, contributed to the general increase of wealth. Paradoxically, Europeans' poverty favored their prospects as competitors and stimulated their exceptionally broad outreach by the standards of the time, compelling them to look overseas because of the dearth of economic opportunities at home, and making their services as shippers relatively cheap.

. . .

From Europe access to the Indian Ocean was well worth seeking. Today, Europe is a magnet for migrants from poorer economies. In Magellan's day, the normal relationship was the other way round: Europe was a relatively poor backwater, despised as barbaric in the richer societies that lined the shores of maritime Asia. European merchants craved a share of the richest trades and most prosperous markets in the world. The widespread assumption that Vasco da Gama was the first to penetrate deep inside the Indian Ocean when he rounded the Cape of Good Hope in 1498 is a vulgar error. Italian merchants often plied their trade there during the late Middle Ages. But the journey was too long, laborious, and hazardous to generate much profit. From the Mediterranean, merchants typically had to travel up the Nile from Alexandria and overland by camel caravan from the first or second cataract to the Red Sea coast, where they awaited the turn of the monsoon before shipping for

Aden or Socotra. It was inadvisable to attempt to join the Red Sea further north because of the formidable hazards to navigation.[24] By an alternative route, Europeans could attempt a dangerous passage through the Ottoman Empire to the Persian Gulf in the rare interstices of war and religious hostility. In either case, they obviously could not take ships with them. This was a potentially fatal limitation because Europeans had little to offer to people in the Indian Ocean basin except shipping services.

Most of the Western venturers who worked in the Indian Ocean before Vasco da Gama's irruption are known only from stray references in the archives.[25] Merchants rarely wrote up their experiences. Two circumstantial accounts, however, survive from the fifteenth century, the first by Niccolò Conti, who had been at least as far east as Java in the 1420s, the second by the Genoese Girolamo di Santo Stefano, who made an equally long trading voyage in the 1490s. Between them, they described the framework in which Portuguese successors, including Magellan, operated.

Conti chose to approach the Indian Ocean overland via Persia to the Gulf, where he took ship for Cambay in the Bay of Bengal. Santo Stefano used the other main route. In company with a business partner, Girolamo Adorno, he traveled up the Nile, joined a caravan bound for the Red Sea, and crossed the ocean from Massawah—a port generally under Ethiopian control at the time.

On his return, Conti sought papal absolution for having abjured Christianity in Cairo in order to save the lives of his wife and children, who traveled with him.[26] He told his story to a Florentine humanist, who made a record of it as a morally edifying tale of changing fortunes. The convention Conti's work established was of "the inconstancy of fortune." When Santo Stefano wrote up his experiences of the Indian Ocean in 1499, he too focused on lamentations against the ill luck he endured "for my sins." Had he eluded his sufferings, he might have retired on the riches that slipped through his hands and avoided the need to throw himself on the mercy of patrons—the obvious subtext of his work. "But who can contend with fortune?" he asked.[27] He and Adorno got as far east as northern Sumatra, where they took ship for Pegu, in Burma, apparently with the idea of engaging in trade in gems. It was a painfully slow business. In Sumatra on the way back a local ruler confiscated their cargo, including the valuable rubies they brought from Burma. Adorno died in 1496, "after fifty-five days' suffering" in Pegu, where "his body was buried in a certain ruined church, frequented by none."[28]

Naturally enough, as they were merchants, Conti and Santo Stefano inventoried trade goods wherever they went and took special interest in spices and aromatics. Santo Stefano described the drying of green peppercorns at Calicut, the profusion of cinnamon in Sri Lanka, the availability of pepper in Sumatra, the location of sandalwood in Coromandel. Conti's description of aromatic oil production from cinnamon berries in Sri Lanka reflects personal observation, whereas he culled other purported observations from his reading. He reported camphor and durians ("The taste varies, like that of cheese")[29] in Sumatra. As specialists in gems, both travelers felt drawn to where rubies, garnets, jacinths, and crystal "grew." Both had antennae for military intelligence. Santo Stefano was interested in elephant breeding for war and confirmed Conti's claim that ten thousand war elephants were maintained in the stables of the ruler of Pegu.

These were hard-headed observations. But the writers seemed to go soft in the head when they succumbed to the lure of exotica. Around the Indian Ocean, they described a topsy-turvy world in which murder was moral, serpents flew, monsters trapped fish by lighting irresistibly magnetic fires on shore, and miners used vultures and eagles to gather diamonds.[30] Some of the tales echo stories in the Sinbad corpus—evidence that the authors really did know the East at first hand. The taste for sensationalism was most apparent in their obsessions with sex. Santo Stefano described how Indian men "never marry a virgin" and hand prospective spouses over to strangers for deflowering "for fifteen or twenty days" before the nuptials. Conti was scrupulous in enumerating the harems of great rulers and commending the sang-froid of wives who committed suttee, flinging themselves on their dead husbands' funeral pyres. In India he found brothels so numerous and so alluring with "sweet perfumes, ointments, blandishments, beauty and youth" that Indians "are much addicted to licentiousness," whereas male homosexuality, "being superfluous, is unknown."[31] In Ava in Burma the women mocked Conti for having a small penis and recommended a local custom: inserting up to a dozen gold, silver, or brass pellets, of about the size of small hazelnuts, under the skin, "and with these insertions, and the swelling of the member, the women are affected with the most exquisite pleasure." Conti refused the service because "he did not want his pain to be a source of others' pleasure."[32]

On the whole, the merchants' reports were of a world of abundance and civility. Beyond the Ganges, according to Conti, in a translation made in the reign of Elizabeth I, people "are equal to us in customs, life,

and policie; for they have sumptuous and neat houses, and all their vessels and householde stuffe very cleane: they esteeme to live as noble
people, avoided of all villainie and crueltie, being courteous people &
riche Merchauntes."[33]

If there was one thing the civilizations of the East lacked, it was shipping adequate to meet the huge demands of their highly productive
economies and active trades. Santo Stefano marveled at the cord-bound
ships that carried him along the Red Sea and across the Indian Ocean.
But while ships were well designed, well built, and ingeniously navigated, there were never enough of them to carry all the available freight.

. . .

As a result, in the 1490s, the Indian Ocean was about to experience a new
future, in which European interlopers would cash in on their advantages.
For that future to happen, Europeans needed to enter the ocean with
ships. Because they lacked salable commodities, they had to find other
ways of doing business: shipping and freighting were their best resources.
Without ships of their own, visitors such as Conti and Santo Stefano were
little better than peddlers. The Indian Ocean region was so rich and productive, so taut with demand and so abundant in supply, that it could
absorb hugely more shipping than was available at the time. Any European who could get ships into the zone stood to make a fortune.

There was only one way to do it: sail the ships in around the southern
tip of Africa. But was such a long and hazardous journey possible? Were
the ships of the time equal to its strains? Could they carry enough food
and water? In any case, it was not even certain that an approach to the
ocean lay along that route. The geographer the age most revered was
the second-century Alexandrian Claudius Ptolemy. His *Geography* was
the favorite book on the subject in the West as soon as the text became
widely available in the early fifteenth century. Readers generally, if inac-
curately, inferred that the Indian Ocean was landlocked and therefore
inaccessible by sea. Maps of the world made to illustrate Ptolemy's ideas
showed the ocean as a vast lake, cut off to the south by a long tongue of
land, protruding from Southeast Africa and curling round to lick at the
edges of East Asia. The fabled wealth of India and the spice islands lay
enclosed within it, like jewels in a strongroom.[34]

Although this was an erroneous view, it was understandable. Indian
Ocean merchants kept to the reliable routes, served by predictable mon-
soons that guaranteed them two-way passage between most of the
trading destinations of maritime Asia and East Africa. There was little

reason to venture to where the belt of tempests girds the sea, or to risk the coasts south of Mozambique, where the storms tear into lee shores. There were no potential trading partners beyond those limits, no opportunities worth braving those dangers for. From within the monsoonal system, the way in and out of it did seem effectively unnavigable.

For anyone who tried to approach from the Atlantic no such inhibitions applied. Other obstacles, however, were equally effective. In 1487 the Portuguese explorer Bartolomeu Dias managed to struggle round the Cape of Storms at the southernmost point of Africa. The king of Portugal is supposed to have renamed it the Cape of Good Hope in a promotional exercise of brazen chutzpah.[35] But the hope was weak, the storms strong. Beyond the cape, Dias found an adverse current and dangerous lee shores. The way to the Indian Ocean still seemed to be barred. Nor had Dias really gone far enough to prove that the ocean was not landlocked. All he had achieved was to demonstrate how laborious was the journey to the southernmost tip of Africa. To avoid the adverse current along the West African shore, his successors would have to strike far into the South Atlantic—farther from home, longer at sea than any voyagers had ever known—to find the westerly winds that would carry them around the cape.

So, while Dias explored the way by sea, the Portuguese crown sent agents overland to the Indian Ocean by traditional routes to gather intelligence and, in particular, to settle the question of whether the ocean was open to the south. Pero da Covilhão led the effort. He was one of many indigent but talented noblemen to cross and recross the permeable border between Portugal and Castile. He spent years in Seville, where he served in the household of the Castilian nobleman the Count (later Duke) of Medina Sidonia. This was probably a useful apprenticeship. The count was an investor in the conquest of the Canary Islands and a major figure in the Atlantic tuna fishery and sugar industry. When war broke out between the two kingdoms in 1474, Covilhão returned to his native Portugal to serve his king. Missions of an unknown nature—perhaps espionage, perhaps diplomacy—took him to Maghribi courts, where he learned Arabic.

At about the time Bartolomeu Dias left to explore the approach to the Indian Ocean from the Atlantic, Covilhão, with a companion, Afonso de Paiva, set off up the Nile and across the Ethiopian desert to Zeila on the Red Sea. His inquiries took him east to Calicut and thence perhaps as far as Sofala on the coast of Mozambique—the emporium from which East African gold was traded across the Indian Ocean. By the end of

1490 he was back in Cairo, from where he sent a report of his findings home. It has not survived. Covilhão then turned to a further aspect of his mission: establishing diplomatic contact with the court of the ruler of Ethiopia, who retained the Portuguese visitor in his service. He was still there when the next Portuguese mission got through in 1520.[36]

To judge from the outcomes, Covilhão's report summarized knowledge gleaned on the spot: the Indian Ocean was open to the south. In 1497–98, a Portuguese trading venture, commissioned by the crown and probably financed by Italian bankers, attempted to use the westerlies of the South Atlantic to reach the Indian Ocean. Its leader, Vasco da Gama, turned east too early and had to struggle around the Cape of Good Hope at the tip of Africa. He managed to get across the Indian Ocean anyway and reach pepper-rich Malabar. The next voyage, in 1500, got to India without a serious hitch. The way to the ocean where Magellan's career took shape was open.

. . .

What awaited him there? South of Ethiopia, trading states speckled the Swahili coast, where Magellan ranged in Portuguese service during a campaign to secure trading posts in 1506: the first piece in a kind of Indian Ocean jigsaw of places that he came to know or hear of in the decade and a half that preceded his great Pacific-spanning voyage. A book of advice on Indian Ocean navigation, formerly attributed to Magellan's own hand, links the pieces of the jigsaw:[37] Swahili emporia where he saw action, such as Kilwa, Mombasa, and Malindi; the Persian Gulf and the island of Socotra that lies beyond the exit from the Gulf to the ocean; the Maldive Islands, a navigational hazard where Magellan foundered and survived in 1510; the Indian mainland, and especially the western coast, where most Portuguese activity was concentrated in Magellan's day; Malacca, which he helped to turn into a Portuguese outpost; the Moluccas—the fabled "Spice Islands" for which he was bound on his great voyage; and the Philippines, where he sought a realm of his own and where he died. We can visit each in turn.

The conventional notion that the Swahili ports housed oceangoing peoples is oversimplified. For generations, the Swahili responded to the racism of Western masters by emphasizing their links of culture and commerce with Arabia and India. After independence some of their hinterland neighbors took revenge, treating them as colonists, rather as the inland communities of Liberia and Sierra Leone resented the descendants of resettled slaves in Monrovia and Freetown as an alien elite. In Kenya, political demagogues threatened to expel the Swahili, as if they were foreign intruders. Yet the Swahili language, though peppered with Arabic loanwords, is closely akin to other Bantu languages. The people who speak it came to the coast from the interior, perhaps thousands of years ago, and retained hinterland links alongside their trade with visitors from the Indian Ocean.[38]

The coastal location of Swahili cities conveys a misleading impression of why the sea was important to them: they were sited for proximity to fresh water, landward routes, and sources of widely traded coral as much as for ocean access. The elite usually married their daughters to business partners inland rather than to foreign sojourners. Few cities had good anchorages. The town of Gedi or Gede, which covered eighteen acres inside ten-foot-high walls and had a palace over a hundred feet wide, was four miles from the sea. Some Swahili traders crossed the ocean,[39] but most plied their own coasts and frequented their own hinterlands, acquiring gold, timber, honey, civet, rhinoceros horn, and ivory to sell to the Arabs and Indians who carried them over the ocean. They were classic middlemen, who seem to have calculated that the risks of transoceanic trading were not worthwhile as long as customers came to their coasts.[40] Kilwa was one of the greatest of Swahili emporia because the monsoon made it accessible to transoceanic traders in a single season. Ports further south, like Sofala, though rich in gold, were accessible only after a laborious wait, usually in Kilwa, for the wind to turn. Merchants from India seem rarely to have bothered to go further south than Mombassa or Malindi, where products from all along the coast as far as Sofala were available and where Indians paid for their purchases with fine silk and cotton.

Visiting Portuguese in the early sixteenth century noticed the love-hate relationship that bound the Swahili to the hinterland. On the one hand, the two zones needed each other for trade; on the other, religious enmity between the Muslims and their pagan neighbors committed them to war. This, thought Duarte Barbosa, who shared a name with Magellan's cousin by marriage,[41] was why the coastal dwellers had "cities well walled with stone and mortar, inasmuch as they are often at war with the heathen of the mainland."[42] There were material causes of conflict too. The Swahili needed plantations, acquired at hinterland communities' expense, to grow food, and slaves to serve them. Coastal and interior peoples exchanged raids and demands for tribute as well as regular trade. When Magellan and his companions arrived in the early sixteenth century, they got the impression that Mombasa, the greatest of the Swahili port cities, lived in awe of their neighbors, the "savage," poison arrow–toting Mozungullos, who had "neither law nor king nor any other interest in life except theft, robbery, and murder."[43]

For Portuguese intruders, Swahili ports were stepping-stones on the way to India, where, in 1505, Magellan got his first recorded experience of war. Along the way, the Maldive Islands, where he was marooned for a while when his ship foundered in 1510, were much-frequented staging posts; seafarers could replace rigging with coconut-fiber ropes and buy cowrie shells, prized by traders in the subcontinent. Low in the water, hazardous to unwary pilots, the islands were frequent scenes of ship-wreck. The most bizarre case, before Magellan's, befell Santo Stefano. In an attempt to head homeward with what little fortune he had salvaged from his adventures, he waited six months at the Maldives for the monsoon to turn. When it did, it unleashed so much rain that his deckless ship sank with the weight of it, "and those who could swim were saved and the rest drowned."[44] After floating on wreckage from morning to evening, the merchant was rescued by a passing ship.

The indefatigable fourteenth-century traveler Ibn Battuta was induced to turn his stage at the Maldives into a stay in 1340, when the locals, impressed with his learning, offered him the job of the chief judge in Male, the islands' largest settlement. The rewards of the job—which included pearls, jewels, gold, slave girls, high-status wives, and the use of a horse and a litter—convey a sense of local prosperity. The cultural level of his new fellow residents, however, disappointed Ibn Battuta, who spent his time enforcing sharia law, having thieves' hands severed, whipping absentees from Friday prayers, and punishing lapses of Islamic discipline in dress—of which there was too little for his taste—and sex—of which there was too much.[45] Perhaps because of large numbers of foreign merchants, jockeying for influence, or the factionalism of nobles, enriched in wealth and encouraged in ambition by opportunities for trade, Maldivian politics were a byword for instability around the Indian Ocean. At the time of Magellan's visit, the sultan, Kalu Muhammad, was kept precariously on a rickety throne, from which enemies periodically ousted him, by the patronage or puppetry of the ruler of Cannanore, for whom the islands were valuable as emporia in the pepper trade.[46]

Cannanore (now Kannur) was part of the comparably turbulent world of the Indian subcontinent. Here too Portuguese intruders could take advantage of political fragmentation that favored newcomers and of tensions between Hindus and Muslims. The once-preponderant sultanate of

Delhi never fully recovered from setbacks of the mid-fourteenth century. Until 1525, when Muslim invaders from Afghanistan reestablished a state capable of reuniting much of the subcontinent by conquest, Hindu states proliferated, some of them sustained by commitment to violence against Islamic neighbors and rivals. There were, of course, regions utterly intractable to Islam. In some circles, Islam met a skeptical reception. Kabir of Benares, for instance, whom Hindus and Muslims alike claim as one of their own, was, rather, a poet of secularist inclinations.

> Feeling your power, you circumcise—
> I can't go along with that, brother.
> If your God favored circumcision
> why weren't you born with the cut?

Hindus fared little better in the face of Kabir's skepticism:.

> If putting on the thread makes you a brahmin,
> What does the wife put on? . . .
> Hindu, Muslim, where did they come from?[47]

Revulsion was more effective than skepticism in setting limits to the spread of Islam. Hindus generally resisted Muslim proselytization with tenacity. The most militaristic Hindu state, perhaps, was Vijayanagar—the name means "City of Victories"—inside its sixty-mile ring of sevenfold walls. Its rajahs called themselves "Lords of the Eastern and Western Oceans." According to the maxims of an early sixteenth-century ruler,

> A king should improve the harbours of his country and so encourage its commerce that horses, elephants, precious gems, sandalwood, pearls and other articles are freely imported. . . . Make the merchants of distant foreign countries who import elephants and good horses attached to yourself. . . . Afford them daily audience and presents. Allow decent profits. Then those articles will never go to your enemies.[48]

In practice, however, the capital was as far from the sea as you could get, and outlying provinces were hard to control. By 1485, the power of Vijayanagar's neighbors seemed not only to have arrested the expansion of the state but to threaten its very existence. Taxation from coastal emporia dried up as the frontiers withdrew inland. Muslim warlords usurped frontier areas. By the time Magellan arrived in India, Vijanayagar had recovered, but only partially. Independent but mutually hostile coastal states provided the Portuguese with opportunities to negotiate or seize places in which to trade.

Along the Malabar coast, where most of the world's pepper grew, Magellan appeared in 1505 in a Portuguese expedition that was designed to finagle or force part of the trade in pepper and textiles. Here, from a few strongholds, Muslim sultans contended for tribute from Hindu subjects and for dominance in trade, perhaps within the "remarkable solidarity" that one historian has detected among the Muslims of Kerala.[49] Calicut (now called Kozhikode), Cannanore (Kannur in current usage), Cochin (Kochi today), and Quilon (now Kollam) handled the bulk of pepper bound for overseas markets, chiefly in China. The origins of the leading families are impossible to reconstruct with certainty from the late chronicles that shroud them in myth; some were indigenous, others Gujarati.[50] But the waning of the power of Vijayanagar was the context that made their independence possible, while the growth of the global pepper trade gave them the resources to sustain their power or dispute it among themselves. Calicut, where Magellan was to receive his first recorded battle scar (below, p. 44), was the most successful in raising armies and fleets. It became the hub of a regional hegemony and the focus of its rivals' resentment. Calicut exercised irksome supremacy when the Portuguese arrived: the newcomers, in consequence, were able to secure local allies and trading partners when the *samorin*, as the ruler was called, tried to exclude them. Here, according to an Italian visitor in the first decade of the sixteenth century, the "sea beats against the walls of houses" crammed together in the center of the city but spreading over six miles of increasingly spacious suburbs. The town's modest appearance belied its wealth: dwellings were low and light, roofed with thatch, because of a water table that inhibited the digging of deep foundations. But there were reputedly one hundred thousand men in the ruler's army.[51]

Further north, the wealth of the Muslim sultanate of Gujarat was conspicuous, thanks to its industrial-scale production of cottons dyed with indigo and the competitive, commercial culture of Muslim and Jain merchant-houses. Mahmud Shah Begarha conquered Champaner from its Hindu masters in 1484 and began rebuilding the city on the grand scale still visible in the sumptuous ruins of palaces, bazaars, squares, gardens, mosques, irrigation tanks, and ornamental ponds. There were workshops producing fine silk, textiles, and arms, and Hindu temples were allowed outside the walls. The sultan's mightiest subject, Malik Ayaz, came to Gujarat, perhaps in the 1480s, as a slave famous for valor and archery in the entourage of a master who presented or sold him to the sultan. Freed for gallantry in battle—or, in another version of the story, for killing a hawk that had besmirched the sultan's head with its droppings—he received the captaincy of an area that included the ancient site of a harborside settlement, just reemerging, thanks to Malik's immediate predecessors, from centuries of accumulated jungle. He induced shippers from the Red Sea, the Persian Gulf, Melaka, China, and Arabia to use Diu as their gateway to northern India. His style of life reflected the value of the trade. When he visited the sultan, he had nine hundred horses in his train. He employed a thousand water carriers and served Indian, Persian, and Turkish cuisine to his guests off china plates.[52]

. . .

The Malay world, where Magellan operated from 1511 to 1513, straddled continental Southeast Asia. In the early fifteenth century Chinese expansion nibbled at the edges; and Chinese rulers were resentful when the Portuguese broke into the region. After the 1430s, however, the Ming dynasty's renunciation of seaborne imperialism had left native states free to try one another's strength. The Thai—founders of what is now called Thailand—had expansionist ambitions. Although in the early fifteenth century they created the region's largest state at the expense of their neighbors, the Burmese, Khmer, Mons, and Malays, the region lacked a dominant empire and remained home to a state system in which a number of polities contended with each other.

Offshore the seas that flowed between the Indian and Pacific Oceans were full of trading states and seafaring traditions. Commercially, Malacca was uniquely important because it dominated the strait at the southern tip of continental Southeast Asia, where the commerce of the Indian Ocean met that of the China Seas. The spread of Islam along the shores of Java in the early fifteenth century made the strait a cynosure for Muslim merchants. But Malacca was also, informally, a Chinese protectorate, perfectly positioned to channel Indian commerce eastwards. A Chinese fleet installed the ruling dynasty in 1402; frequent embassies maintained the illusion that the port was a Chinese outpost. When the Portuguese seized Malacca in 1511, China took revenge by torturing and imprisoning members of the first Portuguese embassy to arrive in the empire. Meanwhile, spices from the islands and textiles from the Indian mainland converged at the strait that Malacca commanded. As the sultan of Malacca observed in 1468, "To master the blue oceans people must engage in trade, even if their countries are barren."[53] In 1502 the text on an Italian map spoke of "cloves, rhubarb, ivory, and precious stones of great value and pearls and scent and fine porcelains and many other products, most of which come from afar on their way to China."[54] Luís de Camões, who ranged the East and celebrated it in verse in the late sixteenth century, bade his readers,

Malacca see before, where ye shall pitch
Your great Emporium, and your Magazins:
The Rendezvous of all that Ocean round
For Merchandizes rich that there abound.[55]

Early in the fifteenth century, Malacca's ruler adopted Islam. From the end of the century conversions multiplied, spread by dynastic marriages, or by a radiating process in which Sufi missionaries fanned outwards from each successive center to which they came. Malacca seems to have provided manpower for the conversion of states in Java, and Java, in turn, around the beginning of the new century, did the same job for Ternate in the Moluccas, from where missionaries continued to neighboring islands. Provincial rulers guaranteed the flow of revenue to the sultans' courts in exchange for the unmolested exercise of power. "As for us who administer territory," said a nobleman in a Malay chronicle, "what concern is that of yours? . . . What we think should be done we do, for the ruler is not concerned with the difficulties we administrators encounter. He only takes account of the good results we achieve."[56]

Beyond Malacca lay two more pieces in the Magellan-related jigsaw of the world: the fabled "Spice Islands" that supplied most of the world's nutmeg, cloves, and mace, and the mass of islands that became "the Philippines"—straddling the way from the Moluccas to China. Magellan did not visit either (see below, pp. 48–50) while serving in "Portuguese India" but was well informed about both from letters he received and interviews he presumably conducted in Malacca. Cloves were probably the most valuable vegetable product in the world per unit of bulk. Five islands specialized in growing them: Ternate, Tidore, Moti, Makan, and Bacan. According to a Portuguese apothecary, Tomé Pires, who was the official in charge of procurement in the early sixteenth century (and whom we shall meet again—below, p. 47), total output was about 6,000 bahars per annum. What that meant is unclear as the bahar was unstandardized, but about 450 pounds weight, or a little over 200 kilograms, is a rough guide to an equivalent in modern standard measures. In any case, purchasers could make dazzling profits: cloves sold in Malacca for up to seventy times their value in Ternate. The trees grow up to forty feet of man-sized trunk, with the cloves clustered at their tips. The oil is most prolific just before the buds open in flower, changing from white to green, then red, and dried or smoked for sale.[57]

Nutmeg and mace, respectively the seed and mantle of *Myristica fragrans*, came overwhelmingly from the Banda Islands, toward the southern end of the region, beyond the island now known as Seram. Both products—"fruits," wrote Tomé Pires, "of which we are so fond"— were equally valued in Europe, but nutmeg was more abundant locally. According to the Portuguese apothecary, Bandanese would sell a bahar of mace only if the purchaser agreed to take seven of nutmeg along with it. Nutmeg was prepared for sale by drying and smoking, whereas mace dried in sun, changing color from crimson to yellow. Before the Portuguese arrived, only Malay and Bandanese traders handled them as far as North Banda, where all comers could acquire them for onward distribution.[58]

The provenance was shrouded in mystery. Chinese sources reveal no awareness of it until the fourteenth century: in previous references, gazetteers always identified the products with the emporia in which they were acquired. The earliest evidence of Chinese penetration of the islands dates from 1349, when the first Chinese text to mention Banda observes that "merchants who . . . trade in the Western Ocean carry out with them such things as cloves, nutmegs . . . and return with goods of ten times greater value."[59]

Riven among rival dynasties, the islands had no sense of collective identity, no prospect of common defense, and no means of discriminating among prospective traders. The name "Moluccas" literally means "many kings," and the political divisions justified it.[60] Ternate and Tidore were locked in rivalry that was almost a ritual, conceived within a notion of a split cosmos, whose unity depended paradoxically on the "dynamism" of divison.[61] The islanders' distinguishing characteristics, according to Tomé Pires, were dark skin, sleek hair, and internecine war. He liked the people of Ternate best—"knights among the Moluccans," barely affected by acculturation with Islam and properly fond of wine, whose sultan and princes wore crowns of gold "of moderate value." Muslims were few—only two hundred out of a population of two thousand in Ternate—and slack in observance: many eluded circumcision. The sultan had, by Moluccan standards, exceptional local authority, dispensing justice and exacting obedience. He was, in Pires's opinion, disposed to submit to Portugal ("He and his lands were slaves of the king") and willing to receive Christian priests "because if our faith seemed to him good he would forsake his sect and turn Christian":

he was, perhaps, desperate for trading partners after the fall of Malacca. Other islands were poorly regulated. In Banda, Pires could not even discern a recognizable polity in islands where no king reigned and village elders monopolized the highest level of authority.[62]

The Moluccas—defenseless and disorganized—were tempting to any merchant community that thought of controlling production of precious commodities, rather than merely trying to dominate trade. They would become the focus of rivalry between Spain and Portugal and the opportunity for Magellan to switch allegiance and lead a Spanish expedition in search of spices. But his own interest, as far as it could be distinguished from those of the merchants and kings he served, seems rather to have focused on the islands that screened the Moluccas to the north. The Philippines—as they later came to be called—had two insuperable attractions: they were close to China and therefore an ideal base from which to control China-bound commerce; and they produced gold—a substance that could always be traded at advantage for almost any other commodity and especially for silver in Japan or textiles in India. Yet the Chinese seem never to have esteemed them highly or tried to conquer them, even in the brief period of overseas imperialism at the Ming court in the very early fifteenth century. Song dynasty annalists had complained of naked, staring, indigent raiders from "Pi-sya-ye" who, on bamboo rafts, plundered villages in Fujian for iron, ripping off door furniture and stripping the armored riders whom they killed with spears, roped for retrieval. The thirteenth-century gazetteer of Southeast Asia by Zhao Rugua included what sound like firsthand observations: "On each island lives a different tribe . . . of about a thousand families" in huts made of rushes, eating copiously of rice, roots, swine, and fowl. Mountain dwellers inland relied on the services of female water gatherers and bartered their cotton, beeswax, coconuts, and fine matting for Chinese imports—silk parasols, porcelain, and rattan baskets. The lowlanders, whom the text calls Han-tai, were described as small, brown-eyed, and frizzle-haired with shining teeth, at home in treetops; there, presumably, they lay in ambush for game and merchant-victims, who, however, could buy impunity with porcelain cups, as the predators, "shouting with joy," "escape with their spoil."[63]

For a prospective conqueror, the possibilities were attractive. Most of the coastal regions of the archipelago housed tiny kin-based communities, known as *barangay*, of between thirty and one hundred families, though on Luzon, Cebu, and Vigan they could number as many as two thousand families, divided among chiefs called *datu*, nobles (or *maharlika* in the prevailing language of Luzon), freemen (*timagua*), and a large servile peasantry. Some chiefs had long evinced ambitions for wider dominion. Embassies had arrived in China at intervals from 1372, attracting gifts of silk from the imperial court, and briefly, in the reign of the Yongle emperor, who was peculiarly susceptible to the idea of founding an overseas empire, the appointment of a permanent imperial representative.[64] When Magellan's expedition arrived in 1521, a leading chief, called Humabon, had evident imperial ambitions of his own, which Magellan exploited to try to coerce neighboring chiefs to form a single state.[65]

The major attractions were economic: in the early fifteenth century, embassies to China from Pangasinan made a display of the wealth of Luzon, tendering tribute in the form of horses, silver, and gold.[66] Tomé Pires, gathering data in Malacca in 1511, confirmed the availability of gold, albeit "of very low quality," and the abundance of foodstuffs and the products of apiculture. He also emphasized the political divisions. His account explains how he and—therefore—Magellan, who was also in Malacca at the time, could learn about the islands: there was a growing immigrant community of islanders, "useful and hard-working," in the city, where they were building increasing numbers of houses and shops. In defiance of the Portuguese occupiers, five hundred Filipino merchants, Pires remarked, had "gone over to the side of the former king of Malacca, not very openly."[67]

Beyond the Philippines, as far as Magellan and his fellow Europeans knew, there was nothing but ocean. On his great voyage he would spot lonely islands and provoke violence in the Marianas (below, pp. 219–20, 227). But the great question was, How much ocean was there? Everyone who thought about it hoped that what Magellan later called the Pacific would prove to be narrow and easy to penetrate and cross from the Atlantic or from Spanish outposts in the New World. Until he made the attempt, no one knew. Even after his death, as we shall see, people in Europe were slow to acknowledge the true breadth of the ocean. We shall see some European geographers' self-deluding calculations and vain hopes in the next chapter.

. . .

Of the world beyond Europe, the last part that concerned Magellan was the Atlantic Ocean. On his great voyage, it mattered to him only as an obstacle he had to cross. The shore he coasted during his search for a strait to the Pacific threatened his enterprise with mutinies, shipwreck, short rations, and devastating cold. There were no European settlements at the time in what is now Brazil or the South American cone—only a few temporary loggers' huts on the edge of the Brazilian forest, where sojourners came to make a quick cull of dyewood. The native inhabitants, reputed cannibals and giants, were, as we shall see (below, pp. 148, 176), capricious and sometimes dangerous. But Magellan's project would have been unthinkable but for a long prior history of Atlantic exploration, establishing exploitable transoceanic routes. In Magellan's day, the Atlantic was Europeans' highway to the rest of the world, bringing formerly sundered cultures into touch, as the divergent, isolated worlds of ancient and medieval times began to resume contact, coming together to form the interconnected world we inhabit today.

How did it happen? How did some European Atlantic–side communities come to exhibit exceptional reach, which enabled them to cross unprecedented distances on previously unexplored routes? It is tempting to seek explanations rooted in notions of "challenge and response"—the theory that tough conditions stimulate human resourcefulness and impel innovation—or "hard times and investment in culture," a concept originally developed to explain the Renaissance as the result of a diversion into arts and learning of funds that a weak economy could not absorb.[68] But the best starting point is the wind-wind map of the ocean. For most of history, winds and currents have played a huge part in conditioning, and even determining, who and what went where in the world.

Europe's only effective access by sea to the rest of the world is along its western seaboard, where fixed winds blow onshore. Instead of changing direction seasonally, as in monsoonal systems, the prevailing winds in the Atlantic tend to be the same; fluctuations are unpredictable and typically short-lived. For most of the age of sail, navigators who wanted to penetrate the ocean had to sail into the wind—which usually resulted in their being blown back without discovering any useful new lands or routes. Norse explorers of the North Atlantic in the tenth and eleventh centuries overcame these limitations by sailing west with the currents that cross the Atlantic below the Arctic, and then picking up the westerlies that took them home. But this route led only to relatively poor and underpopulated regions.

For the Atlantic to become Europe's highway to the rest of the world, explorers had to develop ways to exploit the rest of the fixed-wind system. They had to discover the winds that led to commercially important destinations. There were, first, the northeast trade winds, easily accessed from Spain and Portugal, leading to the resource-rich, densely populated regions of the New World, far south of the lands the Norse reached. The westerlies in northerly latitudes could carry European ships home. The South Atlantic wind system led, by way of the southeast trade winds and the westerlies of the Far South, to the Indian Ocean, or, as Magellan found, by even more laborious means, around South America to the Pacific.

The technology needed to exploit the Atlantic's wind systems only gradually became available during a period of long, slow development in the thirteenth, fourteenth, and fifteenth centuries. Like most technology, for most of history, it developed by trial and error. We know little about the process because the work went undocumented. Humble craftsman labored to improve hull design and rigging—and therefore the maneuverability of ships—and to make water casks secure for the long voyages explorers had to undertake. Historians have traditionally emphasized the contribution of formal science in developing maritime charts and instruments for navigating by the stars. Now it seems that these innovations were irrelevant. No practical navigator of this period in Europe seems to have used them.

In addition to gradually developing technology, fitfully improving knowledge of winds and currents prepared Europeans to explore maritime routes to the rest of the world. The European discovery of the Atlantic was launched from deep in the Mediterranean, chiefly by navigators from Genoa and the island of Majorca. They forced their way through the Strait of Gibraltar, where the strength of the adverse current seemed to stopper their sea, in the thirteenth century. From there, some turned north to the familiar European Atlantic. Others turned south into waters unsailed, as far as we know, for centuries, toward the Madeira and Canary Islands and the African Atlantic. Early efforts were long and laborious because explorers' vision was limited to the small patches of the ocean before them, with their apparently unremitting winds. Navigators were like code breakers deprived of information to work with. Moreover, the Black Death and the economic downturn of the mid-fourteenth century interrupted the effort, or at least slowed it down.

Only the long accumulation of information and experience could make a breakthrough possible. Navigators had no means to keep track of their longitude as they beat their way home against the wind. They made increasingly huge deep-sea detours to find westerlies that would take them home. Those detours led to the discovery of the Azores, a midocean string of islands more than seven hundred miles west of Portugal. Marine charts made not later than the 1380s show all but two islands of the group. Much longer open-sea voyages now became common. From the 1430s, the Portuguese established way stations, sown with wheat or stocked with wild sheep, on the Azores so that passing crews could find provisions.[69]

Several attempts were made during the fifteenth century to explore Atlantic space, but most doomed themselves to failure by setting out in the belt of westerly winds. Presumably explorers chose this route because they wanted to be sure that they would be able to get home. We can still follow the tiny gains in the slowly unfolding record on rare maps and stray documents. In 1427, a Portuguese pilot called Diogo de Silves established for the first time the approximate relationship of the islands of the Azores to one another. Shortly after 1450, the western-most islands of the Azores were reached. Over the next three decades, the Portuguese crown often commissioned voyages of exploration farther into the Atlantic, but none is known to have made any further progress. Perhaps they failed because they departed from the Azores, where the westerlies beat them back to base.[70]

Not only was exploitation of the Atlantic slow, but at first it yielded few returns. One exception was Madeira, which paid enormous taxes to the Portuguese crown thanks to sugar planting in the mid-fifteenth century. The explorers' hope of establishing direct contact with the sources of West African gold proved false, though they were able to get gold at relatively low prices through trade with West African kingdoms. This trade also produced something that could be sold in European markets. From 1440, Portuguese desperadoes obtained increasing numbers of slaves through trading and raiding. But markets for slaves were limited because great slave-staffed plantations, of the sort later familiar in the southern United States, hardly existed in Europe, where most slaves were still domestic servants. The Canary Islands attracted investment because they produced natural dyestuffs and seemed exploitable for sugar. But their inhabitants fiercely resisted Europeans, and the conquest was long and costly.[71]

In the 1480s, however, the situation changed, and Atlantic exploration began to pay off. In the North Atlantic, customs records of the English port of Bristol indicate that quantities of whaling products, salt fish, and walrus ivory from the ocean increased dramatically. In West Africa, the Portuguese post at São Jorge da Mina, near the mouth of the Benya River, was close to gold fields in the Volta River valley, and large amounts of gold now began to reach European hands. In 1484, sugar production at last began in the Canary Islands. In the same decade, Portuguese made contact with the kingdom of Kongo. Although voyages toward and around the southernmost tip of Africa encountered unremittingly adverse currents, they also showed that the Far South of the Atlantic had westerly winds that might at last lead to the Indian Ocean. By the end of the 1480s, it was apparent that Atlantic investment could yield dividends.[72]

The 1490s were a breakthrough decade in Europe's efforts to reach out across the ocean to the rest of the world. In 1492–93, Christopher Columbus, in voyages financed by Italian bankers and Spanish bureaucrats, discovered fast, reliable routes across the Atlantic that linked the Mediterranean and the Caribbean. In 1496, John Cabot, another Italian adventurer, backed by merchants in Bristol and the English crown, discovered a direct route across the North Atlantic, using variable springtime winds to get across and the westerlies to get back. His route, however, was not reliable and, for over one hundred years was mainly used to reach the cod fisheries of Newfoundland. By the time Vasco da Gama found the way into the Indian Ocean in 1497, the wind system of the South Atlantic—dominated by northeast trade winds and strong westerlies in the Far South—had been decoded.

The breakthroughs of the 1490s opened direct, long-range routes of maritime trade across the world between Europe, Asia, and Africa. Success may seem sudden, but not if we view it against the background of slow developments in European chronology and knowledge and the accelerating benefits of Atlantic exploration in the previous decade. Was there more to it than that? Was there something special about European culture that would explain why Europeans rather than explorers from other cultures discovered the world-girdling routes, linking the Old World to the New and the Indian Ocean to the Atlantic? Some European historians have argued just that—that Europeans had something others lacked.

Such a suggestion, however, seems ill conceived. As we shall see in the next chapter, some explorers—including Magellan—shared peculiarly European models of behavior that encouraged deeds of seaborne chivalry; in Spain and Portugal there were monarchs and missionaries with religious notions—variously crusading, evangelizing, and millenarian—that animated outreach to pagan peoples. Compared to the peoples of maritime Asia, however, Europeans were special mainly in being slow to launch long-range voyages. The Atlantic, the ocean they bordered, really was special, because its wind system inhibited exploration for centuries but rewarded it spectacularly once it was launched. Moreover, the breakthrough explorations were the work not of "Europe" but of people from a few communities on the Atlantic Seaboard and in the Mediterranean. What distinguished them was not that they set off with the right kind of culture but that they set off from the right place.

Europe's outreach into the Atlantic was probably the result not so much of science or strength as of delusion and desperation. This was a space race where it helped to come from behind. The prosperous cultures with access to the Indian Ocean felt no need to explore remote lands and seas for new resources. For cash-strapped Europe, however, the attempt to exploit the Atlantic for new products was like the efforts of underdeveloped countries today, anxiously drilling for offshore wealth from oil or natural gas. In some ways, it paid off.

For most people the new opportunities of the increasingly interconnected world seemed to have meant nothing, or to have been at most a source of entertainment in pages of travel and romance, or of prized imports obtained at other people's risk. Some, however, if not born to embrace risk, were brought up to do so. Magellan was among them.

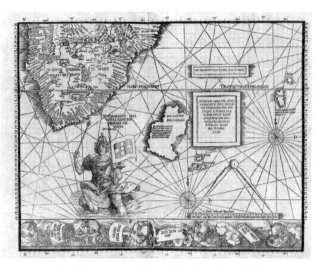

FIGURE 2. Magellan's patron, Manuel the Fortunate, "Most Christian King" of Portugal, bestrides a whale as he dominates the passage from the Atlantic to the Indian Ocean via the Cape of Good Hope, in Martin Waldseemüller's *Carta Marina* of 1516. The king wears armor and tunic of Roman inspiration and an imperial crown, while he waves a scepter and the royal banner on a cross-topped staff. The legend in the large frame, in the style of an inscription, records how "the navigation of the Portuguese in our time" demonstrated the falsehood of the former belief in the inaccessibility of the Indian Ocean by sea. The map combines a grid of longitude and latitude—a device first proposed by Ptolemy, with rhumb-lines intended to help navigators establish a course. Note the scale, shown toward the bottom right, and the putti blowing the winds, which are labeled, with the names of chief points of the compass, in Low German, although Latin is retained for other legends. Courtesy of the Library of Congress.

CHAPTER 2

The Education of an Adventurer

Oporto—Lisbon—the Indian Ocean,
to 1514

Train up a child in the way he should go: and when he is old,
he will not depart from it.

—Proverbs 22:26

Magellan was an Atlantic man, born on the ocean's edge. In childhood he
had plenty of opportunities to stare out to sea and feel the persistent
westerlies blow into his face. We are unreliable witnesses of our places of
birth—too young at the time to perceive accurately or recall subsequently.
Yet bureaucracies bother us for details we can only know, at best, at second
hand. Whenever Magellan had to respond to such requests, he called
himself a native of Oporto—the great gray city at the mouth of the River
Douro, where today barges line the quays to load port for foreign palates.
The consensus among near contemporaries supports Magellan's assertion.[1] But if you travel about ninety kilometers upriver, between granite
hills, to Sabrosa, local cicerones on a street named after Magellan will
point to the sandstone facade and broad windows of the Casa da Pereira
and tell you, confidently but not credibly, that he was born there.

No authentic document supports the tradition, but the house probably belonged to his brother-in-law, João de Silva Telles, husband of
Magellan's sister Teresa.[2] In December 1504, when he made a will prior
to setting out on his first overseas posting, the future explorer declared
an interest in the locality, leaving the income from a modest property for
masses for his soul, in the event of his death, and desiring ardently that
"the little property of Quinta de Souta, which is his possession," remain
in the family.[3] A great Portuguese historian, António Baião, impugned
the authenticity of the document in 1922, chiefly on three grounds: first,

indecisively, that it exists only in a purported eighteenth-century copy; second, suggestively but inconclusively, that not everyone named in it is attested in other sources; and third, and obviously fallaciously, that it is very different from a well-authenticated will Magellan made in 1519.[4] The property is no longer identifiable (notwithstanding antiquarians' claims to have identified it),[5] but the name suggests a modest holding in a rural, wooded location, such as bristle over the hills around the little town. In any case, whether genuine or not, the document says nothing about Magellan's birthplace. He had or acquired property in or near Oporto too, "a farm with vines and chestnut trees and wheat fields," until he transferred it to his sister Isabel in March 1519, when he was preparing for his great voyage.[6]

"My family," he insisted in his first will, of "the very noble and ancient house of the lords of Nobrega," was "one of the most distinguished and best and most ancient of the realm."[7] But the pride Magellan took in his lineage was perhaps a case of excessive protestation. His family belonged to *hobéreau* aristocracy or local squirearchy. He demanded in his will, vainly and in vain, that his brother-in-law add the Magellan arms to the scutcheon of the house of Silva Telles. His anxiety that the name might vanish was understandable, as a Portuguese custom permitted family members who married well to look to the distaff side for more obviously impressive surnames; Magellan's own brother, Duarte, did so, calling himself de Sousa after his mother's ancestors.

In legal documents attested in adulthood, Magellan repeatedly referred to his father as Rui or Rodrigo—they are versions of the same name—de Magalhães and to his mother as Alda de Mesquita, though a scribe at the Portuguese court thought "Pedro" was the father's name.[8] Álvaro de Mesquita, who served alongside the explorer on the great voyage to the Pacific, is always called his cousin, a designation vague but no doubt valid. The names of siblings—Duarte de Sousa, and sisters Isabel, Ginebra, and Teresa—appear frequently enough to be believed. If the evidence of a cousin, recorded in 1563, can be trusted, Magellan had uncles who had "similar tastes" to his own and were "very adventurous, friends of navigation, and ingenious and extraordinary pilgrims." They served in Guinea—which could, in the terminology of the time, be anywhere in sub-Saharan Africa—"and by land and sea."[9] Other bearers of the same surname served around the empire, including one Martín, who was surely a relation, because the children of Diogo or, as he became (below, p. 76) Diego Barbosa, who was to become Magellan's father-in-law, claimed his inheritance in a lawsuit in 1525 and 1526.[10] Magellan's relationship to one Rui de Magalhães, who was custodian of the biggest castle in the vicinity, cannot be established. Long service, however, in the household of the dominant dynasty in the locality, that of the Duques de Viseu, allied almost all Magellan's traceable ancestors and relatives to the great regional powerhouse.[11] It also provided his family with a link to prospects of royal patronage from 1481, when Leonor de Viseu married King João II. Her brother, Manuel, the incumbent duke, ascended the throne in turn in 1495.

No surviving document records Magellan's birth: that is not unusual for the time or place. The assertion that it happened in or about 1480 is a persistent feature of traditions about him, the authors of which copy each other. It seems to have spread in books like red rot: no good evidence supports it. In view of the chronology of the rest of his life, Magellan is unlikely to have been born before the midpoint of the decade. Nor is there any reason to believe that he served Queen Leonor as a page before moving to the king's household in the same capacity.[12] The story can be traced back no further than the early seventeenth century. Equal skepticism seems advisable in contemplating the tale that his parents' deaths, which made him a ward of his Sousa relatives, precipitated his transfer to court; the opportunity created by the accession to the throne of Dom Manuel, a representative of the lineage of his family's ducal patrons, is explanation enough.

The date of Magellan's removal to court is uncertain, but when Manuel became king Magellan was probably of a suitable age: seven was the typical age of an incoming page boy, though more mature recruits were not excluded. No documentary corroboration of the young man's status as a retainer in the royal household appears in surviving records until 1505, when he was designated as such in a list of recruits for the voyage to the east of Francisco de Almeida, Portugal's first viceroy in India.[13] Transfer from household service to an active role in war or exploration was a rite of passage for a royal page—a kind of graduation, marking the completion of one's education at court. One's late teens were a suitable time for such a move.

In Portugal, the royal court was already the focus of political life. The monarch wielded power not by force of arms or the charisma of kingship but by traditional string-pulling. Clout in the aristocratic marriage market counted for a lot, as did disposal of titles of nobility and concessions of fiscal privilege or powers of jurisdiction. So did the king's right of disposal over offices of profit in the administration, the chivalric orders, and, to some extent, the church.[14] The thugs and desperadoes who dominated aristocratic affinities were kept from rebellion—somewhat precariously—by the opportunities of war and overseas adventures.

The poverty of the treasury kept royal pretensions modest. João II was an almost landless king. Like Roger Miller in the famous vagabond song, he was, in his own jape, "king of the road,"[15] because predecessors had alienated the rest of the royal patrimony. Before Manuel's accession, expenditure always exceeded income.[16] In the 1480s the opening of the Guinea trade (above, p. 27) enriched the realm of King João and furnished the court with some elegance. Hieronymus Münzer, a critically alert physician of Nuremberg, who visited Portugal on a tour of Iberia—undertaken out of a mixture of piety, curiosity, and the desire to elude plague at home—was delighted to see imports from Germany alongside such dazzling exotica as gold, slaves, elephant ivory, and Malaguetta pepper—grains of paradise. He found the king's dinner-table conversation learned and brighter than the gloomy décor of the palace—"which he always kept badly since the death of his son, Afonso, in a fall from a horse."[17] The traveler's hopes that so admirable a king would enjoy long life were unfulfilled. João died shortly after Münzer's departure.

The accession of Manuel I changed everything. The poet Garcia de Resende, who had served João II as a page and who became intimate with his successor, summed up the transformation:

> Ennoblement brought
> The Portuguese court,
> Which once was small,
> To outshine all
> Of Christian sort,
> Or so I thought.[18]

The new king was, by most calculations, next in line of succession, which, however, his predecessor had done everything possible to frustrate; a mixture of foreign support and judiciously placed bribes secured it. He was called "Manuel the Fortunate," and fortune attended him in both senses of the word. A series of dynastic incidents—including battlefield tragedies, sporting accidents, and political assassinations—had eliminated rival heirs to egregious riches and made him the most prosperous man in Portugal.[19] His private wealth and the skill with which he managed it augmented royal coffers. I recall R. W. Southern's anecdote of his boyhood: as a schoolboy he wrote an essay arguing that Henry VII was "the first businessman to rule England. . . . Of course," he added, "I was wrong, but at that moment I became an historian."

Manuel was, if not a businessman, a comparably businesslike king, squeezing revenues, cutting costs. He gave himself the title of "Lord of Commerce," which a traditionally minded monarch might have regarded as derogation. The new king literally lived above the shop. Royal apartments surmounted the Casa da índia, the combined warehouse and countinghouse through which all Portugal's trade from the Indian Ocean was funneled. His personal estate included a fifth of the yield of the booming sugar industry of Madeira.[20] His inherited wealth from the duchies of Beja and Viseu and his share, as Grand Master from 1487, of the income of Portugal's richest order of chivalry made the crucial difference that allowed him to expand his household. Numbers and the attendant costs leaped. Pensions to personnel he replaced helped boost costs in the first three or four years of his reign by over 50 percent to 16 million reais. Over the reign as a whole, the number in his retinue of young men rose from 140—close to the official limit that the cost-conscious taxpayers' deputies in the Cortes, the assembly of the realm, prescribed—to over 400. In the most expensive category, that of household knights who were members of the royal council and typically received pay in excess of 50,000 reais a year, numbers roughly quadrupled.[21] The king could afford it.

It was a strategy much affected by European monarchs at the time. By accumulating retainers and gathering the youth of the kingdom's nobility into the king's embrace, where they were under observation and dependent on the crown, monarchs could exercise a form of control. Noble scions brought up alongside princes and princesses were guarantors of future collaboration and even of present submission, as they were, in a sense, hostages for their parents' good behavior. The arrangement suited both parties: it spared noble families expense and placed their children in line for royal favor. It gave monarchs what political mavens today call leverage. Young nobles at court received stipends, called *moradías,* which bound them closely in reliance on the crown. Amounts were assigned roughly in proportion to what royal officials thought the stipendiary's family's support was worth. The beneficiaries ate in the king's company, arrayed in order of rank in parallel tables, like diners in an old-fashioned school or college refectory today. As partakers in the king's bread and salt they incurred a symbolic duty of deference. They also received arms and apparel worth more than their stipends. Financial dependence made their bonds of service hard to break.

. . .

Magellan entered this world, probably just as Manuel began to expand the household. What did the future explorer do in the undocumented period of up to a decade or so between his arrival at court and his departure for the East? Hints of what it was like to grow up in King Manuel's entourage survive in the sources. Education at court was more fastidious than ever before—even than under João II, whose humanist table talk so impressed Hieronymus Münzer. Greek lessons were available. The schools of Lisbon self-transformed, thanks to the alchemy of royal gold, into what could respectably be called a university. In 1500 Manuel I made the payment of young nobles' *moradías* dependent on their success in an examination in grammar.[22] Letters were beginning to rival arms as aristocratic accomplishments.[23]

Emphasis on upbringing for war, however, was irreplaceable. Battlefields still needed knights, equipped with costly appurtenances—horses, weapons, armor—and prepared in protracted training for a highly specialized form of combat. The aim of noble youths—*moços fidalgos,* such as Magellan—in the royal household was to proceed to the rank of squire and take part in warfare, which would open, for those sufficiently skilled and successful, chances of elevation to knighthood. For those well enough qualified in letters, membership of the royal council beckoned as what one might nowadays describe as a retirement option.

Magellan learned chivalric principles along with the exercise of arms. Chivalry was more characteristic than humanism in the values and education of most members of the elite. It could not, perhaps, make men good, as it was supposed to do. It could, however, win wars. In 1492, for instance, the monarchs of the Spanish kingdom of Castile extended the frontier of Christendom by conquering Granada, the last Muslim kingdom in Spain, in what the Venetian ambassador called "a beautiful war. . . . There was not a lord present who was not enamored of some lady," who "often handed warriors their weapons . . . with a request that they show their love by their deeds." In 1504 Queen Isabella II of Castile died uttering prayers to the Archangel Michael as "prince of the chivalry of angels."[24]

Chivalry was the aristocratic ethos of the day.[25] It had been gaining familiarity and currency for four centuries, ever since priests and members of the warrior class combined to formulate it in the twelfth century, when crusades turned violence into a source of sanctification for participants who fought for ground hallowed by the footsteps and relics of Christ and the apostles. For knights, fighting was a full-time vocation. It required training that left little time for anything else and commitment in proportion to the incidence of warfare. The privileged access to heaven that monks and nuns enjoyed seemed inaccessible to laymen whose job—to put it crudely, but not excessively so—was to kill people. Chivalry was a means of outflanking the obstacles to heaven that a military life interposed.

Those who followed its "code" imitated religious life in secular occupations. Like religious, they took vows, preceded by vigils. They practiced, in theory, virtues that exceeded the traditional seven works of mercy in the demands they made: material largesse, generosity of spirit, reproachless courage. They could love outside marriage, as long as they did so chastely; much tragic chivalric literature concerns flirtations that got out of hand, like Lancelot's with Guinevere. Many knights belonged to "orders" of chivalry that sometimes, like religious communities, required elements of common life or diverted to the orders concerned the wealth required to support a knight, who managed it on the organization's behalf, rather than retaining it as a burden of the kind that kept a rich man from entering the Kingdom of Heaven. In the only order to have survived to the present day from the early crusading era, that of the Knights of St. John, the obligation to care for the poor and sick persisted even when most of the manpower was diverted to the desperate effort to keep infidels at bay on the frontiers of Christendom.

The knights' saints were holy warriors, like St. Michael the Archangel, whom the prayer of the church still hails as "princeps militiae caelestis"—chief of the chivalry of heaven—in war against Satan, or St. James "the Moor-slayer," whose battlefield apparitions started, allegedly, in the tenth century and continued during the wars of Christian invaders of new worlds in the sixteenth. If the aim of chivalry was to make sinners good, it can hardly be said to have succeeded: medieval wars were as disfigured by ruthlessness, cruelty, butchery, and massacre as those of every period. At least, however, the notion prevailed that war was potentially hallowing, as well as a proper occupation in which a nobleman could take part without social slippage.[26]

The same notion shaped naval warfare. Ships dueled with each other. Their fighting crews closed in mêlées that individual prowess might resolve. No text better illustrates the influence of chivalric traditions at sea than the chronicle of the deeds of Count Pero Niño, written by his standard-bearer in the second quarter of the fifteenth century. A treatise of chivalry, as well as an account of campaigns, *El victorial* celebrates a knight never vanquished in joust or war or love, who fought his greatest battles at sea—and "to win a battle is the greatest good and the greatest glory of life." When the author discoursed on the mutability of life, his interlocutors were Fortune and the Wind, whose "mother" was the sea, "and therein is my chief office."²⁷ A maritime milieu had an important advantage for the teller of chivalric tales: where rapid cycles of storm and calm succeed, the wheel of Fortune revolves most briskly. The sea is God's own element; the wind bloweth where it listeth, and the rover's awareness of dependence on God is ineluctable. Magellan knew that and would recall it shortly before he died (below, p. 256).

There was something Homeric in the descriptions of seaborne battles of the time, in which ships contended like participants in single combat. In the Middle Ages most lay people, if they could read, knew Homer's tales only through romantic versions, in which Greeks and Trojans were recast as knights in the equivalent of airport-bookstall pulp fiction.

As chivalry infused seafaring, it made maritime service attractive for more than prize money. The sea became a field fit for kings. In the thirteenth century, one of the great spokesmen for the chivalric ethos in the Iberian peninsula was Jaume I, king of Aragon and Count of Barcelona. When he described his conquest of Majorca in 1229 he revealed that he saw maritime war as a means of chivalric adventure par excellence. There was "more honor" in conquering a single kingdom "in the midst of the sea, where God has been pleased to put it," than three on dry land. A metaphor quickly established itself, which was to be a commonplace for the rest of the age of sail. A ship, in the words of King Alfonso X "the Wise" of Castile, was "the horse of them that fight by sea." St. Louis planned to create the Order of the Ship for participants in his Tunis crusade. The Order of the Dragon, instituted by the Count of Foix in the early fifteenth century, honored members who fought at sea with emerald insignia.²⁸ Pirates and cutthroats composed the entourage of the Portuguese prince Dom Henrique; they staffed his campaigns to win a kingdom in the African Atlantic, colonized the Madeira Islands and parts of the Azores, and explored the coast of West Africa as far as Sierra Leone. They called themselves knights and squires and gave themselves storybook names, such as Lancelot and Tristram. Henrique himself—traditionally misrepresented as "the Navigator" motivated by scientific curiosity—imagined himself a romantic hero, destined by his horoscope to perform great deeds and win a kingdom of his own. The truth is that he never went exploring, and his desperate efforts to make enough money to pay his retainers included slave raiding and a soap monopoly.²⁹ His followers included the father-in-law of Christopher Columbus, a weaver by training who reinvented himself as a "captain of cavaliers and conquests" and who took to exploration to escape the restricted social opportunities of home.[30]

By the time of Columbus, the Portuguese poet Gil Vicente, in a poem he wrote in Spanish, could liken a ship at once to a warhorse and to a lovely woman without incongruity, for all three were almost equipollent images in the chivalric tradition.

> Tell me, tar,
> who dwell at sea,
> Can ship or sail or star
> be fair as she?
> Or you, knight, say,
> Who armor wear,
> If horse or arms or fray
> Can be as fair?[31]

The resemblance seemed strong to various authors from the thirteenth-century royal bard, Alfonso the Wise, onwards.[32] If you look at late medieval pictures of fighting ships, caparisoned with pennants as gaily as any warhorse, you can grasp how, in imaginations of the era, the sea could be a knightly battlefield and the waves ridden like jennets.

. . .

Fictional reading supplemented aspirants' image of knightly life. The typical chivalric hero in books of the time took to the sea, conquered an island, married a princess, and became a ruler. Explorers—often men of humble social origins—tried to embody these fictions in real life. Strange as it may seem to those who know Arthurian legends only from Victorian and later versions, the Arthur cycle—the most influential of all knightly tales—included many seaborne adventures, albeit in texts that have been lost. The most widely read, to judge from surviving allusions, was the *Gesta Arthuri*, a romance about the conquests planned around the round table.[33] Arthur's realm, as a late sixteenth-century reader put it, was "too small for him"; so he embarked on conquests in Iceland, Greenland, Norway, Lapland, Russia, "and the Arctic Pole."[34]

A fourteenth-century romance, *The History of Melosina,* which Columbus's son bought for his famous library, illustrates a typical adaptation of chivalric narrative to a seaborne setting. It tells of the adventures of Melosina's sons as they seek conquests in Cyprus and Ireland. One of the boys perfectly expresses the vocation of an explorer when he tells his mother,

Lady, if you please, it seems that the time has come for us to undertake a voyage so that we may learn of strange lands, countries, and places in order to achieve honor and good renown in far-off frontiers. . . . So shall we learn of the various things that there are in distant lands and how they differ from what is commonly found here. And then, if fortune or good luck befriend us, we shall surely want to conquer lands and countries.[35]

Taking leave of her daring offspring, Melosina gives them leave to "do as you wish and see to whatever is to your honor and advantage." She advises them to follow all the rules of knightly conduct, adding some points that seem to anticipate the imperialism of the sixteenth century:

And if God so favors you that you are able to conquer lands, govern yourselves and your subjects according to each one's nature and rank, and if you see that any rebel against you, take care to humble them and act as their lords. Do not let any of your rights of lordship lapse. . . . Take dues and taxes from your subjects without charging more, save for a just cause.[36]

In one aspect, at least, Melosina's counsel went unheeded. "Do not," she said, "make report unless reasonable and true." Writers of books of romance filled them with fables, marvels, and magic. So did the narrators of explorers' adventures. Among stories popular in Magellan's day, the hero of *Huon of Bordeaux* sails in a ship that bounds like a warhorse to the "Gulf of Perils" and the "Isle of the Castle of Adamant." The protagonist of *The Book of Count Partinuplés* boards an enchanted ship that takes him to a castle where he enjoys the embraces of a Byzantine empress. Oliverio de Castilla's fictional adventures happen in Ireland—a disappointingly unromantic destination, but accessible only by sea. In the *History of the Love of Paris for Viana,* the lover sails to India and learns falconry. In a Spanish version of the romance of Alexander, the Macedonian makes his journey to India by sea.[37] In the best-known of the knightly fictions—*Tirant lo blanc,* the world's best book, according to Don Quixote—a king of the Canary Islands launches an invasion of England.[38] The usual plotline of such books inspired one of Cervantes's jokes: Sancho Panza asking his master to make him "governor of some island, with maybe a bit of the heaven above it."[39]

The chivalric tradition blended at times with hagiography. After all, to complete Christ's command to evangelize the corners of the world, saints had to travel. St Paul's seaborne adventures were models. Legends of respectable antiquity attributed comparable journeys to St. Thomas and St. Mary of Magdala. Some saints' lives were indistinguishable from tales of knight errantry. St. Eustace, for instance, underwent terrible trials at sea, searching for his lost family. The *History of Sir Placidas, Who Later Became a Christian and Took the Name Eustace* was an explicitly secular rewrite, as was the *Book of Zifar, the Knight,* which omitted the saint's conversion and martyrdom.

. . .

We should not expect abundant evidence that Magellan read such stuff. But he did. Unlike Columbus and Vespucci, who wore their hearts on their sleeves and poured self-revelations through the nibs of their quills, Magellan has left very little work in his own words. References in the sources to anything he may have read is sparse. Unlike Columbus, he left no annotated library of works from his own shelves. His unique reference to work he read is, however, revelatory. As we shall see, he recalled the *Books of Primaleon* on his great voyage, when he was coasting Patagonia (below, p. 182). Primaleon, in a sixteenth-century translation into English, was a paragon of every knightly virtue prized at King Manuel's court, who

> in his yongest yeares exercised himselfe so well in vertuous Disciplines (whereof according to his verie naturall inclination and desire, he declared himselfe a studious louer) as one might iustly name him the true miracle of perfection: So among other endowments, for the spirite of wisedome he might be compared to *Salomon,* for beautie both exteriour and interiour, to the gentle Greeke *Alcibiades,* and for magnanimitie, councell, millitarie strategemes and such like, to valiant *Scipio* the *African,* or the subtill *Hanniball* of *Carthage,* especiallie in this time of his youth, when he had not receiued his order of Knighthood.[40]

The fictional life laid down part of the course of Magellan's real one. Primaleon, like Magellan, experienced marginalization—not, like Magellan, as a result of being an orphan but because of the illegitimate birth of his father. Both protagonists suffered accusations of treason. Both endured long exiles at sea, battled (as, in Magellan's case, we shall see) giants, and aspired to vindicate themselves by seaborne adventures. Both were knighted. Both declined help in combat. Primaleon's life was dominated by his rivalry with an English prince and his struggles to win the hand of a bride whose mother had declared implacable enmity, with an oath that none should marry her daughter but the man who brought her Primaleon's head. Magellan could not parallel those episodes of his hero's trajectory, but his perusal of the romance is certain and its influence on him inferable. No one who read *Primaleon*—which was hot off the press in 1509, shortly after Magellan graduated, as it were, from his formation in the Portuguese court—would have done so without reading other chivalric fictions. The work was part of a series that opened with the adventures of Primaleon's father and included numerous cross-references to other romances.

Alongside chivalry, millenarian fantasies may have influenced overseas expansion and colored Magellan's outlook. Manuel's imagination was alive with images of messianic kingship and expectations of the end of the world. Ever since the twelfth century, anticipations of an imminent "age of the Holy Spirit," preceded by the cosmic struggle of a "last world emperor" against Antichrist, had dominated the prophetic tradition in western Europe.[41] The first king of Portugal's ruling dynasty was actually called "Messiah of Portugal." Columbus and other courtiers flattered Fernando, king of Aragon, that God might have cast him for the starring role in precipitating the millennium, with a mission to lead a crusade, capture Jerusalem, and lay Mecca waste. Columbus claimed that the profits of his discoveries could be used to conquer Jerusalem and help complete God's plans for a new age. Franciscan friars who supported Columbus believed that an "Age of the Holy Spirit," which would precede the final rapture, was coming soon, and some of them came to see the New World as the place where such an age might begin.[42] King Manuel was equally susceptible. "God wishes to choose your Highness," urged one promoter of Oriental exploration, "to subject and receive and reduce to tribute the barbarous kings and princes of the Orient."[43] Images of ancient heroes from Alexander to Solomon were invoked to urge the same program. The Augustinian superior Fr. Gil de Viterbo saw Manuel as "the messianic king who shall overthrow Islam, reform the church, and establish a universal empire of peace."[44]

When the reign began, doubt about how best to fulfill the king's destiny divided factions at court between proponents, on the one hand, of a policy of extending the reconquest of Iberia into North Africa, and on the other of seeking to break into the commercial world of the Indian Ocean. Islam could then be assailed, as it were, from the rear with the advantage of enormously enhanced resources, and perhaps of alliances with the Christians who were known to live on the shores of the Indian Ocean, but with whom contacts were fragile or severed: Vasco da Gama, when he reached India, announced that he was looking for "Christians and spices."[45] While Christian allies were not forthcoming, and although Manuel never lost sight of objectives in Morocco, success in taking a share of India's pepper trade gradually drove the king's ambitions eastwards. In a map of 1516, Martin Waldseemüller, the great cartographer of St Dié in Lorraine, showed him carrying the flag of Portugal on a whale's back into the Indian Ocean.

. . .

Magellan's dispatch eastwards was part of the king's strategy. The training of the postulant for knighthood was over. His practical education now began. He was neither personable nor imposing. "His person," according to the only observer who left a physical description of him, "did not carry much authority, since he was of small stature and did not look as if he would amount to much, so that people thought they could exploit him for his want of prudence and courage."[46] The appearance was deceptive: as events would show, he rarely displayed prudence but never lacked courage. Gonzalo Fernández de Oviedo— theorist of chivalry and chronicler of Magellan's career—detected the ambition that shaped Magellan's life from this point or soon after: "He thought of becoming a great lord"[47]—the same ambition that Columbus's crewmen attributed to their leader.

Magellan sailed with Francisco de Almeida, leaving Portugal on March 25, 1505. It was Portugal's most ambitious venture yet: over twenty ships, including at least seven merchant vessels, a thousand crewmen, an army about 1,500 men strong, and a fourfold mission: to show the flag; to cow rivals and opponents; to secure bases for sharing and if possible controlling trade; and to adumbrate, if not establish, an empire—a framework of outposts under Almeida in charge.

Hans Mayr, who was aboard to represent German commercial interests, wrote a swashbuckling account of the outward voyage that makes it sound like a piratical expedition along the East African coast, with casual descriptions of banner-hoisting ceremonies to the cry of "Portugal! Portugal!" preceding massacres and frank rapine.[48] Previous Portuguese expeditions had disturbed and divided the elites of the Swahili coast—some seeing the newcomers as a threat, others appraising them as potential allies and arbiters. Almeida arrived determined to take advantage of the divisions and coerce cooperation. He or his subordinates imposed a new ruler and Portuguese garrisons on Kilwa and Sofala. At Mombasa, gunfire greeted the Portuguese, and a Christian renegade resident warned them that "they would not find people with hearts that could be eaten like chickens as they had done in Kilwa." Almeida responded by looting and burning the town. Malindi, awed, perhaps, by Almeida's violence, opted to collaborate.

On arrival off the Indian coast, the Portuguese commander tried the same shock tactics. Why he made a show of force that he would be unable to sustain is baffling. If his intention was to intimidate Indian merchant-states, he was only partly successful. Cannanore, where Almeida proclaimed himself viceroy, was biddable, at first, if unreliable. So was Cochin. Calicut was intractable. The outcomes were not the result of Almeida's bullying but reflections of local politics: the elites of the cities concerned hated each other more than they feared or resented the Portuguese. At Calicut, said Mayr, Almeida "did zilch" (fizeram nichil).[49] But some action seems to have taken place: it was perhaps here that Magellan acquired a limp; a few years later a collector of local tales, Gaspar de Correia, claimed that the future explorer was badly wounded in an otherwise unrecorded action.[50]

In 1506 Magellan returned to the Swahili coast on an expedition to reinforce the garrisons Almeida had left and to arbitrate a succession dispute at Kilwa.[51] No data survive on his role, but according to more of the stories that Gaspar de Correia picked up, Magellan took part in Almeida's most celebrated campaign: the battle of Diu in February 1509—a spectacular Portuguese success against a numerically more impressive fleet. Egyptian ships, sent to succor coreligionists, had combined with the navies of Calicut and Gujarat to put the Portuguese in their place; but Malik Ayaz (above, p. 19) managed to shift the odds in Portugal's favor by impeding the Gujarati fleet.[52] Almeida's victory helped induce Malik Ayaz at Diu to become an ally and confirmed Portugal's presence as an established part of life on the western coast of the subcontinent.

. . .

Magellan was back in Cochin in July of that year, when a surviving document includes his receipt for part of his salary, compounded in wheat.[53] In or about the same year, his friendship with a shipmate, Francisco Serrão, began—or began to be noticed—on a cruise from Cochin, perhaps to investigate inadequately explored reaches of the ocean and look for reputed gold along the coast of Sumatra. From there the flotilla was to proceed to Malacca to negotiate a treaty that would allow Portuguese ships to trade. It was Magellan's first taste of exploration: no Portuguese ship had yet got that far east. In Malacca, according to the unreliable consensus of early historians, unconfirmed by surviving records, he detected, or at least reported, some Malay deckhands' plot to seize the ships. In the ensuing fracas, he saved Serrão's life.

The story sounds like a romanticization, but Serrão's importance is undeniable. His provenance is unknown, but his future, as we shall see, lay in the Moluccas, as the islands' first Portuguese resident, the confidant of the sultan of Ternate, a promoter and handler of trade in cloves, and a source of information on the wealth of the islands that helped to ignite or fan rivalry between Spain and Portugal. The official line of Portuguese court historiography in the mid-sixteenth century came to condemn Magellan's friendship with Serrão as "a cause of great harm to our country" (sucedeo muito damno a este Reyno) on the grounds that the latter encouraged the former's concupiscence.[54]

By the time the Malacca reconnaissance got back to the Malabar coast, Almeida had reluctantly yielded power to his appointed successor as governor (from 1515, viceroy), Afonso de Albuquerque. The new incumbent encouraged or commanded the departure of leading figures in the prior regime. Magellan hardly seems important enough to count in that category, but he took advantage of the opportunity to try to cash in on the profits of his service so far and, like a typical conquistador with a lucky strike to his credit, to head home to Portugal to enjoy the proceeds.

Surviving documents about his business activities in India date only from October 1510, when he described himself as "a man who trades in this city [Cochin] in partnership with merchants to whom he gives money at lawful interest by mutual agreement and they generally agree with him to pay him half the profit of the deal at the rate of 10 percent per annum."[55] It is unlikely that he spent his first four or five years in the Indian Ocean without exploiting the commercial prospects that were available, without dishonor, to the servants of the king of Portugal. The records of his dabblings in trade were part of court proceedings he launched in Cochin and Lisbon, demanding a return on his investment in two cargoes of pepper, one of which he was to accompany back to Portugal. The inference that the cargo in question was the motive of his homeward voyage of 1510 is hard to resist. For whatever reason, shortly after Almeida left in December 1509, Magellan was aboard a homebound ship from Cochin. It went aground on shoals off the Maldives.

According to one of the most famous anecdotes about him, Magellan took charge of the wrecked ship, while a boat he sent for help struggled to Cannanore with all the other officers and gentlemen aboard. Having exacted his departing comrades' promise to send relief as soon as possible, he remained on the shoals to quell the ordinary seamen's fears that they were being abandoned.[56] He managed rations and nerves with equal acuity for several weeks before rescuers arrived. The story is unverifiable but not unbelievable, since João de Barros—a chronicler whose hostility to Magellan will become increasingly obvious as our story unfolds (below, pp. 77–8)—endorsed it. "If only," he remarked, Magellan "had shown equal loyalty to his king and country."[57] The feat seems to have recalled the explorer to his duty, reconciled him to the new governor, and obliged him to revert to military service. Rather than resume his journey home, he returned to India: presumably the loss of the cargo in which his capital and hopes were invested left him no choice. The fact that the governor presented him with a secondhand suit of armor in September 1510, just ahead of the next major campaign, suggests that Magellan had lost equipment, as well as cargo, in the disaster on the shoals.[58]

The following month he took part in a conference the governor summoned on a proposal to take over a port on the Malabar coast. Cochin and Cannanore had been serviceable refuges so far for Portuguese ships and personnel, but a stronghold wholly under Portuguese control would be preferable. Goa was the target. It is not clear why; but it was less defensible than Calicut, where Albuquerque was repulsed and wounded, and had many of the same natural advantages for potential development as a globally important emporium. The first attack failed; but the assailants seem to have been encouraged by what they saw. Goa looked ripe for plucking. The city and hinterland had changed hands between increasingly avaricious rulers four times in the previous half century. Taxes—paid mainly by Hindu subjects to Muslim masters—were punitive, charged directly on the produce of peasants. Expropriations of land were common in favor of the Turkish warriors the sultans employed.[59] Most of the population would probably welcome a further change of regime. In the event, Albuquerque "pressed into service all from local dancing girls and musicians to war elephants and mercenaries," while Hindus collaborated by cutting off retreat or support for the defenders to landward.[60]

Magellan went unmentioned—as it were—in dispatches: unlisted, that is, among the noblemen Albuquerque singled out for commendation in his reports home. In the council meeting that preceded the campaign he argued against the governor but with the majority of captains that merchant ships should be omitted from the attacking fleet so as not to disrupt commerce—perhaps because investments of his own were at stake.[61] The governor had to accept his officers' advice, but not to approve of those who gave it. The success of the campaign transformed Portuguese prospects. For Tomé Pires, the official apothecary of the Portuguese Empire, Goa was or became the "most important kingdom in India." Its conquest was a sign that God was preparing the final rout of Islam and the end of the world.[62]

Albuquerque's focus shifted to Malacca, where "more ships arrive than at any other place in the world."[63] The sultan had rebuffed previous Portuguese overtures. Where diplomacy had failed, coercion was worth trying. Tomé Pires, who lived and worked in Malacca for a while after it fell into Portuguese hands, knew exactly why the port mattered. "The smallest merchandise here is gold," he observed, but it was not merely wealth that made Malacca "more prized than the world of the Indies." It was the strategic hub of the Orient. It commanded the strait through which all the nutmeg, mace, and cloves of the Spice Islands passed on the way west. Venice's traditional privileges in the spice trade, already eroded, could be further curtailed to Portugal's advantage. "Whoever is lord of Malacca," Pires continued, "has his hand on the throat of Venice. . . . Who understands this will favour Malacca; let it not be forgotten, for in Malacca they prize garlic and onions," dirt cheap in Portugal, "more than musk, benzoin, and other precious things."

So diverse a throng of merchants gathered there, each with the right to be judged according to his own law, that "a Solomon was needed to govern Malacca, and it deserves one."[64] The incumbent ruler was no Solomon. The siege took only six weeks. Malacca fell in mid-August 1511. Magellan and Francisco Serrão were among the besieging force. Pires became, in a sense, a victim of Portugal's success. Chosen to be the Portuguese ambassador, he went to China—the richest and superficially the most attractive of the destinations that now came within Albuquerque's purview, which the capture of Malacca extended eastwards. In China, however, Portuguese behavior in seizing a former Chinese protectorate seemed a typical piece of barbarian impertinence, for which Pires and his companions were tortured and imprisoned, never to return home.

. . .

As usual up to this point in his life, Magellan was not important enough for his deeds to attract detailed attention. In Malacca, however, his life can be said to have entered a new phase, focused on destinations further east. Albuquerque's policy, like Almeida's, seems in retrospect riskily aggressive. If it had been relatively easy for him to seize Goa and Malacca, it would be even easier to lose them if indigenous powers chose to eject the intruders. He hoped to fulfill his king's chiliastic fantasies by turning west, attacking the heartlands of Islam, and threatening Mecca. His professed justification for taking Malacca was not to enrich Portugal but to strangle Muslim trade. "I am very sure," he wrote, "that, if this Malacca trade is taken out of their hands, Cairo and Mecca will be completely lost."[65] He had already tried to bottle up the Persian Gulf and Red Sea by establishing forts in the narrows that led to the Indian Ocean. In 1513 he would burst into the Red Sea with a fleet, boasting that he would burn Mecca, seize the Prophet's remains, and hold them to ransom. He kept up remorseless war against Calicut until he finally forced the sultan off his throne, completing Portuguese hegemony over most of the west coast of India. It seems a wildly ambitious program for a small number of Portuguese—of whom no more than a thousand or so could be in arms at any one time—with a modest fleet, half a world from home, tenuous supply lines, poor prospects for reinforcement, and a vast ocean to police.

In his underlying policy, however, Albuquerque appreciated his own weakness. East of Malacca, wherever valuable trade was available, and wherever he thought he could do business without resorting to banditry or blackmail, he sought to establish peaceful relations and offer good terms to native merchants to use Portuguese shipping and ports. The possession of Goa and Malacca gave the policy a chance of success. In some ways, the fact that Portugal obviously lacked resources to reach beyond Malacca made Albuquerque seem an unthreatening partner to most powers beyond the strait. Overtures to China failed. In parts of Sumatra and Indochina, however, Portuguese emissaries were well received.

Offshore, the most inviting destination for the governor's diplomacy was the Moluccas, where production of the world's most valuable spices was concentrated (above, p. 21). Late in 1511, probably in November, Albuquerque took advantage of a trading venture there by Nehodá Ismael of Malacca, in a junk laden with goods of Malay and Javanese merchants to trade for cloves in the islands that produced them and for nutmeg and mace in Banda. The governor sent an envoy with a flotilla in which Francisco Serrão commanded a ship. The mission would be "well received," Barros suggested, "because our name was fearsome among those peoples."[66] But difficulties bedeviled the voyage. Historians have expended much inconclusive energy on the problem of what, if any, was Magellan's role in the expedition. Early chroniclers, whose lives overlapped with Magellan's and who, in at least one case, probably knew him personally, knew or assumed that he was aboard.[67] Abreu himself spoke of Magellan as "able to give testimony about the whereabouts of the islands" (fôra testemunha, e tendo certeza onde aquelas ilhas jaziam) if he had formed part of the expedition.[68]

Contradictory reports make events hard to reconstruct. Malay and Javanese pilots led the expedition along the northeast coast of Java, hauling Serrão's ship from shoals off Sapudi on the way. The weather was so bad that they wintered at Guliguli, where Serrão's ship had to be scuttled because "she was already old" or perhaps because the grounding had damaged the hull.[69] At Amboina they acquired a junk to replace Serrão's vessel and loaded so much nutmeg, plus cloves imported from the Moluccas, that they decided to turn back for Malacca without completing the proposed itinerary, especially as the weather-beaten ships all needed repair.

A storm separated Serrão from the rest of his fleet and drove his junk ashore on Luca Pino, according to one account, in the Tartarugas islands, about thirty-three leagues—an unstandardized measure, equivalent to between three and four land miles—from Banda. A rival narrative places the castaways in Mindanao. This seems unlikely but is worth mentioning with subsequent events in mind (below, pp. 93, 222) because it would give Serrão, and therefore his friend and correspondent, Magellan, access to direct knowledge of the Philippines—the islands which, as we shall see (below, p. 94), came to dominate his plans for his future and to which his ambitions, toward the end of life, would be mysteriously diverted.[70]

Without food or water, prospects were bleak, especially when wreckers arrived by ship to strip the stricken vessel. But Serrão turned the tables on the pirates, captured their ship, and returned to Amboina, where the locals at his point of landfall immediately recruited him to help fight their neighbors from Veranula, who were also, in the manner of neighbors, traditional enemies. They rewarded his part in their victory with a wooden house for him to inhabit, which he decorated with the arms of Portugal and the cross of Christ. His renown, supplemented by Nehodá's recommendation, drew the attention of the ruler of Ternate, Bayan Sirullah (or Cachil Boleif, as the Portuguese called him). Portuguese sources claimed that prophecy influenced the chief. Having heard that men of iron would come "from very distant parts of the world," the potentate "raised his arms, giving thanks to God for having shown him, before his death, these men of iron, in whose strength was the security of his kingdom."[71] Such stories were topoi of conquistador literature, serving to justify intrusions or invasions of other peoples' countries. Even so, Serrão seems genuinely to have become a court favorite in Ternate, where he opened commerce with Malacca and proposed the erection of a Portuguese fort.

Serrão arrived in Ternate in 1512 and stayed till perhaps 1521, when he died. Letters he sent to Malacca to officials and friends, including Magellan, have not survived, but early references and abstracts make it clear that they extolled the islands and recommended active Portuguese participation in the production of cloves or at least in the purchase of them directly from the growers.[72] His reports, though greatly overstating the islands' attractions, gave his compatriots in Malacca a new and clearer idea of what the Moluccas were like. Barros was probably summarizing Serrão's letters when he updated the information in Ptolemy's *Geography*—the second-century Alexandrian text that was still Westerners' starting point for picturing the world. Ptolemy, Barros remarked, knew little of what lay beyond Sumatra, but "we know thanks to our explorations that the land there is shattered into many thousands of islands, covering a great part of the circumference of the world . . . and that in the midst of that great number of islands are those we call Maluco."[73] He noted that the Moluccas lay under or around and close to the equator.

Magellan had access to similar data. The questions of how and how much he knew of the Moluccas, and when he came to know it, are of enormous importance for understanding the circumstances in which the idea of promoting an expedition to the archipelago arose in his mind. They are not, however, easy to answer. Barros, who had access to copies of Serrão's letters, was unequivocal: Magellan, as the writer's friend, was among the recipients, indeed, the principal addressee. Barros repeatedly and insistently stressed how greatly Magellan prized Serrão's friendship "from the time they served together in India, especially in the taking of Malacca." What was worse, from the point of view of the Portuguese court, was that Serrão sought "to make his feat seem more admirable, exaggerating the manner and hardship of it, and doubling the distance to the islands." Serrão gave his readers to understand "that he had discovered another world greater, more distant, and richer than that which Admiral Vasco da Gama had discovered" and "amplified his account with so many words and mysteries, representing the Moluccas as so far from Malacca as to augment his gallantry in the eyes of King Manuel, inasmuch as those letters would seem to have traveled farther than from the antipodes or from that other New World."[74] By the time the humanist historian Francisco López de Gómara wrote his compendium of the history of Spanish explorations and conquest in 1552, the notion that Serrão's letters attracted Magellan had become entrenched. He referred to "a letter from Francisco Serrano [*sic*], written in the Moluccas, in which he asked him to go there if he wanted to get rich quick."[75]

An unintended consequence of Serrão's exaggerations was to make it seem that the islands were accessible from Spain's American possessions. The farther they were from Malacca, the nearer they would be to America and to access from the Atlantic. As a result, "That same Fernão de Magalhães began to formulate new notions, which would lead him to his death and this kingdom to a certain disadvantage."[76] In other words, Serrão's distortions put the idea of trying for the Moluccas via the Atlantic—and therefore, by implication, in Spanish service and disloyalty to his home country—into Magellan's mind.

No doubt Serrão did exaggerate, but so, I suspect, did Barros. Magellan had other sources of information about the Moluccas: the large number of traders in Malacca who did business with the islands; the data Tomé Pires deployed in the account that we have already summarized (above, p. 47), which he wrote between 1512 and 1515 in Cochin and Malacca, where Magellan was a fellow resident until 1513; and perhaps further insights from Magellan's slaves, including the much-celebrated and occasionally reviled Enrique, whom he described as "a mulatto, native of Malacca." Enrique's provenance is not known. Pigafetta thought he was from Sumatra, while others, perhaps because of the easily confused place-names, traced his origin to the Moluccas.[77] But he was expected to be useful on the great voyage, ostensibly to the Moluccas, in 1519. In a last will and testament, just before they sailed, Magellan promised Enrique freedom and a substantial legacy of 30,000 maravedíes "because he is a Christian and that he may pray for my soul."[78]

It seems, in any case, doubtful whether Magellan heard from Serrão before leaving the Indies.[79] It is impossible to track and trace all the letters. The first reached Malacca sometime in 1513, but in January of that year Magellan had left for Cochin. As we know from the declaration of Jorge de Albuquerque, the messenger who carried the correspondence from Amboina, where he had an interview with Serrão, other letters arrived only in 1514, when Magellan had already left India for Portugal.[80]

Whatever part Serrão's missives played, questions arise about what Magellan may have learned of cosmography and current affairs—and therefore of the whereabouts and disputed status of the Moluccas—in his boyhood and youth. The origins of the idea of reaching the Moluccas via the Atlantic lay in a broad confluence of influences that came together

MAP 1 *(next spread)*. Magellan in Iberia.

1. **Oporto**—probable birthplace, perhaps in late 1480s; owned a small property nearby; probably also owned a property at Sabrosa, 90 km inland; his sister and brother-in-law also lived there

2. **Lisbon**—educated at the court of Manuel I, perhaps from about 1495; leaves for the Indian Ocean, March 1505; visits again 1515 and 1516, when starts lawsuit over proceeds of investments in pepper, and for the last time in 1517, when, on April 15, the king rejects his application for an increased stipend

3. **Burgos**—home city of the Haro brothers; hub of merchants with interests in the spice market of Lisbon and of backers of Magellan at the Castilian court

4. **Valladolid**—interview with Grand Chancellor, March 1518; contract with crown for a voyage to the Moluccas signed March 22

5. **Medina del Campo**—visited on the way to Valladolid, Janu-ary-February 1518

6. **Zarauz**—shipyard where *Victoria* was built

7. **Málaga, Cádiz, Huelva**—recruitment for the fleet, 1519

8. **Puente del Duero**—negotiates deal with Juan de Aranda, January 1518

9. **Seville**—place of self-exile from Portugal, October 1517; marries late in 1517; assembles fleet, February 1518- August 1519; dictates will August 24, 1519, having shifted the fleet to Sanlúcar

10. **Sanlúcar de Barrameda**—sets sail, bound ostensibly for the Moluccas, September 20, 1519

11. **Barcelona**—visits May 1519; makes depositions relating to dispute with Juan de Aranda

12. **Bilbao**—Magellan's father-in-law buys gunpowder for Magelan's fleet, 1519

13. **Triana**—takes oath of loyalty to Castile, August 1519

14. **Ciudad Rodrigo**—place of refuge of Magellan's runaway slave, Antonio, 1519

15. **Morocco**—campaigns for Portugal, 1513-17; perhaps wounded; exonerated of charges of peculation

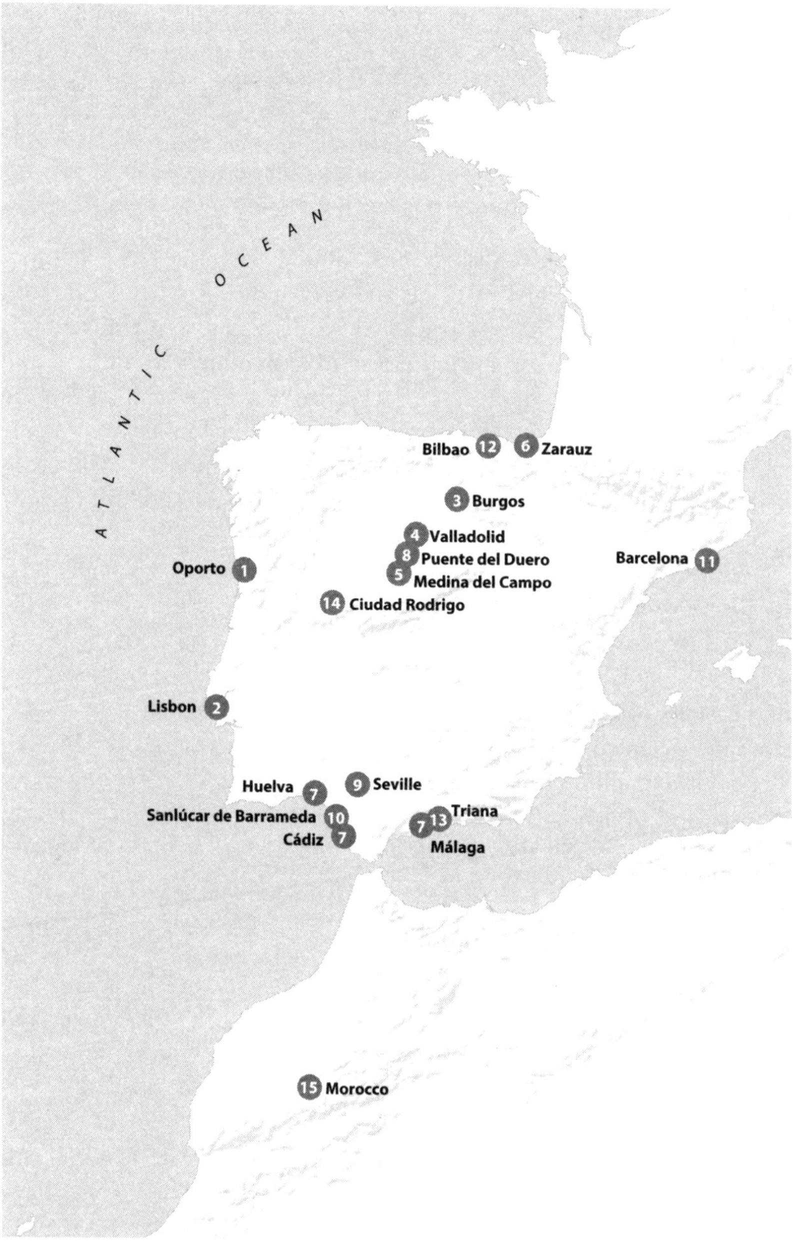

ATLANTIC OCEAN

Bilbao **12** **6** Zarauz

3 Burgos

4 Valladolid
8 Puente del Duero
Oporto **1** **5** Medina del Campo Barcelona **11**
14 Ciudad Rodrigo

Lisbon **2**

Huelva **7** **9** Seville
Sanlúcar de Barrameda **10** **7** **13** Triana
Cádiz **7** Málaga

15 Morocco

at court in Lisbon during his time there. They included geographical lore about the size of the globe, diplomatic disputes about the extent of Spain's and Portugal's proper spheres of navigation, and the personal ambitions of more than one man. The story of how these circumstances combined now claims our attention.

. . .

The story we have to tell takes us back to the Portuguese court in 1493, before Magellan's arrival there, at the very moment when Columbus brought the news of the discoveries he made on his first transatlantic voyage. If we can trust the information of Columbus's first editor and early biographer—the equivalent of what we might now call his literary executor—King João II reacted with immediate cupidity, expressing "annoyance over the loss of such valuable opportunities."[81] The problem the king now had to face was of how to divide the spoils of oceanic exploration with his neighbor in Spain.

Portuguese and Spanish rulers were familiar with the chore of allotting, by agreement, rights of reconquest of Iberian territories supposedly usurped by Muslim intruders from the last Visigothic rulers of the peninsula. From the Treaty of Tudilén of 1151 through periodic revisions, it was normal for Christian Iberian kingdoms to limit quarrels over their respective spheres of expansion by defining potential victims among the Muslim kingdoms and allocating them accordingly. The precedent was imperfect for purposes of dividing the New World, because—save in unconvincing fantasies about ancient Iberian dominion in America, such as one chronicler's claim that the New World had belonged to the mythical King Hesperus or the nonexistent hero-blacksmith Tubal Cain or Hercules, whose legacy descended to the kings of Spain[82]—rights of conquest based on the claim to recover usurped territory could not extend across the ocean. One principle, however, applied in the ocean as in the peninsula. Spain and Portugal could agree on a line of demarcation, of the kind traditional in the Reconquista, between their respective zones of expansion.

The size of the world was known with some accuracy, thanks to calculations made by Eratosthenes, the librarian of ancient Alexandria, and communicated to Latin Christendom by late antique texts.[83] As Columbus's son Hernando pointed out in 1524, the ancients' relevant experiments and computations had not been repeated in modern times, but their results were well known and widely trusted.[84] One of the great attainments the Spanish cosmographer, Pedro de Medina, claimed for his profession toward the middle of the sixteenth century was the ability to measure the globe.[85] Until then, cosmographers relied on ancient authorities, modified by explorers' always-impressionistic reports, guesswork, and wishful thinking. Fifteenth-century estimates of the value of a degree of latitude along the great arc of the Earth are hard to give precisely, but the most commonly cited figures—sixteen and two thirds or seventeen "leagues"—can be read to suggest a total circumference of about twenty-five thousand miles or so, in modern imperial measures, or some forty thousand kilometers: impressively close to the correct length.

The effects of Columbus's voyages on geographical received wisdom were paradoxical. The world seemed, in uneasily compatible ways, at once bigger and smaller than before. On the one hand, Columbus's discoveries and those by explorers in his wake seemed to add vast *terrae incognitae* to the map; on the other, they made the world seem a lot smaller than it really is, shrinking it to comprehensible proportions and putting the whole of it—or, at least, more of it than ever before—within reach. Cosmographers faced the dilemma of the indecisive student in one of Rafael Dieste's short stories: he thought the world was small but "not as small as they say."

Even in antiquity, the size ascribed to the Earth, according to scientific orthodoxy, had begun to shrink in some perceptions. Eratosthenes's model yielded 252,000 stadia (usually reckoned at a little under two hundred meters each in modern terms), which other ancient geographers reduced—Posidonius to 240,000 and Strabo to 180,000.[86] The same tendency to revise the figures downward continued under Columbus's influence. The smallest estimate was his own—about 20 to 25 percent below the real value.[87] Paradoxically, the hemisphere Columbus discovered would not have fitted into a world as small as he claimed ours to be.

Other contemporaries' calculations of the girth of the globe were not wildly different, notably those of the Florentine academician Paolo Toscanelli and the maker of the Nuremberg globe of 1492 attributed to Martin Behaim (below, p. 85), both of whom seem to have made an underestimate on the order of 13 percent.[88] The apparent success that attended Columbus's first voyage persuaded many experts that his calculations were correct—at least at first, before sober reflection affirmed that Columbus could not really have gotten anywhere near the extreme Orient.[89] Among his followers, Vicente Yáñez Pinzón and Amerigo Vespucci evidently thought their own voyages took them to the proximities of Asia.[90] The pope at the time echoed the claim that Columbus had approached India.

Columbus's image of the world, therefore, captured contemporary imaginations, in spite of the absurdity of supposing that he could have neared Asia after only some four thousand kilometers—to put his own figures into modern terms—under sail. Wishful thinking continued to make explorers underestimate distances traversed and cartographers to underrepresent them. As late as 1524, when Magellan had made the vastness of the Pacific known, Hernando Colón could still unabashedly echo some of his father's calculations, reckoning the circumference of the Earth at 5,100 leagues, or about 36,000 kilometers.[91] The persistence of the image of a relatively small globe, inexplicable as a reflection of reality, becomes intelligible against the background of negotiations between Spain and Portugal to divide the spoils of the world. As Hernando Colón remarked, to divide the world effectively, "We must first consider and establish how big it is."[92] Cosmographical disputations dominated and bedeviled the diplomacy.

When Columbus's first voyage made the matter seem timely, Pope Alexander VI, born a subject of the king of Spain, was the only possible mediator to whom the Spanish and Portuguese rivals could appeal. The previous pontiff, Innocent VIII, had assigned to Portugal a monopoly in the African Atlantic beyond the Canary Islands, and the limits of that grant had to be fixed in order to liberate Spaniards to follow up Columbus's initiative. In the Bull *Inter Caetera,* on May 4, 1493, Alexander granted to the Spanish monarchs "all islands and mainlands found and to be found, discovered and to be discovered towards the west and south" of a line drawn one hundred leagues from "any island" of the Azores and the Cape Verde archipelago, "by drawing and establishing a line from the Arctic pole, namely the north, to the Antarctic pole, namely the south, no matter whether the said mainlands and islands are found and to be found in the direction of India or towards any other quarter, the said line to be distant one hundred leagues towards the west and south from any of the islands commonly known as the Azores and Cape Verde." Despite the opacity of the text, *Inter Cetera* clearly envisaged that the dividing line the pope proposed would reach right round the globe, not merely across the Atlantic. The next bull, *Dudum Siquidem,* of September 26, was even more explicit, defining the grant as covering "all islands and mainlands whatsoever, found and to be found, discovered and to be discovered, that are or may be or may seem to be in the route of navigation or travel towards the west or south, whether they be in western parts, or in the regions of the south and east and of India."[93]

This was landlubbers' geography. As a bureaucrat in Hispaniola pointed out a few years later, the pope's division of the world "as if it were an orange" was inexact, the demarcation lines "imaginary," and the task of pilots sent to determine them impossible.[94] The point from which the one hundred leagues were to be measured was unclear. Nor was it clear at what latitude the measurements were to be made. There were, in any case, no reliable means of measuring distance along the surface of the ocean. The study of longitude was in its infancy, and no accurate method existed for reckoning it on the open sea. As the size of the globe was unknown, the extension of the dividing line across the Eastern Hemisphere could not be drawn. If and when it could, how much of the Orient would be on the Portuguese side and how much on Spain's? The smaller the world, the more would accrue to Spain. Wild imaginations or unchecked self-interest on either side assigned every-thing from India to Japan to one or other competitor-kingdom.

At first it hardly seemed to matter in practice, as all the prizes of the Indian Ocean were beyond reach. Once Portuguese ships penetrated the ocean, however, the unresolved problems became increasingly urgent. When Malacca fell, it hardly seemed worthwhile for Spain to dispute Portugal's rights there. But the Moluccas were another matter. The rich-est prize of all, in terms of concentrated value, lay waiting to be assigned by computational cleverness, cosmographical jiggery-pokery, or preemp-tive strike.

The need for Spain and Portugal to come to an accommodation that would supersede the bulls was obvious, for locating meaningfully the boundaries between the Castilian and Portuguese zones on both sides of the world and for avoiding conflict between the crowns. A further pur-pose underpinned the negotiations that followed: to eliminate the pope from influencing matters at issue between the two powers. Paradoxically, both states continued to rely on papal legitimation of their overseas acqui-sitions of territory. No other source of validation was available at the time, apart from the fantasy, never convincing to any but the most obtuse or obstinate, that the New World was an inheritance from supposititious ancient rulers of Spain or, in the Portuguese case, medieval refugees from the Moors. Yet monarchs of neither kingdom could ever feel happily beholden to Rome. Kings rarely want to compromise their sovereignty.

There was a precedent for capping and thereby overriding papal donations: in 1479 and 1482 Portugal and Castile had fixed their respective spheres in the African Atlantic, assigning the Canary Islands, long coveted by Portugal, to Castile, with a small coastal outpost on the mainland known as Santa Cruz de la Mar Pequeña. Portugal, meanwhile, retained rights of conquest over the rest of the West African shore and the other archipelagoes in the region.[95] The settlement marked a new departure in diplomacy. In the Treaty of Tordesillas, of June 7, 1494, the parties effectively tore up the papal concessions to Castile and made their respective rights depend on a bilateral agreement. They sidelined the papacy and found a way of grounding the legitimacy of their imperial ventures by secular means.

The Treaty of Tordesillas was more political than practical—hardly more helpful than the papal bulls it displaced. It satisfied Portuguese sensibilities by pushing the line of demarcation far out into the ocean, like a net cast speculatively in the hope of catching a miraculous draft. It named the Cape Verde Islands as the starting point for the calculation of the line of demarcation but failed to specify which island was in question or from where exactly the measurements should be taken. The text specified that the line would run from pole to pole 370 leagues west of the starting line, without saying whose league would be used—and leagues were rough measures of which the value varied, according to custom, from place to place.

The treaty deepened, if anything, the obscurity over the question of the "antimeridian"—the whereabouts of the continuation of the line on the far side of the Earth, where the two Iberian powers would soon clash again and contend for control of lucrative markets, with no satisfactory means of determining which islands and ports lay in the zone assigned to which crown. Implicitly, and in the opinion of the Mallorcan cosmographer and admirer of Columbus Jaume Ferrer, the location of the antimeridian was an open question, to be decided later. Above all, the Tordesillas agreement was self-stultified by the negotiators' strictly impossible aspirations: to fix a line on the unstable surface of the sea, and to separate the zones of contending states with a line the position of which could not be established with the technology available at the time.[96]

The treaty, in short, solved few problems and raised many. Ferrer supposed at first that the intention was to take the vital measurements along the equator from a starting point along the meridian of one of the Cape Verde Islands, but he revised his reading to suggest that the distance in question should be measured along the line of latitude fifteen degrees north. In consequence, in terms of longitude, the demarcation line might lie, by Ferrer's own calculations, anywhere between eighteen degrees twenty minutes and twenty degrees west of the starting point.[97] Even the latitude of the Cape Verde Islands was in doubt. Columbus reckoned it on average at nearly six degrees further south than Ferrer. According to cosmographer Gian Battista Gesio, who reviewed the terms of the Treaty for Philip II in 1580, the demarcation line could fairly be assigned to fifty-four distinct locations without violating the text.[98]

It is unsurprising, therefore, that the Spanish and Portuguese cosmographers and negotiators never succeeded in locating the whereabouts of the Tordesillas line. Indeed, the requisite data remained unavailable until 1743, when a Franco-Spanish expedition (part of a project to determine how far the shape of the world differed from that of a perfect sphere) made reliable observations of the longitude of Grão Pará in Brazil.[99] Shortly thereafter, a new treaty between Spain and Portugal discarded the Tordesillas line once and for all.

. . .

Meanwhile the consequences of the treaty shaped Magellan's life. Whatever the future of negotiations between Spain and Portugal for access to the spices of the Moluccas, the starting point would be the location of the antimeridian. A small world suited Spain. By shrinking estimates of the size of the planet, geographers could shift the antimeridian westwards and justify Spanish claims to coveted territories. As three of the king of Spain's pet experts avowed in 1524, "We must assign a value to the distance occupied by a degree of the circumference of the Earth, and make it as small as we can, because the smaller the amount, the smaller the world—which will be very much to the service of our king and queen."[100]

While the cosmographical squabbles wore on, adventurers from Portugal and Spain mounted practical efforts to preempt the issue by getting to the Moluccas first. For a while, Portuguese efforts seemed unpromising: the Indian Ocean was full of hostile powers; the monsoonal system, despite its advantages (above, pp. 13–14), imposed a time-consuming routine on ships that had to await the turn of the winds. From 1499, repeated—and repeatedly disappointed—Spanish ventures tried to find a way through America. Columbus scoured the central American coast, searching for a strait and for proof of the smallness of the globe, in 1502. The following year, three of his former henchmen, Vespucci, Juan de la Cosa, and Vicente Yáñez Pinzón, were among a committee in Spain that proposed a further venture in the same direction in the expectation that the Moluccas were approachable from the east. Another meeting three years later decided to seek an opening to the Pacific "west of the Antilles and north of the equator."[101] Failure to get through hardly deterred continuing efforts.

Francisco Serrão's establishment in Ternate might have clinched the race in Portugal's favor. So might the first appearance of the Moluccas in Portuguese maps, presumably after Javanese originals. In effect, however, Portuguese success only served to whet appetites and excite efforts in Spain. By the time Magellan got back to Portugal in 1513, Spanish exploration overland across the Isthmus of Panamá had reignited expectations, as it demonstrated that in parts, at least, the American continent was narrow. By the time Magellan departed on his own quest for the Moluccas, predecessors' efforts had got at least as far as the River Plate, where the leader of the most advanced expedition died, reputedly eaten by cannibals. Magellan did not need to rely on any single source for the idea of trying a westward approach to the Moluccas. The notion was in the air. Other people's plans were already unfolding or unraveling.

FIGURE 3. Dom Jaime I, Duque de Bragança, from a seventeenth-century Italian engraving of the family tree of the kings of Portugal. Jaime was Magellan's nominal commander-in-chief during his campaigns in Morocco in 1513–14, which preceded and contributed to the explorer's renunciation of Portuguese nationality and defection to Spain. The Bragança dynasty, descended from a bastard son of King João I, were the ruling house's rivals for the Portuguese throne. Under King Manuel's predecessor, members of the family fled to Spain to escape royal anger or revenge. Manuel attempted reconciliation: the Italian genealogist mentions in the scroll that accompanies Jaime's portrait that the subject was the king's nominee to succeed him, should there be no direct heir. Some Bragança exiles, however, remained in Seville, where they welcomed Magellan and created the circle of Portuguese exiles among whom he settled, married, and found supporters. Courtesy of the Biblioteca Nacional de Portugal.

CHAPTER 3

The Trajectory of a Traitor

Morocco, Portugal, and Spain,
1514 to 1519

And whosoever will not receive you, when ye go out of that
city, shake off the very dust from your feet for a testimony
against them.

—Luke 9:5

Prophets without honor in their own countries seek it in exile. In late
medieval and early modern Latin Christendom, explorers—like church-
men, artists, poets, rhetoricians, scholars, alchemists, and potential
royal consorts—crossed frontiers in search of preferment with little or
no sense of betraying their homelands. Some of the crowd of aspiring
knights and squires of the Infante Dom Henrique came from Italy,
including Columbus's father-in-law, who was from Piacenza, and the
Venetian Alvise da Mosto, who supplied the first accounts of the Gam-
bia and the Cape Verde Islands.[1] Columbus, who hawked his services
around European courts before settling on and in Spain, came from
Genoa. John Cabot, as he was called in his adopted England, was Venetian. Vespucci and Verrazano were Florentines. Juan Díaz de Solís, in
the Castilian version of his name, was also known, in Portugal, as João
Dias de Solis. He served both crowns indifferently: it is not known in
which kingdom he was born. He left Portuguese service, allegedly for
the same reason proposed on Magellan's behalf—to be able to undertake a voyage to the Moluccas, supposedly on the grounds that they
were in Castile's sphere, but apparently as a fugitive from justice, suspected of murdering his wife.[2] Estevão Gomes, a pilot who, as we shall
see, defected to Spain at about the same time as Magellan—before joining the great voyage, mutinying, and deserting—had a long career in

Spanish service, exploring much of the Atlantic coast of North America.[3] Vasco da Gama allegedly threatened desertion to Spain when the king seemed to want to dishonor his promise to make the explorer a count.[4]

Magellan was one of literally scores of recruits Spain got from Portugal. Sometimes they incurred blame or vengeance. An undated document, of which Afonso de Albuquerque had a copy, specified forfeiture of all property and exile to St. Helena for seamen who defected.[5] In other cases, individuals could switch back and forth with impunity. Yet when Magellan abjured Portuguese service and shifted to Spain, reproaches and alienation were among the results. The family he left in Portugal repudiated him as an unpardonable turncoat. His great-nephew, Francisco da Silva Telles, ordered future generations never to heed Magellan's wish for his scutcheon to adorn his house near Sabrosa,

> since I desire that it should for ever remain obliterated, as was done by order of my lord the king, as a punishment for the crime of Fernão de Magalhães, in that he entered the service of Castile to the injury of this kingdom, and went to discover new lands, where he died in the disgrace of our king.[6]

Magellan had been equally vehement in renouncing his former allegiance. In August 1519, on the eve of his great voyage, he ordered that the king of Spain's standard be placed in the friary of Santa María de la Victoria, Triana, where he hoped his body would rest, and where his would take "his oath and pledge of homage, according to the laws and customs of Castile, that he would undertake the voyage with all loyalty as a good vassal of His Majesty."[7] Two months later he entailed his estate on condition that his heirs transfer to Castile and change their names to his. "It is my wish," he explained,

> that my brother, Diego de Sousa, who at present dwells with the most Serene Lord, the king of Portugal, have all the aforesaid property if he comes to live in these kingdoms of Castile and marries therein and on condition that he take the name of Magallaes [sic] and bear the arms of Magallaes, in the same form as I bear them, which combine those of Magallaes and Sousa.[8]

The same conditions were to apply to the next person in line of succession, their sister Isabel.

The testator can hardly have meant these demands seriously. He had a baby son of his own to think of; and the likelihood that his brother or sister would desert Portugal could be reckoned at zero. The demand that they change their name was surely a calculated act of defiance, bravado, or insult. The inference is inescapable: in Portugal, Magellan came to be seen as a renegade; in Castile he treated his tergiversation as a matter of pride. "Oh, Magalhães," exclaimed Camões,

Truly, Portuguese-born,
Faithlessly, then, forsworn![9]

What turned the former page into a labeled traitor?

Some of Magellan's contemporaries shared the assumption or believed the claim that his move to Spain was an act of exasperation, motivated by a desire to be free to attempt to execute the design of sailing to the Moluccas via the Atlantic. The Portuguese king's refusal to sponsor or license such a voyage was, according to this theory, an intolerable provocation. It is not an incredible theory, but association with partisan traditions taints it. We have seen how Francisco Serrão importuned his friend to join him in Ternate. In the only purported fragments copied from Magellan's replies, which, if authentic, must have been written while he was in Portugal or Morocco between 1514 and 1517, "he said," in Barros's digest of the correspondence,

> that if God pleased they would soon meet in person, and that if it were not by way of Portugal it would be by way of Castile, such was the state of his affairs, wherefore Serrão should await him there. . . . And the devil, as he always moves the minds of men for evil ends and to destroy them therein, so arranged matters that this Fernão de Magalhães should become discontented with his king and with the kingdom.[10]

When Magellan failed to get what he wanted in Portugal, Barros continued, "he put into effect the plan he wrote to his friend, Francisco Serrão, while he was in Malacca, so that it appears that his departure for Castile had been in his mind for some time, rather than arising suddenly as a result of his dismissal."[11] Antonio Pigafetta, who, as we shall see, became a close confidant of Magellan's, confirmed that "when our captain was at Malacca, [Serrão] had written to him several times" from Ternate and "was the cause of inciting . . . the enterprise," without saying when the recipient got the letters or whether they induced his desertion to Castile.[12] In any case, the Moluccas were never far from Magellan's awareness. Correspondence from the sultan of Ternate had been received and pondered at the Portuguese court at intervals from 1514.[13] Magellan had been aboard the Molucca-bound ships that got as far as Amboina in 1512 (above, p. 48).

It is more likely that the plan to discover the Atlantic route to the Moluccas was a consequence rather than a cause of Magellan's defection. He had returned to a country where everyone was complaining of inflation and where copper was replacing silver at the royal mint. His discontent, whether or not diabolically induced, arose from financial embarrassment, from which neither the king's patronage nor the Portuguese courts could relieve him, combined with an aggrieved sense of honor, ignited when the king refused the rewards Magellan expected for his service. Magellan's growing links with other malcontents in Portugal contributed.

. . .

The immediate context was an expedition to Morocco. It left Lisbon in August 1513 to relieve the Portuguese stronghold of Azamor, on the Atlantic coast of the Maghrib, and to extort tribute due from, but withheld by, Moroccan chiefs. Magellan, newly arrived home, was part of the task force, under the nominal command of Dom Jaime, Duke of Braganza. Effective leadership seems to have been delegated to subordinates, but the connection with the duke is the first evidence of links between Magellan and a lineage that could plausibly challenge King Manuel for the Portuguese throne.

A royal bastard, elevated to be Duke of Braganza in 1442, had founded the house. João II responded to the growth of Braganza wealth, power, and ambition by executing the third duke, confiscating his property, purging the family, and driving most of the survivors into exile in Spain, where the opportunity to take part in the conquest of Granada provided an honorable pretext for absence from home; some of them formed the nucleus of a community of disaffected Portuguese, whom Magellan was later to join, in Seville.[14] Dom Jaime's restoration to favor was part of King Manuel's reversal of his predecessor's policies, but it was conditional; by murdering his first wife the duke put himself at the king's mercy, and the Azamor expedition was part of his expiation of his guilt.

According to Barros, it was in Morocco that Magellan received the lance wound that caused his limp[15]—contradicting the story (above, p. 44) that linked the affliction to a wound sustained at Calicut. What Magellan gained, however, was worth a wound: the prospect of recovering his fortune. As we have seen, his destitution at the time of the conquest of Goa was such that Albuquerque had handed him a suit of old armor as a gift. He had lost one of the pepper cargoes in which he had invested (above, p. 45); he was hoping to recover something from the debacle by suing his former partner's family, but that would take time and cost more money. Proceeds from the second cargo had not yet reached him, though he had placed a lien on a disputed consignment, valued at sixty cruza-dos.[16] To fight in Azamor, he had to buy a horse on credit and, when it died under him, to petition the king to make good his loss "because they killed it while I was in your service in an honorable place, with great peril to my person, where I had to reach safety on foot."[17]

Success in the Azamor campaign improved his prospects.

The Portuguese took 890 captives and two thousand head of cattle. Jointly with one Álvaro Monteiro, Magellan took official charge of the bovine spoils. The role, which included management of the herd, its grazing, stabling, fodder, and disposal, provided opportunities for mar-ginally honest enrichment and temptations to major corruption. Magel-lan's proposal to sell four hundred head of cattle to native purchasers evidently came into the latter category. Allegedly, he handed the beasts over by night, on the pretext that they had been rustled. His return to Portugal without leave in July 1514 looks like an attempt to evade his accusers, though he justified it on the grounds that he had urgent busi-ness with the king and the courts.[18]

Presuming, perhaps, on his limp as evidence of self-sacrificial service, and on a sense of entitlement to which his financial misfortunes may have added a measure of resentment, Magellan began to make demands of the king, among which, said Barros, "it is said that he asked for an increase in his stipend—a matter that always causes difficulties for the noblemen of this realm and seems to be a source of a kind of martyrdom for the Portuguese and, for their kings, a common cause of scandal."

Manuel's investment in a large household (above, p. 35) was a strain even on resources as great as his. The duties recipients of stipends owed were poorly defined. Princes' graces, as Barros explained, are recipro-cal. They must be proportionate to services rendered and the quality and ancestry of the recipient. When requests are refused, they must be borne with patience. Magellan, however, felt that he was the victim of unequal treatment and that less qualified rivals were better favored. That, at least, is the implication of Barros's observation that

> when, however, one sees an equal favored, especially in cases where jobbery and friendship, rather than genuine merits, are responsible, that is when all patience is spent, wherefrom indignation arises, and thence hatred, and at last despair, until those concerned come to actions that damn them, and oth-ers with them.[19]

The pending case of the purloined cattle overshadowed Magellan's claims. Accusers suggested that his protestations of poverty were feigned.

When a letter arrived from his superior officer to say that Magellan was absent without leave, the king declined to hear his explanations and ordered him back to Azamor to face his denouncers. The accused returned to Morocco in July 1515, where he succeeded in having the charges against him dismissed or deferred, "either," in Barros's nicely calculated version, "because he was clear of blame or (as most affirm) because it was inopportune to do anything to prejudice the security of Azamor. The king however, remained ever suspicious of him."[20] Historians have fantasized about irrational character flaws or psychological pathologies that might have made Manuel hate Magellan. Samuel Eliot Morison, the battle-scarred and battle-ready admiral who was a professor at Oxford and Harvard, put it with characteristic pithiness: "Nobody knows why."[21] There is no evidence of hatred—only suspicion, justified in the circumstances and amplified, as we shall see, by subsequent events.

By March or April 1516, Magellan was clear of afflictions in Morocco and able to return to Lisbon and to his financial problems. Funds from the second of his shipments of pepper from the East at last awaited.[22] Getting hold of them was another matter.

The terms of Magellan's contract with his partner in Cochin, Pedro Anes Abraldez, seem, from the records of the lawsuits their agreement generated, to have made little sense:[23] the terms can hardly have suited either party; so considerations now lost to us must have been in play. The details are tiresome but important to understand for their bearing on the restlessness, discontent, and shortness of temper Magellan displayed for the rest of his time in Portugal.

His initial investment in the shipment of pepper, made in 1510, was of 100 cruzados; 200 cruzados were due to him in Portugal—a return of 100 percent, equivalent, according to the terms of the original contract (above, p. 45) to 50 percent of the profits, and payable in installments. Presumably Magellan expected to get his due after the sale of the next consignment, as the profit would be impossible to calculate until then. Abraldez, however, had died in debt in 1513. No doubt the grounding of one of the ships involved in the deal had played a part in his ruin.

From the treasurer of the Casa da India, at an uncertain date some time in 1515, Magellan received a substantial payment in compensation for pepper and cash, "which came to him in the ship as a partner with a half share."[24] In continued litigation, however, he cited his deceased partner's father. The suit prospered in the end. In 1516 the plaintiff's brother, Duarte de Sousa, received 80,000 reais on his behalf: that is, 277 cruzados at the prevailing official rate, but only about 205 if a recent fall in negotiable values is taken into account. The source of the balance is unclear: presumably interest and costs account for it. Magellan, however, continued to claim, unsuccessfully, that he had been insufficiently reimbursed.[25]

In May, 1517, he acknowledged that he had received most of what was owed.[26] He still needed money. And, presumably, he still craved status. He therefore continued to petition for an increase of his standing as a member of the king's household, and of his stipend (which stood at 1,250 reais monthly in April 1516).[27] The amount in dispute is unrecorded in any reliable document, and historians seem to have resorted to guesswork. But Sebastião Alvares, the king's representative or "factor" in Seville, might be expected to know how much was involved, as he later had the thankless task of trying to turn Magellan back to Portuguese allegiance after defection to Spain. He put the sum at a mere 100 reais.[28] Honor, as well as cash, was at stake, and on a point of honor a true nobleman would cavil to the ninth part of a hair. The king's refusal, definitively issued on April 15, 1517, precipitated a crisis.

The chronicler Gaspar Correia, writing a couple of decades later, probably made up the dramatic scene he described, but there is a sort of poetic truth about it: faced with the king's intransigence, Magellan retorted by asking "leave to go to some place where they would grant him favor, whereat the king tersely replied that no one would take him." Magellan responded by "rising and leaving the house where the king was, ripping up his commission and flinging the fragments from his hand."[29]

. . .

Spain was the obvious place of exile for any Portuguese malcontent. The Iberian peninsula was like political Lego, with kingdoms coupling and uncoupling according to the vagaries of war or dynastic accident. The inexorable tendency had been toward progressive unification, as the crown of Castile gradually incorporated or absorbed most of the others; the creation of a kingdom "of Spain"—or rather of a combination of realms whose sovereign was loosely called king of Spain—was a novelty: Granada succumbed only in 1492, Navarre in 1513. Aragon, which included or was combined with other historic entities—the kingdom of Valencia, the Catalan principality, the Balearic islands—became definitively part of the king's dominions only in 1516, when Carlos I, whom most English readers know as Charles V because he later became the fifth Holy Roman Emperor of that name, ascended the throne. He surely thought of himself as Charles, in French—the language he reputedly chose to "speak to men" rather than "Italian to women, German to my horse and Spanish to God";[30] but he was Carol in most of his Netherlandish dominions. He would be Karl or Karel to his future German and Bohemian subjects, Carlo or Carolo in most of Italy, and Carolus in the common language of Latin Christendom. In Spain he was Carlos, and that shall be his name (and therefore, for consistency, Spanish the language for rendering the names of Spanish monarchs) in these pages.

There were no common institutions throughout Spain except the Inquisition, no uniformity of laws or administration, and only at best a weak sense of common identity. The process of adhesion by way of royal marriages and rights of inheritance would bring Portugal into the fold in 1580. Meanwhile, the notion of "Hispania"—the group of provinces that had shared a single name since the Roman Empire—included the whole peninsula.

The culture of the kingdoms overlapped. We have seen (above, p. 39) that Portuguese poets might choose to write in Castilian—which functioned as an elite lingua franca—instead of their native tongue; that Spain had been the resort of the Braganza exiles and their Portuguese allies in the reign of João II; that the exchange of personnel across the frontier happened frequently and often unrancorously; and that the recruitment of Portuguese for Spanish service was especially intense in the context of maritime adventures.

The commercial sector that helped to back overseas exploration and imperialism in Spain was looking for new opportunities. Bankers had begun to see the enormous returns that were possible: the gold that flowed into Portuguese coffers from the 1480s as a result of trade with West Africa; the slaves the same region supplied; the sugar and marine products that accumulated in the same decade in the Atlantic archipelagoes or the waters around or beyond them; the pepper that Portuguese outreach to India brought from 1498; the tantalizing prospects of a western route to Asia that Columbus had broached. Marginal noblemen, shut out from advancement at home and imbued with chivalric ideas, were willing to take amazing risks.

So were merchants, whose eagerness to finance adventures was based on gamblers' thinking: huge price differentials between markets meant that one could make a profit from a few successful voyages, even if most cargoes were lost. In Spain one firm above all became fixed—perhaps fixated—on the prospects of infiltrating the spice trade. The Haro family were long-standing wool merchants in Burgos, in northwest Castile, with an increasingly international portfolio. One brother, Diego, was in Antwerp, where cloth merchants paid more for wool than their counterparts anywhere else in Europe. His sibling, Cristóbal, had moved to Lisbon in order, at first, to buy Brazilian logwood and Madeiran sugar for onward trade to richer markets. In Lisbon his ascent to the rank of a sort of merchant-prince began, which led at last to the sumptuous grave in the Church of San Lesmes in Burgos, where he kneels in effigy, alongside his wife, praying, as if in concert with the five masses a week that he endowed at his death in 1541.

Acquaintance with senior figures in the foreign merchant community seems to have introduced Cristóbal de Haro to the attractions of speculating in spices. He collaborated with Jakob Fugger, the representative of the richest German merchant house in the city, in fitting out two of the more than a score of ships in the expedition that took Magellan to India in 1505. He earned a return of 300 percent and was said to have wept because he had not invested more.[31] According to a royal secretary who married his niece, Cristóbal had traded with Oriental markets via Lisbon "for many years." He had reputedly helped to finance a voyage in search of a strait across America in 1514.

Fellow natives of Burgos shared his interest. In 1511, for instance, one of Haro's associates, Martín Alonso, bought 5,500 cruzados' worth of pepper in the Casa da índia in Lisbon, while the king of Portugal ordered the Casa to make a donation of 500 ducados (each worth perhaps about half a cruzado) to the Church of San Cosme, Burgos, presumably as part of a thank-offering or a deal on the side. In partnership, Cristóbal de Haro and Diego de Covarrubias (who would finance the Loaisa expedition in Magellan's wake in 1525 (below, p. 000) bought the enormous quantity of fifty quintals (each typically reckoned at a hundredweight) of pepper in the same year.[32] The merchants cultivated the court of Spain. They lent money to Ferdinand of Aragon. They paid retainers, which were bribes in effect, to Juan de Fonseca, bishop of Burgos, whose responsibilities in the king's service included the supervision of trade passing through Seville from the New World. Fonseca was an arachnoid bureaucrat, whose web covered the realm.

In Burgos, the policy of cutting out Portuguese intermediaries by opening a direct route to the Spice Islands was obviously attractive. The possibility or, by some estimates, likelihood that the Moluccas lay in Castile's zone of navigation encouraged the notion. So did the inconclusive results of the search for a westward route to Asia in the wake of Columbus. By the end of the first decade of the sixteenth century, explorers had scoured the region from the bulge of Brazil to the Gulf of Mexico so thoroughly as to be confident that no strait existed there. On March 27, 1512, King Fernando, who at the time ruled Castile as regent as well as Aragon as king, commissioned Juan Díaz de Solís—as we have seen, another recruit seduced form Portuguese service—to go, via Ceylon (that is, Sri Lanka), to "the island of Maluco, which is inside the limits of our demarcation, and you will take possession of it"[33]—adding, along the way, Sumatra, Pegu in Burma, "the land of the Chinese and the land of the junks" (la tierra de los chinos and la tierra de los Jungos). The ambition was proportional to the ignorance that made it seem possible. The king's reason for fixing Ceylon as a point of departure was that "according to very learned men" (por personas muy sabias)—cosmographers, that is, in Spanish pay—the antimeridian of the Treaty of Tordesillas passed through the center of the island: an absurd claim by any standards.

The plan, it seems, involved sailing through waters assigned by agreement to Portugal's monopoly; but the king salved his conscience by adding further strict instructions: "You must take great care therein, such that in God and in your conscience you measure the demarcation line as exactly as you can."[34]

Fernando aborted the enterprise when Manuel objected. Haro, undeterred, joined his Fugger friends in sponsoring an expedition from Portugal via the Atlantic and along the coast of South America, in search of a strait, the following year under a Portuguese pilot, unnamed but unequaled in eminence, as far as forty degrees south. The evidence comes from a document published in *Newe Zeytung auss Presillig Landt* (New tidings from the land of Brazil) in Augsburg in 1514. The *Zeytung* described skin-clad natives and snow-covered peaks, which the explorers glimpsed before passing through a strait between the southernmost point of America, or Brazil, and a land to the southwest, referred to as *vndtere Presill* or *Brasilia inferior*—that is, lower Brazil. The description is vague enough to refer to Tierra del Fuego, where the combination of skin and snow alerted Darwin, on the *Beagle* voyage, to the way biology adapts to environment.[35]

The text might have been part of what we should now call a campaign of disinformation, designed to lure rivals into fruitless explorations in the same direction. Since, however, no one could confidently dismiss the possibility of such a strait, the report is more likely to have been an optimistic exaggeration of what the expedition achieved. "They sighted land on the other side," it declared,

> when they had gone sixty miles beyond the cape at the far end—in the same way as, when you pass the Strait of Gibraltar, you see Barbary. And when they had rounded the cape, they headed northwest in our direction. There they met such a terrible storm and the wind grew so great that they could sail no further. And so they were obliged thereby to turn back to the hither side of the cape and proceed northwards along the coast—that is, of the land of Brazil.

The *Zeytung* adds that, in the opinion of the pilot of the vessel concerned, "from this cape there are no more than six hundred miles to Malacca." where Chinese merchants could be found.[36]

. . .

The claim that there was a means of circumnavigating South America had at least one and perhaps two effects on the cartography of the time.

Johannes Schöner of Karlstadt was a mapmaker and astronomer of exceptional gifts, both technical and scholarly. In his youth he felt so close to Regiomontanus, the compiler of the most widely used astronomical tables of the day, that he used the margins of the book to write an intimate diary, including details not only of his scientific observations but also of life with the concubine and child that, as a priest, he should not have had. His map and globe of 1515 derive most of their outline of America from a more famous prototype, made in 1513 by Martin Waldseemüller of St. Dié (who is famous for having drawn the first map to include "America" as the proposed name for the New World). It shows two straits: one in central America, copied from Waldseemüller, is purely speculative or derived from traditions connected with Columbus's search for a westward passage to Asia. The other, at about forty-six degrees south on the map and about seven degrees beyond on the globe, is an obvious allusion to the *Newe Zeytung*, which Schöner roughly quoted in the book *Luculentissima quaedam terrae totius descriptio,* or "A Certain Very Bright Description of All the World," which the map accompanied: "The Portuguese," he wrote,

> sailed in this region, Brazil, and found there a strait very similar to that of our own Europe, and which runs from east to west. From one bank the opposite shore is visible and the farther neck is about sixty miles away, as if we were navigating the Strait of Gibraltar that separates Seville from Barbary. Beyond, the distance is not very great from this region of Brazil to Malacca, where the apostle St. Thomas was martyred.[37]

A further map, attributed to an accomplished pilot, João de Lisboa, and bearing the date 1514, also shows South America tapering to a tip. The purported author, who died in 1525, is perhaps identifiable with a shipmate of the same name who took part in the conquest of Malacca with Magellan and was in Terceira in the Azores in October 1514. The map forms part of his treatise of navigation, but other maps in the same volume show signs of later emendation. As we shall see, Magellan claimed to know for certain of the existence of the strait he sought and found—albeit not where he was expecting it. He also had, by some accounts (below, p. 84), a map that illustrated its whereabouts. Whether a work by Schöner or João de Lisboa was in question, the maps help to situate Magellan's voyage in context: as part of a long and increasingly frantic series of voyages in the race for the Moluccas in the years preceding his attempt, and as evidence of the persistence of efforts to seek a route via the Atlantic.

The attempts continued almost to the eve of Magellan's venture. On November 24, 1514, King Fernando enlisted Solís for another foray, perhaps in response to the latest Portuguese initiatives and with evident encouragement from the overland explorations of Central America—or "Castilla del Oro" in the promotional nomenclature of the time. Here, in September 1513, Vasco Núñez de Balboa—former stowaway and pig farmer, who had imposed himself as governor of the region by dexterity, determination, and diplomacy—brandished sword and flag while he waded knee-deep into the Pacific and claimed the entire "South Sea" as he called it, for Castile. Ferdinand's commission to Solís was to go "to the rear of Castilla del Oro" (a las espaldas de Castilla del Oro). As we have seen, the explorer got as far as the mouth of the River Plate before cannibals supposedly devoured him. His failure hardly seems to have dampened the continuing excitement at the thought of western access to the sea Balboa had identified.

If, before leaving Portugal, Magellan had formulated a plan along similar lines, the idea could have been implanted or encouraged by letters from Francisco Serrão, which at last caught up with him by the vagaries of interoceanic forward mail. But the company and conversation of Cristóbal de Haro were also available to him.

Haro considered himself the begetter of Spanish involvement in the Moluccas. His record of devotion to the quest for a route there started long before Magellan's voyage and continued long after it. In 1520 Bishop Fonseca and Cristóbal de Haro together commissioned another failed expedition to the Moluccas via Panama, with Andrés Niño, pilot of the Casa de Contratación (the Spanish crown agency charged with overseeing the trade of the New World), and Gil González Dávila, a servant of the bishop's. Balboa was to provide ships. Dávila explored Lake Nicaragua, reportedly engaging in conversation about the reality of the soul with native chiefs and sponsoring thousands of baptisms, but getting not much further with the task of exploration.[38] In 1522 Haro sponsored another vain attempt via the Northwest Passage that cartographers depicted along the Arctic coast of America.[39] Finally, in 1529, when Portugal settled the dispute over access to the islands by buying out Spain's claims at the bargain price of 350,000 ducats, Haro wrote to his king demanding a cut. "This sum," he averred,

> constitutes the benefit and interest on the cost of fleets equipped and sent from Spain to the Spice Islands, and I beseech His Majesty to make me the heir thereof, on equal terms with His Majesty and other investors, because we have been the cause of the discovery of the Moluccas. . . . I have suffered great losses at Portugal's hands for having served as an instrument for the said discovery and for having organized it.[40]

Haro left Lisbon after an obscure dispute with King Manuel in 1517—roughly at the time Magellan's career and country of residence underwent a similar twist. He had a concession for trade with Portuguese outposts on the African coast, where local commanders cast doubt on his credentials and seized or destroyed some of his goods.[41] It is not clear, however, whether the outrage occurred just before or just after his break with the Portuguese king: it may have been a consequence rather than a cause.

In any case, in introductions to other merchants of Burgos, who were influential in Seville, Haro's hand is perceptible among the connections Magellan subsequently made at the Spanish court, where the trader's friends and relations were among the auditors and supporters of the Portuguese renegade community. As we shall see in the next chapter, a thread led from Portuguese exiles who welcomed Magellan, via Burgos merchants who helped to finance them and Spanish bureaucrats with whom some of them worked, to decision-makers at court who could influence the king in Magellan's favor.

In Seville Magellan benefited from the friendship of Portuguese, many of whom enjoyed extraordinary levels of trust and had powerful jobs in the administration of the Casa de Contratación. Arriving in Seville on October 20, 1517, he slotted into the community that had gathered around the Braganza exiles of the previous reign (above, p. 66). The welcome he received from one of the most prominent of the local Portuguese, Diogo or, as he was known in Spain, Diego Barbosa, was of transcendent importance. Barbosa had taken over the charge that King Fernando had given to Jorge de Braganza: the post of warden of the complex of buildings that housed the Casa de Contratación, and of the shipyards. He was a member of Spain's most coveted Order of Chivalry, the Order of Santiago. While in India Magellan had perhaps gotten to know Duarte, Barbosa's nephew (the relationship the term designated was vague): Duarte joined him in exile and sailed on the great voyage.

The foot Magellan had in the door of the Barbosa household soon slid over the threshold. He married Diego's daughter, Beatriz, before the year's end. It was what people now call a whirlwind romance, consummated without waiting for a contract, which was not drawn up until June 4, 1519, perhaps prompted by the birth of the couple's son, Rodrigo, at about that time. The historian Juan Gil recently discovered that a Greek survivor of Magellan's great voyage, known as Nicolao de Nápoles, testified to having dandled little Rodrigo in his arms before the departure of the fleet.[42] The contract promised a dowry of 60,000 maravedíes (enough to pay the wages of fifty common seamen for a year in the Castilian currency of account), half in cash payable immediately; the rest would be compounded for in the customary payment in kind: jewels and furniture and ornaments and household goods over a period of three years.[43] Magellan acknowledged receipt of the cash on August 23, 1519, just before his fleet departed. In his will, however, Barbosa said he had not paid the balance, nor had Magellan handed over the customary thirteen coins that were the symbolic bride-price due from a groom.[44] These irregularities did not affect the legitimacy of the union or imply regret or reluctance on either side. Magellan had married in haste; there was no time for repentance at leisure.

A modest migration of fellow exiles from Portugal enlarged the size of Seville's Lusitanian colony at about the time of Magellan's arrival. He of course brought slaves, including Enrique (above, p. 51), with him. A well-informed mid-sixteenth-century history by the elegant humanist Francisco López de Gómara (best known for his adulatory account of Cortés's deeds in Mexico), claimed that Magellan's wife and her father maltreated Enrique, stoking resentment that led him, after Magellan's death, to abscond and stay in the Philippines rather than return to Spain.[45] The story is believable: Magellan had another slave, called Antonio, who fled the family home and was recaptured in Ciudad Rodrigo, in March 1519,[46] and Duarte Barbosa, Beatriz's kinsman, is said to have threatened to scourge Enrique for idleness when Magellan's expedition was in the Philippines and its leader dead (below, p. 258). Evidently, though the image of a dandled baby on a common seaman's knee may evoke a sense of charming domesticity and unsnobbish hospitality, life at home with the Magellans or Barbosas was not endurable for all inmates.

A party of friends and relations also joined Magellan in Spain: as well as Duarte Barbosa, there were his mother's kinsmen, Álvaro and Francisco de Mesquita, and Martín de Magallanes, whose relationship we have presumed (above, p. 77, note 72). Barros complained that he "took with him certain pilots who suffered from the same disease" (levando alguns Pilotos tambem doentes desta sua enfermidade) of disloyalty.[47] Estevão Gomes (above, p. 63), if he was not one of them, arrived at about the same time, as did Pedro and Jorge Reinel, learned in cosmography and cartography, and Diogo Ribeiro, who was to play the role of official cartographer, supplying most of the expedition's maps, but who did not accompany the fleet—"nor," according to the king of Portugal's agent in Seville, "does he wish for more than to earn his living by his art."[48]

. . .

A nucleus of experts to support a project for a new voyage was taking shape. Its most important member, by common agreement among commentators at the time or soon afterwards, was Rui Faleiro, another Portuguese exile, whose work on navigation and geography impressed some, at least, of those acquainted with it, and supported the Spanish case for placing the Moluccas in Spain's zone of navigation. He was an astrologer; at the time astrology and astronomy were closely allied arts. The horoscope of the Infante Dom Henrique inspired his patronage of explorers.[49] The same knowledge of the stars fed navigation, cosmography, and divination—all almost equally unreliably. Astrology was not perfectly respectable: paganism tainted fortune-telling. But links between the configurations of the stars and other natural phenomena were presumable and, in the science of the time, worthy of study.

Faleiro's special skill, real or imagined, was in calculating longitude by the lunar distance method, which he described accurately and in detail: differences in longitude can be reckoned from comparative measurements of the distance between two celestial bodies (commonly including the moon) at different places on the surface of the Earth. The technique held out the prospect of at last being able to fix the lines of demarcation between Spain's and Portugal's zones of navigation. As we shall see, however, the method was easier to advocate in theory than to execute in practice.

Testimony divulged in later lawsuits shows that Magellan and Faleiro made solemn promises to each other at an early stage of their collaboration: that they would work together, that they would not divulge their plans to a third party, that either could abrogate the agreement, on giving due notice, and that they would submit any dispute to arbitration. They agreed "to be equals and that whatever either learned, whether of Portugal or Castile, that might be relevant to their voyage, he would communicate it to the other within six hours."[50] The nature of the joint plan is not specified in the surviving evidence. However, Faleiro's place in Magellan's circle in Seville shows that the Moluccas were by this stage the focus of the expedition that was under consideration, since his picture of the world placed the islands on Castile's side of the antimeridian and was probably among the reasons for his flight from Portugal. According to Barros, Magellan brought Faleiro into the team specifically to encourage the Spaniards to believe that the Moluccas were theirs by right. "He joined to his side Ruy Faleiro, a Portuguese by birth and an astrologer, who also had a grudge against the king of Portugal, who did not want to employ him, as astrology was a business of which he had little need."[51]

Faleiro arrived in Seville a month and a half after Magellan in December 1517, with his brother Francisco and perhaps his wife, Eva Afonso.[52] He expressed annoyance that his associate had already initiated contacts with the Casa de Contratación in pursuit of support for an expedition; but that may have been merely because Faleiro wanted to make an independent proposal himself or represent himself as the senior partner. Magellan's proposal, as Spanish officials understood it at this stage, was that he "offered himself to discover a very great source of spices and other riches in the Ocean Sea within the limits assigned to Castile."[53] Barbosa smoothed over the collaborators' differences, with help from another leading light of the Portuguese community, Rui Lopes, who had served Dom Jorge de Braganza as a man of business and was something of an amateur cosmographer himself: his library included the *Geography* of Ptolemy and a treatise on the astrolabe.[54]

As we shall see, Faleiro was designated joint commander on the projected expedition, with responsibility for navigation and specific orders to take measurements of longitude in situ, using both lunar distance and the standard method, by timing eclipses, while allowing for magnetic variation—all techniques that feature prominently in his own writings.[55] But Faleiro never sailed. It is tempting to suppose that his exclusion followed further dissensions, but it seems rather to have been the result of his own increasingly acknowledged mental instability. In September 1518 Portugal's ambassador, Álvaro da Costa, who was working to dissuade the fellow countrymen from working for Spain, wrote to his king, reporting that Faleiro was "almost out of his mind" (quasi fora de seu seso).[56] There is a taste of sour grapes about the remark, but the Portuguese representative in Seville—the "factor," who was rather like a modern consul—had similar misgivings (below, p. 91). Retrospectively, Spaniards concurred in dismissing Faleiro as mad. Gómara, a generation later, suggested that awareness of the impossibility of the planned voyage drove him over the edge.[57]

. . .

Faleiro's support was not, in any case, needed to confirm Spaniards' convictions that the Moluccas were properly theirs. We have seen (above, p. 59) that the cosmographers employed to negotiate with Portugal wanted to reduce the commonly accepted figure for the girth of the globe and that Spanish sources put the antimeridian as far west as Sri Lanka. In the year Magellan's fleet set sail, the *Suma de Geographia*, by perhaps the best informed of all Spanish experts, Martín Fernández de Enciso, appeared in Seville: it made the line bisect the Bay of Bengal, through the mouth of the Ganges.[58] Spanish negotiators continued to make the same claim, or similar ones, in talks with Portugal after the survivors of the Magellan expedition returned home and the mariners' tales of their long sufferings had revealed the vast extent of the Pacific. Enciso placed Malacca in the Spanish sector, whereas in reality the city lay less than 150 degrees east of the Tordesillas line, as most authorities place it. The true antimeridian was probably just beyond the Moluccas, on the way to New Guinea.

Where did Magellan place it in his own mind? No surviving declaration from his hand is earlier than a memoir dated September 1519, just before he set sail, and addressed to Carlos I. His avowed purpose in writing was to provide the king with the true facts in case the writer's death should leave Spain ill informed in refuting any future Portuguese claims, for "no one understands them as well as I."[59] He advised the king to keep his memorandum "very well guarded, for the time may come when it shall be very much needed and serve to resolve disagreement; and I say this with a clear conscience, without respect to any other purpose than to speak the truth."[60]

His understanding does not seem to have justified his confidence, as he suggested that the Portuguese might try to underrepresent the true extent of the intervening seas, or "gulfs" as he called them: that would have favored, rather, the Spanish case. Presumably Portuguese charts, perhaps those Francisco Rodrigues made on or after Abreu's expedition (above, p. 48–9), were among those he deployed for calculations "agreed by the Portuguese pilots who discovered" the islands (asentadas por los pilotos portugueses que las descubrieron).[61] It was not unreasonable for Magellan to assume that Portuguese sources doctored the figures in their own favor; his own computations, though not as wildly unreasonable as most of the Spanish calculations, were equally partial.

Magellan's other assertions were surprisingly tentative, although advanced on the basis of figures to which he assigned overconfidently exact values. The basis of his calculations, for what they are worth, was that the Tordesillas line should be measured from the westernmost point of São Antão in the Cape Verde Islands. He placed it twenty-two degrees west of that point. The Portuguese sector, he averred, began "in Portugal's own land of Brazil" (en la misma tierra del brasil de Portugal). What he called Cape Santa María—that is, the mouth of the River Plate—he placed six and a quarter degrees further west.

The three spice islands closest to the antimeridian were, he claimed, two and a half degrees east of it. He located the easternmost point of the Moluccas a degree and a half further east. It is worth noting that he was aware that the Moluccas straddled the equator, since—as we shall see—he did not behave accordingly when he reached the broadest part of the Earth at an advanced stage of his voyage across the Pacific in 1521 (below, p. 221).

His figures suggest that he thought that from the River Plate estuary he would have about 170 degrees of the surface of the Earth to traverse in order to reach the meridian of the Moluccas—an underestimate by some 20 degrees. His opinion that Malacca was 17.5 degrees West of the antimeridian seems inaccurate to an extent consistent with his other errors. We have to take into account the fact that in any case he seriously underestimated the size of the world. We do not know by how much: the nearest clue he provides is his suggestion that Malacca lay 1,600 leagues from the Cape of Good Hope—a real difference in longitude of nearly 84 degrees, or perhaps something over 75 in Magellan's mind; but as we know neither the value he assigned to a league nor the route over which he was reckoning the distance, the datum does not help much. Also to be borne in mind is that he probably thought the strait he sought was at a lower latitude than he in fact had to reach. By combining these observations, we can get some inkling of the extent to which he underestimated the task he faced. He made what was almost impossible with the technology at his command seem relatively easy.

To understand the image of the world Magellan had in his mind, one must think one's way into the geographical uncertainties of the time. The small world of Columbus was still a possibility: indeed, it looked more convincing than ever, now that awareness of the narrowness of America at the Isthmus of Panama shrank the dimensions of the hemisphere in most imaginations. The breadth of the Pacific was unknown, and to those who were emotionally invested in a viable westward route to the Pacific, almost inconceivable. The relationship of the New World to the Old remained unestablished. America might be a vast, previously unknown hemisphere, detached from Asia; but it was at least equally likely that it was a prolongation of the world known to Ptolemy and the other ancient giants of geographical wisdom: another great peninsula or salient, like those of Indochina and Malaya, displaced somewhat toward the east.[62] Many maps of the time showed it as such. In Magellan's mind, he had only to round or pierce such a protuberance—not cross a hemisphere to face an ocean wider than any known.

. . .

However well advanced his plans were when he arrived in Seville, they had four months to gestate or mature while he dallied with Beatriz, squabbled with Faleiro, plotted with Burgos merchants, and pestered the officials of the Casa de Contratación. When the petitioners were at last allowed to put their case before the king's chancellor, Jean de Sauvage, at Valladolid in March 1518, an eyewitness was at hand: Bartolomé de Las Casas, Dominican priest and loquacious lobbyist, was the most persistent spokesman at court on behalf of the king's native subjects in the Americas, unremittingly seeking respect for their traditional polities and hierarchies, exemption from exploitation by Spanish colonists, and relief from excessive tribute or demands on their labor. He was also, in effect, Columbus's literary executor and the self-appointed historian of the inception of Spain's overseas empire. His observations, the notes he made, the questions he put to Magellan, and the replies he recorded all deserve respect, albeit not unquestioning credulity.

Las Casas helps us answer a nagging question: What materials did Magellan deploy in support of his argument that he could reach the

---

MAP 2 (*next spread*). Magellan in the Indian Ocean.

Diu—took part in battle, February 1509

Goa—took part in Almeida's campaign, 1511

Cannanore—visits with Almeida and at intervals, 1505-13

Calicut—campaigns with Almeida, 1505; perhaps wounded

Cochin—resident, 1509-10 (at least); trades in pepper

Maldives—shipwrecked, with probable loss of part of his pepper investment, December 1509

Malindi—campaigns with Almeida, 1505

Mombasa—campaigns with Almeida, 1505

Kilwa—campaigns with or for Francisco de Almeida, 1505 and 1506

Sofala—campaigns with Almeida, 1505

INDIAN

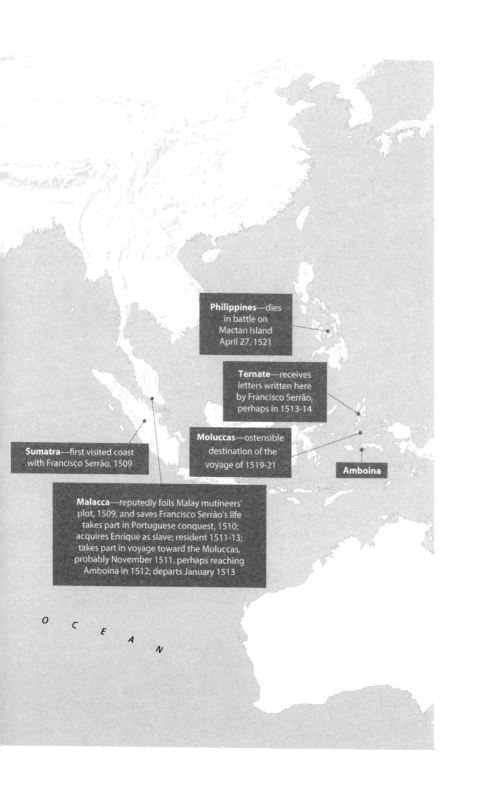

Philippines—dies in battle on Mactan Island April 27, 1521

Ternate—receives letters written here by Francisco Serrão, perhaps in 1513-14

Sumatra—first visited coast with Francisco Serrão, 1509

Moluccas—ostensible destination of the voyage of 1519-21

Amboina

Malacca—reputedly foils Malay mutineers' plot, 1509, and saves Francisco Serrão's life takes part in Portuguese conquest, 1510; acquires Enrique as slave; resident 1511-13; takes part in voyage toward the Moluccas, probably November 1511, perhaps reaching Amboina in 1512; departs January 1513

O C E A N

Moluccas via the Atlantic? The would-be explorer's saddlebags, according to Barros, were full of maps and documents, calculated to appeal to the king of Spain, who "was in love with mariners' charts and globes." Barros, as usual, placed most of the blame for what he considered an act of treachery on the correspondence of Francisco Serrão, on which Magellan "principally rested his case and also relied for his reasons for it." He also condemned Faleiro's input.[63] But what else had contributed to the formation of the plan?

We know from a later revelation of Francisco, Rui Faleiro's brother, that Magellan had access to maps supposedly by Vespucci—whose reputation as a mapmaker seems to have been entirely the result of self-recommendation—and Nuno Garcia, who was a reputable professional (below, p. 111). Faleiro reworked their products, adding lines of longitude.[64] He might also have had maps in which, as we have seen, Schöner and Lisboa recorded rumors of straits arising from recent voyages. But no such maps seem to have been to hand in the crucial encounter with Sauvage.

Las Casas's most startling revelation—which there seems no reason to doubt—is that Magellan demonstrated his plans by exhibiting a globe on which no means of exit westward from the Atlantic appeared: he "brought with him a well painted globe showing the entire world, and thereon traced the course he proposed to take, save that the strait was purposely left blank so that nobody could anticipate him." It is as if an applicant for a patent declined to describe his product, or a supplicant for investment concealed part of his plan on the grounds that his potential investor could not be trusted. When a king or a king's representative is the object of the distrust, the exhibitor's chutzpah seems breathtaking.

By Las Casas's own account, he did not know the basis for Magellan's confidence until the voyage was over, when he was able to read an account by Antonio Pigafetta, one of the expedition's survivors, who claimed that the explorer had seen "a well-hidden strait" depicted on a map "by that excellent man [*quello exelentissimo huomo*], Martin of Bohemia":[65]

According to what an Italian gentleman named Pigafetta of Vicenza wrote in a letter, Magellan was perfectly certain to find the strait because in the treasury of the king of Portugal he had seen on a nautical chart made by one Martin of Bohemia, a great pilot and cosmographer, the strait depicted just as he found it. And because the said strait was on the coast of land and sea, within the boundaries of the sovereign of Castile, he therefore had to move and offer his services to the king of Castile to discover a new route to the said islands of Malacca and the rest.

Martin Behaim, often called "of Bohemia," was a merchant by vocation, based in Nuremberg, who knew Portugal well. Whether he deserved Las Casas's and Pigafetta's compliments may be doubted. One of his trips abroad, in 1483, probably had an ulterior motive: to postpone or avoid a sentence of three weeks' imprisonment for dancing during Lent at a Jewish friend's wedding. He was in Lisbon in 1484 and seems to have caught the geography bug in that city of Atlantic explorers, where coastal surveying voyages down the west of Africa were under way. His claim to have accompanied those expeditions is unsupported by any other evidence and seems incompatible with errors he made in his representation of the lay of African shores. His ambitions exceeded his knowledge.

When he got back to Nuremberg in 1490, his tales excited expectations that he could not honestly or perfectly fulfill. Still, although he had little or no practical experience of navigating or surveying, he was a representative armchair geographer of his day, who conscientiously compiled information of varying degrees of reliability from other people's maps and from sailing directions recorded by real explorers. The data he brought to Germany from Portugal were bound to arouse the enthusiasm due to shards of insight from the cutting edge of the exploration of the Earth.

A globe of his survives, made in 1492. The most conspicuous feature Martin incorporated from the latest Portuguese discoveries was his depiction of the Indian Ocean as accessible from the Atlantic, around the southern tip of Africa. He shows the African coast trailing a long way eastwards—a relic of an old mapmaking tradition that represented the Indian Ocean as landlocked and effectively barricaded to the south by a great arc of land, stretching all the way from southern Africa to easternmost Asia. No surviving work of his shows America; so in the globe of 1492 the way westward to the Moluccas is open. If Magellan displayed a later globe by the same maker, it must have shown Columbus's discoveries, blocking the route. Either Pigafetta misattributed the map, or an unknown work of Behaim's included information similar to that which Schöner and Lisboa used in their versions of an opening or openings from the Atlantic to the sea beyond America.

In any case, if Las Casas is right, Magellan's emphasis must have been on the attractions of the Moluccas, rather than on the location of a strait that might lead to them. Serrão's letters may have helped him. According to Gómara, as we have seen, he flourished missives from the Moluccas in which Serrão "asked him to go there if he would like to get rich quick."[66] He also brandished a printed account of the islands: a copy of an account of travels in the East, published in 1510, which was an immediate hit in the busy market for travelogues—so much so that the work went through at least twenty-seven editions in four languages by the middle of the century.[67] The author called himself Ludovico Varthema and claimed to travel for no more probable purpose than "for my pleasure and to know many things."[68]

. . .

Varthema is so elusive and practically undetectable in other sources that it is tempting to dismiss his identity as suppositious, his work as spurious, and his astonishing gift for accuracy in description and fidelity in narrative as the product of perverse genius, rather like that of George Macdonald Fraser's justly famous "Flashman" novels—improbable historical fictions so meticulously researched that it is almost impossible to catch the author in an anachronism or factual slip. Like Flashman, Varthema is a picaro whose adventures lend themselves to the episodic structure of a traveler's narration. He resembles Flashman too in his conviction of his own sexual irresistibility and displays a canny sense of the book market in every lubricious scene of seduction and sexual hospitality.

There is a Sinbad-like timbre in his tales, which take us through a series of hairbreadth escapes from disaster that beggar belief. His most egregious escapade leads our hero disguised as a Muslim to Mecca, and thence to Aden, where, after detection as a Christian, he escapes the punishment he deserves in the arms of the sultan's wife. The story, if true, qualifies him as the first recorded Christian visitor to the Muslim holy city, which he describes with such circumspection and infallible detail that it is hard to disavow his credibility. He was a gifted linguist, whose transcriptions of exotic languages fortify his claims to have been able to speak them.

His existence is impeccably corroborated—not least by the most skeptical of his denouncers, the physician and renowned herbalist García de Orta, who had it on the authority of common friends that Varthema was a great liar.[69] At least, to lie he had really to exist. Magellan had the opportunity briefly to know the liar in person because Varthema, by his own account, took service with Almeida in Cannanore, where he brought stories of his alleged visits to the Spice Islands. Corroboration may be seen in a reference to one "Ludovyco," cited as a source of intelligence in a report about Aden in late 1506.[70] Orta denied that Varthema had ever got beyond the Malabar coast, but Magellan seems to have placed some faith in Varthema's claims about twenty or so islands east of Java, including Bantam, Borneo, Tidore, "and other spice islands."[71]

Did Varthema sucker Magellan, or was the latter merely willing to call any text, however dubious, into service? The availability of Varthema's book, which was published in a Spanish edition in Seville shortly after Magellan's departure, and the reissue of the *Book of Marco Polo* in May 1518 suggest that there was some interest in works on the Orient in Seville at the time.[72] On the face of it, however, Varthema's text seems hardly serviceable to Magellan in Spain because of its extravagant praise of Portugal and of Portuguese valor and invincibility— clues, perhaps, to Varthema's undisclosed agenda as a Portuguese propagandist and to his true identity.

Moreover, the writer was at such pains to vindicate Portuguese claims to sovereignty over the Spice Islands that he denigrated the inhabitants in order to impugn their right to protection under natural law. The effect was to make the archipelago seem frankly repellent. In Banda, "ugly and gloomy," where the search for nutmeg focused, life was nasty and brutish, law was unknown, and the people were "so stupid that if they wished to do evil they would not know how to accomplish it." Those of Ternate, which produced in cheap abundance cloves even more valuable than nutmeg, were even worse and equally savage in their manner of life.[73] Magellan must have known from his correspondence with Francisco Serrão that Varthema's text traduced the sophisticated culture of the islands and the commercial canniness of their denizens. The implication, however, of islands that it would be lawful to seize, easy to conquer, and profitable to exploit was as appealing to King Carlos as to King Manuel. The truth as told by Francisco Serrão and confirmed by Tomé Pires was that Ternate was a legitimate polity, such as where "powers that be are ordained of God," dispensing justice according to law, and therefore exempt, according to the jurisprudence of the day, from conquest by Christians. Lands of lawless savages, however, were up for grabs.

Las Casas felt immediate anxiety about the viability of the proposed venture. "I asked him," the Dominican recalled,

> what route he proposed to take. He replied that he intended to take that of Cape Santa María, which we call Río de la Plata, and thence to follow the coast along until he found the strait. I said, "What will you do if you find no strait to pass into the other sea?" He replied that if he found none he would follow the course that the Portuguese took.[74]

The admission showed a potentially fatal flaw in the plan. Having betrayed his former master, the king of Portugal, Magellan now showed himself willing to disregard international treaties and infringe Portugal's rights of exclusive navigation around the Cape of Good Hope. The king of Spain could not lawfully authorize such an outrage, even had he been willing to do so.

. . .

Doubts of Magellan's good faith dogged him. Throughout the Atlantic phase of his great voyage, as we shall see, subordinates would question his intentions; some would simply assume that he never intended anything other than a covert operation around the southern tip of Africa. Spaniards who observed his preparations for the voyage, once it was authorized, or who were assigned to his command never ceased to suspect that he would betray them, as he had betrayed his native country, or that his entire course of action was a subterfuge designed to hoodwink the king of Spain in Portuguese interests. Some mariners refused to ship with him on the grounds that he was a foreigner. Officials of the Casa de Contratación recorded repeated insults Magellan endured from those who questioned his goodwill, while his father-in-law complained of the constant hostility, although "the more obstacles and burdens there were to frustrate his voyage, the stronger the will for it Magellan and Faleiro showed."[75] Xenophobic rumor-mongers kept up their campaign of vilification throughout the voyage (below, pp. 116, 170–2).

An incident in October 1518, while Magellan was at work in Seville on preparations for the voyage, illustrates the tensions that arose from the mutual suspicions of Spaniards and Portuguese. Magellan hoisted flags bearing his own coat of arms from the yardarm of one of his vessels—a gesture typical of his self-advertisement and pride of ancestry. He wrote to the king of Spain, explaining that he was unable to fly the monarch's banners because "the painting of them was not yet finished and because, what with the work of getting the ship afloat, I did not have time to see to it."[76] A mob gathered, indignant at what they saw as Portuguese devices, though Magellan explained to the police officials who came to investigate that "the arms displayed were not the king of Portugal's but mine, and that I am your Highness's vassal."[77]

The incident escalated. Dr. Juan de Matienzo, a high official of the Casa de Contratación, ordered Magellan to remove the offending symbols. The official report of the incident confirms that Matienzo then attempted to have Magellan and his henchmen arrested, "calling up officers to seize the Portuguese captain who hoisted flags for the king of Portugal."[78] Allegedly, they dispersed the workmen who were helping to prepare the fleet, stabbed a pilot, and disarmed members of Magellan's crew. The aggrieved commander and his men resisted, assaulted Matienzos, and drove off the intruders. "It seems very much at variance with your Highness's nature," Magellan concluded, "that men who leave their own kingdom and homeland to serve you in so important a matter should be so ill treated."[79]

Magellan went unpunished, despite what must have seemed a dangerously veiled threat in his letter to King Carlos. He was willing to remove the flags, he said, "because there was present a knight of the king of Portugal's service whom his king had ordered to come to Seville to treat with me to return to Portugal."[80] In other words, Magellan could renege on his commitments if he did not get what he wanted. The threat was serious. Portuguese representatives were indeed hard at work on two fronts: trying to detach Carlos from Magellan and Magellan from Carlos.

Toward the end of September 1518, the Portuguese ambassador, Alvaro de Costa, reported home on how hard "I have worked in the matter" (sobre el negocio de Fernam de Magalhees [sic] he trabajado muchísimo). He had told the king frankly "how ugly it looks for one king to take vassals of another into service" and "that it was something unwonted among knightly persons."[81] To these considerations of propriety he added others more practical. "In a matter of such slight prospects and so uncertain," the king of Spain "had enough vassals of his own to undertake exploration without reaching for the help of men who came to him because they were discontented" in Portugal.[82] Carlos, the ambassador continued, made a show of dismay but referred the matter to the bishop of Burgos to resolve: as we have seen, the bishop was already deeply implicated in Magellan's dealings with members of his flock. Advisers, the ambassador complained, "persuaded the king of Spain that he must continue with what he had started and that the planned destination for the voyage lay within Spain's zone."[83] Carlos's final reply included one point intended for reassurance—Magellan and Faleiro were "of little account" (de poca sustancia)—and one of countercomplaint: Portugal made similar use of many Spanish personnel. Costa summed up his report by suggesting that his master should recall Magellan to Portugal: "It would be a serious blow to these Spaniards" (que seria gran bofetada para estos). Faleiro, on the other hand, the ambassador opined, was negligible—"sleepless and half out of his mind."[84]

At the end of February 1519, Carlos wrote in person to reassure Manuel of his good intentions: "I have heard," he disingenuously explained, "that you have some suspicions that some prejudice to your rights in those regions may arise from the fleet that we have ordered to assemble to got to the Indies, under the command of Hernando de Magallanes and Ruy Faleiro. . . . The first item and command of ours," he added, "that the said captains have in writing is that they respect the line of demarcation, and, under grave penalties, that they in no way infringe on the regions and lands and seas that are assigned to you according to the demarcation agreement." The language of the king's instructions to his prospective commanders reflects as much his fear that they might exercise their own initiative as his anxiety to oblige his Portuguese counterpart.[85]

Portuguese agents' efforts to detach Magellan from Spanish allegiance continued through the summer—first in Barcelona in May,[86] then, while the fleet was in preparation, in Seville, from where in July the Portuguese representative, Sebastião Alvares (above, p. 69) wrote a circumspect account of his efforts. He had begun by taunting Magellan for derogation from gentleman to chandler. "Seeing a favorable opportunity," he wrote,

> for doing what your Highness had told me, I went to Magellan's lodging and found him gathering victuals and preserves, etc., and I said to him that such was the outcome of his evil purpose; and as this was the last time that I should speak to him, as his friend and as a good Portuguese, he should think well on the error he was about to commit."[87]

Alvares still hoped for a change of heart on Magellan's part, but the ensuing dialogue, which can be reliably reconstructed from the letter's *oratio obliqua* (in accordance with my threat or promise above, p. xiii), disabused him.

"My purpose," said Magellan, "is to continue with what I have begun."

"There is no honor in improper gains," Alvares countered. "Even the Castilians regard you as a reprobate and a traitor against your homeland."

"My intention in my voyage is to do nothing against his Highness's service and to infringe on nothing that belongs to him."

"It will be a great cause of harm to Portugal," insisted Alvares, "if you find the riches you have promised within the Castilian zone. But if you do your duty, I promise that his Highness will consider himself well served."

"But I know of no reason for abandoning the king of Spain, who has shown me such great favor."

"The best reason is to do your duty and avoid dishonor. It is shameful that you should have left Portugal and broken your allegiance, merely because the king omitted to give you a paltry 100 reais' increase in your stipend."[88]

Despite the show of resolve, Magellan seemed impressed with how much his interlocutor knew of his affairs and evidently was as mistrustful of his new employers as they were of him. Alvares played on his fears, intimating that, if intelligence reports were reliable, Faleiro would not adhere to the planned route but would head south—by implication, for the Cape of Good Hope. Other subordinates had secret orders, countermanding Magellan's, and the bishop of Burgos was untrustworthy.

The warnings struck home. As we shall see, Magellan was in constant fear of mutiny when the expedition was at sea. If Magellan wanted to succeed, Alvares insinuated, he had better make sure he was in sole command. "I said," he wrote,

> that it would better behoove him to see the rival orders, and check on Rui Faleiro, who said openly that he would not follow his lantern and that he rather intended to sail southward or he would not go in the fleet. And I warned him that he should take care to sail as commander-in-chief, and that I knew that there were other orders contrary to his own of which he would remain unaware until it was too late to redeem his honor. And that he should pay no heed to the honey with which the bishop of Burgos's lips were coated.[89]

Magellan began to shift his ground, saying that he might desist from his undertaking if the Spaniards defaulted on any matter that had been agreed. What favors would the king of Portugal grant him, he asked, if he should change sides again?[90] Some dust, it seemed, still clung to Magellan's feet. Alvares pressed his advantage. The time had arrived, he continued,

> when, if Magellan would give me a letter for your Highness, I would undertake to deliver it for love of him and would do my best for him; for I had no message to give him from your Highness but was merely telling him frankly what I understood, just as I had on many previous occasions. He said to me that he would make no reply before seeing whatever message from your Highness the post might bring.[91]

Alvares also approached Rui Faleiro, finding that his mental condition contributed to his intractability. "He would say nothing," Alvares reported, "except that he would not go against the king his lord who did him so much honor. . . . It seems to me that he is like a man whose reason has gone awry [*como homem torvado do juizo*], but if Magellan changes his mind I think Rui Faleiro will follow where Magellan leads."[92] Suspicions of Faleiro's sanity were not confined to self-interested Portuguese. Gonzalo Fernández de Oviedo reported that he "lost his reason and was very mad."[93] After his condition led to dismissal and confinement, rumor had it that his renown as a magus was a sort of Faustian infirmity, contrived by the same diabolic spirits as caused insanity.[94]

. . .

Alvares's account was of course designed to excuse his failure. But the picture he provides is as convincing as it is disquieting: Magellan was infinitely flexible in his loyalties; willing to trim if he caught a whiff of a favorable wind; disposed to tack between rival courses; beset by anxieties about the good faith of his partners and collaborators; unwilling to commit himself, and susceptible to suspicion. He was glib in the language of honor he had learned from his chivalric readings but unreliable in adhering to principles.

What was his real objective? Was he double-crossing both his masters? The king of Spain's confidence in him was shaky. As we shall see in the next chapter, he got the commission to make the voyage he proposed less because of his own merits and arguments than because of the representations on his behalf that merchants of Burgos made. The king's concern—justified in the event—was that Magellan, once afloat, would disobey orders. It seeps between the lines of the many documents ordering strict compliance with the agreed plan. On April 19, 1519, for instance, Carlos wrote to Magellan and Faleiro, referring to the corroboration merchants' evidence provided of the wealth of the Moluccas:

> Inasmuch as I have it for certain, according to the ample information I have received from persons who have relevant experience and have seen it with their eyes, that in the islands of Maluco there is the spicery that it is your principal objective to seek with that fleet, my will is that you go directly on the route to the said islands . . . so that, first and foremost and before any other destination, you go to the said islands of Maluco, absolutely without fail.[95]

What other destination might have lured Magellan? There are no means of knowing where, apart from the Moluccas, he might have contemplated before he set out. But there are reasons to wonder whether he was as single-minded as Carlos demanded, or even whether the Moluccas were his main priority. Their location, in relation to the Tordesillas antimeridian, was doubtful. Francisco Serrão had already got there and was acting in Portugal's interest. Magellan knew from his friend's letters of the proposal, at least, to build a Portuguese fort there. The islands were already in touch with Malacca.[96] Any Spanish interlopers would be vulnerable to attack from Portuguese bases relatively near at hand. When choosing between hypotheses in the absence of decisive evidence, the most appealing is the one that fits all the facts. That the Moluccas were Magellan's objective is consistent with some evidence but hard to reconcile with the fact (below, pp. 221–2) that, when he had the chance, he did not go there.

Magellan knew of another archipelago, beyond the Moluccas, still unpreempted by Portuguese rivals and more likely to yield easy riches. The attractions of the Philippines (as they would later come to be called) were obvious, especially the gold—the commodity that surpassed all others in value (above, p. 24). The colony of Filipinos in Malacca made Magellan, like other inhabitants of the city, aware of the location and relative vulnerability of the islands, as well as of their wealth. As we shall see, one of the most astonishing decisions he made in command of his fleet was that when they got to the equator—where, as he well knew (above, p. 80), the Moluccas were located—he passed the islands by in order to proceed to the latitude of the Philippines. The change of destination defied the wishes of his officers and men and imperiled the welfare of shipmates. who were already, by that time, after traversing the Pacific, weak, sick, and starving.[97]

When he got there, he found the expectations of abundant gold to be entirely justified. The log of one his pilots, whose descriptive style was always austere and who rarely mentioned anything outside his strict professional ambit, supplied unaccustomed detail:

> At a large island, . . . which is inhabited and has gold in it, we sailed along the coast and made for the southwest, where we made landfall at another, small island . . . and it is inhabited, and the people of it are very good, and there we raised a cross on top of a hill, and from there they showed us three more islands toward the west-southwest, and they say that there is much gold, and they showed us how they gather it and they find nuggets as big as chickpeas and others the size of lentils. And this island lies on nine and two-thirds degrees north.[98]

Magellan took extraordinary measures to prevent the gold from escaping his control. He forbade his crew to exchange iron, clothing, or other truck for the precious metal. The testimony his unruly lieutenant, Juan Sebastián Elcano, gave after all the disasters of the voyage were over and he had led the expedition's handful of survivors back to Spain was never likely to flatter Magellan, but it was backed by the witness's authority as bookkeeper of the expedition at the relevant stage. "Let no man dare," Magellan ordered in Elcano's account, "on pain of death, to trade gold, or take gold, because the commander wanted to keep the price of gold low." Every transaction, Elcano continued, was recorded in the accounts he kept as treasurer.[99] López de Gómara's narrative, from data gathered nearly a decade later, recorded the rumor, rife among the crew, that Magellan had all along been seeking "mines and sands of gold." When they first beheld it on arrival in the Philippines, he let slip a declaration that Ginés de Mafra, who was an eyewitness, recorded. "Now," he told his men, "I am in the land that I have desired" (Ya estaba en la tierra que había deseado).[100]

The contract Magellan had made with the king of Spain before departure contained a mystifying clause—mystifying, at least, if the Moluccas were Magellan's only objective. If he were to find more than six islands, he could select two for his own benefit and receive a fifteenth of any profit or product they might yield (below, p. 104). The received wisdom, confirmed in the books of Pires and Varthema, was that there were only five Moluccas (though nowadays we count far more as belonging to the region). So, irrespective of the priorities of the king of Spain, of the merchants of Burgos, and of the men who sailed with him, Magellan had an objective beyond the Moluccas in mind. Where can it have been but the Philippines? Biblical noises rattled in heads. Sacred imagery speckled maps and works of geography. The realm of Tarshish, distant Ophir, the mines of Solomon, the provenance of Sheba and the Magi, the isles that would fear the Lord and the ends of the Earth that would draw near—these were real places in the pages of scripture and the minds of those who heard readings from it. Columbus sought them across the Ocean. They might lie only a little way beyond it.

FIGURE 4. The harbor of Sanlúcar de Barrameda, where Magellan's fleet left the Spanish mainland on September 20, 1519, as imagined by Theodore De Bry in 1594. He made the engraving to illustrate a German translation of Girolamo Benzoni's *Historia del mondo nuovo* (History of the New World), a work adversely critical of the Spanish Empire and therefore agreeable to De Bry, a propagandist on behalf of Protestantism and resistance to Spanish power. Though the image is meant to be evocative rather than realistic, many elements recognizable to Magellan appear: the thirteenth-century walls; the spire of the Church of Santa María de la O; on the hill in the background the fort of Santiago, newly built by the local overlord, the Duke of Medina Sidonia; the Torre del Oro in the foreground; the beach still used for lading; the bustling port, with diverse types of shipping; and the tunny fishers, shown mending their nets at bottom right. From De Bry, *Das sechste Theil Americae* (Frankfurt, 1594). Courtesy Library of Congress.

CHAPTER 4

The Making and Marring of a Fleet

Seville and Valladolid,
1517 to 1519

The ships of Tarshish did sing of thee in thy market: and
thou wast replenished, and made very glorious in the midst
of the seas.

—Ezekiel 27:25

That an overdose of aphrodisiacs killed him is a plausible myth. King
Fernando of Aragon was in want of an heir when a painful seizure overtook him on the road to Guadalupe in Extremadura on January 23,
1516. A weak heart, strained by the exertions of the road and the ill-
advised temptations of the hunt, were probably responsible for his
plight. He was hurriedly installed in a handy hermitage—the only
accommodation available, "comfortless and badly appointed" (des-
guarnecida e indecorosa)—to die.[1] Though with his first wife, Isabel the
Catholic, he shares a reputation as the founder of Spanish unity, he
intended to split the kingdoms of Castile and Aragon after his death, as
so many temporary unifiers of Iberian states had done in the past.
Whatever the effect of the aphrodisiacs, his second marriage was infer-
tile. His grandson, as his only heir, scooped the pool; the whole of Spain
became the inheritance of Carlos I.

Carlos arrived in Spain from his previous home in the Netherlands
only about a month before Magellan crossed the frontier from Portugal.
Everyone with a plan to propose or a petition to present must have
thought the moment propitious. After King Fernando's failures with
previous expeditions in search of a westward route to Asia, Magellan

was presumably among those who thought that the change of regime enhanced their projects' chances of adoption. The route, however, to royal attention—let alone approval—was long and arduous, even at the best of times. Outsiders remained excluded without inside help. The royal court was a labyrinth: you needed guides, and skill in unraveling complex threads. To reach the guides, you needed *enchufe*, as Spaniards call it—personal connections.

In Magellan's case, the obvious route led through the Casa de Contratación, if only he could attract the favor or interest of an official there. He had introductions from two sources: first, his father-in-law, who, as we have seen, was in close professional touch with the people who ran the Casa; and second Cristóbal de Haro. As a Burgos merchant, Haro was a fellow citizen and collaborator of a figure who was on what we might now call the board of the Casa and who proved to be key: Juan de Aranda. Both men belonged to a network that bound Seville's Burgos-born colony in ties of marriage, godparentship, and commercial partnerships. Aranda had lived and worked for a time in Magellan's native city.[2] His associates numbered other Burgos merchants who had made profitable investments in the Indies trade, including Cristóbal de Haro. One of his Lisbon correspondents, who sent information about Magellan, was Diego de Haro, Cristóbal's brother.

Aranda, as Magellan avowed in testimony recorded in November 1518, was uniquely qualified to help him. "No one," he said, "was as well fitted to understand" what he and Faleiro wanted.[3] Aranda's estimate of his own importance was no lower, but he emphasized, by implication, Diego de Haro's data as well as his own skill. "I helped," Aranda stated, referring to Magellan and Faleiro, "in reporting what I had heard concerning their abilities and the feelings their defection aroused in Portugal."[4]

The Casa, however, was a nest of vipers. Other officials detested Aranda. He complained to the king at how they excluded him from their counsels.[5] It is hardly surprising, therefore, that Casa officials were reluctant to give Magellan and Faleiro much credence; more surprising was the confidence Aranda showed in the adventurers—a function perhaps of commercial judgment but more probably of his obligations to the likes of Haro and Barbosa.

Every adventurer needed a royal patron, albeit not necessarily to put money into any particular venture: Magellan's voyage was unusual in relying, in the event, on royal financial sponsorship. The normal pattern was for private investors to carry all or most of the risk. Without backing, however, from a great potentate or sovereign authority, no explorer could hope to hold onto whatever he might find or to get rewards commensurate with his service: without princely protection, his finds would be up for grabs by rivals or interlopers.

Eventually, Aranda received the king's leave to bring the two Portuguese supplicants to court. They left Seville on January 20, 1518. Perhaps to boost their credentials as conners of the Orient, they took with them Enrique de Malacca and an unnamed "Sumatran" woman— exotic exhibits to enrich their pitch.[6] The so-called Sumatran was another of Magellan's acquisitions in Malacca and, like Enrique, a useful linguist. Presumably her provenance could be traced as far as Sumatra, but she may have come from further afield. Though she is not named in the crew lists, López de Gómara, who had no discernible reason for inventing or misrepresenting her, reported that she sailed on the expedition and was able to interpret the speech of islanders in the Ladrones, as Magellan called the islands now known as the Marianas.[7] If that were true—and no other source corroborates it—it would make her the only person identifiable as a woman aboard Magellan's ships. Faleiro's brother, Francisco, made another of the party that traveled along on the road to court. Aranda or his messenger caught up with them when they were about twelve miles out of the city, with confirmation of their rendezvous at court, but soon left them to follow him as he hurried ahead for undeclared reasons—apparently to communicate with the royal chancellor, Jean de Sauvage, and prepare the would-be explorers' reception.

After meeting up and spending a night at the great crossroads of Medina del Campo, the party reconvened at an inn at Puente del Duero, about fifteen kilometers south of Valladolid. We can reconstruct the events that followed in some detail, thanks to the record of the interrogation of the main participants in Barcelona, more than a year later, when litigation between the partners threatened. In the inn at Puente del Duero, according to Magellan, Aranda set out the terms on which he expected recompense for his help. "You," he told his protégés,

> will have now no cause to complain of what I have written to the Grand Chancellor. On the contrary, in recognition of it and of what I shall do in laying before his Highness the information I have concerning you from Portugal, you must give me part of whatever gains, God willing, you make as a result.[8]

His initial bid was for a fifth of the profits. Francisco Faleiro, who claimed, when the case came to court, that he did not remember the amount in question, wanted to reject Aranda's demand outright. Aranda, in his turn, testified that Francisco offered a tenth. He broke off negotiations and withdrew to Valladolid, where the chancellor awaited.

Aranda was in a strong position. The project might never reach the chancellor or come to the king's attention without his help. He threatened and cajoled the Portuguese with the extent of his influence. "I have many friends at court," Faleiro recalled hearing him say. The sentence conveys a nice mix of menace, promise, and boast. Faleiro responded with "annoyance," Magellan with anxiety. The danger that Aranda would use his influence against them was real. The negotiations ended in a compromise that, as Aranda recalled, Magellan proposed: Aranda would receive an eighth of the proceeds of the voyage, if and only if he was able to negotiate a deal with the crown for the entire cost of the proposed expedition. Otherwise he would get nothing. Compromise was not a typical recourse for Magellan: not with the king of Portugal, or with litigants, or with rivals. In the case in point, however, he chose to impose his will on the Faleiro brothers rather than on Aranda.[9]

Aranda's brinkmanship worked, at least initially. On February 23, Magellan and Faleiro visited him while he was staying in the home of a fellow merchant, Diego López de Castro. According to Magellan's testimony, Aranda continued to pester his interlocutors, demanding to know "whether they did not want to give him some satisfaction for the trouble he had taken and the assistance he had given, and saying that he would take it as a favor."[10] The language of Magellan's evidence seems calculated to imply that Aranda had no right to what he asked. He invited his interlocutors to sign a notarized agreement, which he had ready to hand, granting him an eighth of their prospective gains:

> May you have the eighth part of all the profit and interest that we may have from the discovery of the lands and islands, which, please God, we have to discover and find among the lands and within the limits and line of demarcation of King Don Carlos, our lord.[11]

They signed reluctantly.

Why did Aranda propose this settlement, let alone agree to it? It presented, from his point of view, two obvious advantages. First, it put the job of negotiating with the king in his hands. Second, it meant that at no cost to himself he had the chance of cutting out potential rivals.

Haro was the obvious target. According to the royal secretary who married his niece, Haro would have been willing with the help of "friends" to fund the expedition if the king was unforthcoming. Indeed, he signaled his interest in a letter to the king as early as March 5. Faleiro was the dissatisfied party: he later claimed that he had never endorsed the agreement with Aranda, which Magellan negotiated without reference to him. In any case, the Portuguese were left to languish in Puente del Duero, or at inns in other spots on the Valladolid road, while Aranda returned to court and worked on the susceptibilities of the chancellor.

He found Valladolid seething with intrigue. The venerable court humanist Peter Martyr of Anghiera, an inveterate gossip who was writing his own history of Spanish explorations and conquests, complained of the rapacity of the Flemish hangers-on who clung to the young king—"denizens," as he called them, "of the sea of ice," who looked down from the north on the south and despised Spaniards "as if they were the offspring of their stews."[12]

Nonetheless, Aranda's proceedings prospered, perhaps because large sums were arriving from the Indies. Seventeen thousand ducats poured into the treasury within a few days of the new king's arrival; petitions from the Indies were holding out fabulous prospects, promising revenues of 400,000 ducats a year and rising.[13] Investment in exploration looked worthwhile.

Aranda procured an audience for Magellan and Faleiro with the grand chancellor and Bishop Fonseca; evidently, he was sufficiently satisfied with the progress of the deal to allow an interview to go ahead. The would-be explorers arrived in Valladolid at his invitation on March 17. He expected them to stay in his house. They did so for one night, but their haste in decamping to lodgings, which they found for themselves, shows the tensions in the partnership or, at least, their unwillingness to be under their mediator's surveillance.

. . .

The interview took place on March 17 or 18. We have seen it through Las Casas's eyes (above, p. 84): the unfolded mystery-map with the crucial part of the route unmarked; Serrão's letters and Varthema's pages, flourished and flashed to prove the wealth of the Moluccas; the skeptical interrogation. The decision was deferred to the king, who in turn referred the matter to the royal council. The members' opinion was unfavorable. They did not believe there was any strait or any navigable way out of the Atlantic into eastern seas. God

had perhaps for his good reasons left the eastern and western moieties separate and closed off from one another, in such a way that it was impossible to

navigate or find a passage between them. Or it might be that that great mainland . . . was so continuous and endless that it parted, limited, and separated
the eastern from the western seas, in such a way that by no means could
anyone pass or navigate toward the Orient in that direction.[14]

It sounds like a description of the old Ptolemaic-inspired world maps
that showed a world without a Pacific Ocean, and an Indian Ocean
enclosed, as the tips of Africa and the easternmost peninsula of Asia
curled inwards to meet. The Moluccas, in this version of how the world
looked, were inside the enclosed ocean. If the councilors' rejection were
not well attested, it would be tempting to see it as a literary topos—a
peripety of fortune, designed to enhance the drama, an echo of a similar
change of fortune in Columbus's story, when, after the dismissal of his
suit at court, he was suddenly recalled at the behest of the king and
queen—as he disconsolately rode his donkey in the direction of exile—
to receive a royal commission, after a sudden change of heart on the
monarchs' part, in January 1492.

The councilors, of course, were no geographical experts. But others
saw the attractions of the scheme. "If it succeeds," wrote Peter Martyr,
"we shall wrest the trade in spices and precious stones from the Portuguese."[15] Fonseca had to be satisfied: it seems unlikely that the council
would have issued its demurral in defiance of his wishes: either he
changed his mind, or the council's dictum was a negotiating ploy. In any
case, the king, with his notorious love of maps and charts, trusted his
own judgment. He made up his mind in favor of the scheme with astonishing rapidity. He arrived back in Valladolid on March 20, after escort-
ing his mother to a place of retirement in Tordesillas. He countermanded
the council and approved Magellan's and Faleiro's plan on March 22.

The likelihood that Cristóbal de Haro would collar the profit by
putting together a consortium of his own may have influenced the king's
decision to finance the entire fleet himself; but the option of exploiting
the financier's help for extraneous or marginal costs remained. The document Magellan and Faleiro had submitted for the consideration of
king and councilors allowed for the finance for the voyage to come from
either a private company or the crown. Although annotations by royal
officials show that the king had already decided not to alienate the
project to the private sector, there was scope for private investment in
items of cargo that could be exchanged for spices or other trade goods.[16]
Royal instructions ordered the Casa to provide the truck but did not say
that it had to be acquired at the king's expense. On March 10, the king

had asked Fonseca to find financiers willing to lend money for the voyage, in exchange for the right to load goods to exchange for spices in the event of success.

How much cargo was wanted and in what it would consist were moot questions. Magellan wanted at least a million maravedíes' worth of trade goods. The king thought in terms of at least a million and a half. The commodity most likely to be in demand in the Far Orient was silver, but none was available. Mercury, however, which was vital in most processes of extraction of silver from ore, was abundant in Castile and became the most important item of cargo, together with dyestuffs for the cloth that India was known to produce for export to the Moluccas and other destinations in the region.

Haro put together a consortium that included none of the other great merchant houses: his named co-investors were Alonso Gutiérrez, a royal councilor and city councilor of Seville with a reputation as a financial fixer, who in the end supplied only 157,500 maravedíes, and Juan de Cartagena, who, in disregard of potential problems of conflict of interest, would serve aboard as financial overseer and captain of a ship. His contribution was negligible: less than 50,000 maravedíes.[17] Evidently, Haro wanted to keep in his own hands, and for as long as possible, any commerce the venture might inaugurate. Where, in turn, Haro got the money he put up is uncertain. The Fugger later claimed that he had borrowed some of it from them, but never proved their case.[18]

Haro's coup cut Aranda out of the deal. His goodwill was worth more than his rival's good offices: it was measurable in cash. Aranda did, however, have a negotiable asset: the agreement he had signed with Magellan and Faleiro. Royal officials took a long time to ponder how best to stifle his claims. On June 25, 1519, Fonseca informed him that his contract with the Portuguese was null and void, on grounds of conflict of interest in his capacity as an official of the Casa. Aranda accepted defeat but continued to petition instead, as "inventor and first cause" (imbentor y primera causa) of the voyage, for an increase of salary and for his job at the Casa to be made heritable.

. . .

The terms on which the king would provide his support for the voyage had still to be settled. A record of the witness statements taken in Barcelona the following year shows that Juan de Aranda advised his partners to moderate the demands they proposed to make of the crown.[19] He advised them to ask for a twentieth instead of a tenth of the revenues that might accrue from the expedition ("de lo que rentase todo lo que se descubriese").[20] That would have been consistent with typical Spanish practice. Magellan and Faleiro admitted to the official interrogators that they felt annoyance when their demands were met with caviling. They blamed Aranda.

The crown also rejected their demand to be designated "admirals." The term implied more than command of a fleet; in Spanish tradition, it was a hereditary office, which conferred some rights of jurisdiction. It was contrary to Carlos's policy to alienate such rights. The example of Columbus, moreover, who had been given the title Magellan and Faleiro wanted, was unencouraging: the lawsuits Columbus had launched against the crown were still being pursued by his heirs. Instead, Magellan and Faleiro received the titles of "governors," which implied that the administration of any colonies they founded would be in their hands, at least at first. The style of hereditary *adelantados* was added, but it conferred no necessary political responsibilities, signifying rather the honor due to leaders of conquests, colonization, or defense on the frontiers of the monarchy.

Magellan and Faleiro were more successful in the most puzzling of their demands: they asked—in a prefiguration of Sancho Panza's request to Don Quixote—to be given an island each, if they were to discover more than six,

> your Highness having the right to make first choice of any six, and that thereafter, out of all the others, we may be allowed to take the two that seem best to us, whereof your Highness shall grant us lordship, with all their revenues, present and future, and all their trade and that of what we gain thereby your Highness shall not take in taxes more than 10 percent.[21]

The request was granted, with the proviso that the crown would have one-fifteenth of the total yield of the islands concerned.[22] As we have seen (above, p. 94), the clause raises a presumption, or at least a query, about Magellan's plans: it confirms what we have already seen or at least suggested (above, p. 49)—that he had more than the Moluccas in mind and that an undisclosed objective guided his thinking.

The contract, which was signed at once, without bickering, on March 22, 1518, also made provisions for the conduct of the fleet. Magellan and Faleiro were to be styled joint captains, in command of the whole expedition. That was important because, as we have seen, sharing supreme command was a limitation on his power that vexed and worried Magellan. Equally significant, as we shall see, for the way the voyage would unfold—with mutinies met by rough justice—was the proviso that all crew and all residents "in the lands and islands you shall discover . . . shall defer to you and comply with your orders, under pain of such punishment or punishments as you on our behalf shall impose or order to be imposed." The right to resolve summarily all disputes on board was specified.[23] These sound like plenary powers, but they had to be understood as governed by existing laws and qualified by the stipulation that they would be exercised on the crown's behalf: that is, neither arbitrarily nor inconsistently with law and justice.

The crown undertook to provide five ships, with supplies for two years, and 234 crew to man them. Royal nominees would oversee financial matters and keep a record of all events, transactions, and agreements made on the voyage: in effect, evidence to be used against the captains should they exceed their authority, abuse their power, or infringe the rights of the crown. "We shall name a factor, treasurer, accountant, and scribes for each ship, and they shall make an account and tender an explanation of everything, and whatever shall be taken on board or bought must be authorized by them, as must everything that shall be aboard the said fleet."[24]

The appointment of officials in charge of accounts was an absolute priority. It was also where the preparations began to go wrong, introducing the potential—perhaps determining the inevitability—of conflicts between rival sources of authority. It began on March 30, 1518, when Luis de Mendoza was granted a salary of 60,000 maravedíes a year as treasurer and Juan de Cartagena was appointed inspector general, with a retainer of 70,000 maravedíes for the whole voyage, on which he was commissioned to "exercise the duties of that trust in accordance with the instructions given him under the king's hand." Cartagena and Mendoza were part of the affinity of Bishop Fonseca—his dependents or, from Magellan's point of view, the bishop's creatures: the nominees of the Casa de Contratación. They were there to keep the two Portuguese under surveillance. Cartagena was also to have command of a ship, after Magellan and Faleiro had made their choices. The fact that he had private instructions from the king was not calculated to enhance Magellan's and Faleiro's sense of control. Nor was the fact that the emoluments of the new appointees seemed to dwarf the captains', whose pay was set initially at 30,000 maravedíes a year—later increased to 50,000, perhaps in order to redress some of the apparent disparity.

On April 6 followed the nomination of other associates of Fonseca's—Gaspar de Quesada as captain of the fourth ship, with an annual salary of 48,000 maravedíes, which must have seemed to Magellan like a calculated insult. On the last day of the same month Antonio de Coca became the expedition's accountant, "who shall have account of everything contained in the ships, giving note of everything to the treasurer." His salary was to be of 50,000 maravedíes annually.[25] On April 19, another of Fonseca's nominees, Gonzalo Gómez de Espinosa, said to be thirty-two years old, was commissioned to join the officers of the fleet, though without command of a ship. To Magellan, Espinosa, who shipped with him on the same vessel, would become close and loyal, but Coca remained on the side of Fonseca's nominees, suspect and, as events would prove, hostile.

Meanwhile, as if to offset any ill feeling these appointments might cause, the king made an extraordinary gesture of confidence in Faleiro and Magellan, elevating each to the rank of knight in the Order of Santiago. They were addressed as such in all documents from April 19 onwards. No record of Magellan's response survives, but his satisfaction can be read obliquely in his request, which the king granted, that all the captains and pilots should be knighted in the event of the expedition's success.[26] At last he had attained the accolade he had trained for at court in Portugal, and the honor his reading had taught him to covet. It came in exile at the hands of a foreign prince: more a mark of Cain, from one point of view, than a sign of honor—or consolation for a context in which control of his enterprise was slipping from his grasp.

A grant of 125,000 maravedíes accompanied the bestowal of the accolade. Magellan had not received the money by the time he left Spain; so he petitioned the king to give it to the shrine of Nuestra Señora de la Victoria in Triana, as a life interest, with reversion to his estate after his death "because of the great devotion I have for the friary" (por la mucha devoción que yo tengo al monasterio).[27] The sanctuary played an important part in Magellan's last days in Spain as the place where he received the royal standard on the eve of departure and where the fleet's entire complement heard their last mass on home soil.

Finally, the contract emphasized that Magellan and Faleiro must respect the limits of Spain's zone of navigation and not infringe on Portugal's.

> You must undertake the discovery in question on the understanding that you must not explore or do anything inside the lines of demarcation and limits assigned to the most serene king of Portugal, my very dear and well-beloved cousin and brother monarch, nor do anything to his disadvantage, but must stay within the limits assigned to us.[28]

The lines, in hindsight, make the fatuity of the whole enterprise clear, as the Moluccas were within "the limits assigned to Portugal." Even had the Spanish case been valid, and had the islands lain where Spanish cosmographers claimed, other defects, which mounted in the months ahead, foredoomed the expedition.

. . .

The deal looked definitive. But the hard part—assembling, equipping, and manning the fleet—was still to come. The terms of the king's understanding with Magellan would change as circumstances unfolded.

The costs were staggering; it is not clear whether royal officials realized in advance that private sector partnership was inescapable, but the expenses mounted alarmingly, while the staff of the Casa de Contratación, whether in prudence or malevolence, did their best to keep Magellan short of the money he needed. The funds emerged painfully and slowly, as if by titration; restraint rather than urgency seems to have been the officials' priority. On September 24, 1518, for instance, Casa officials allocated a miserly 5,000 maravedíes for Magellan's and Faleiro's expenses, prompting them to implore the king a month later to call for more funds, especially for merchandise, on the grounds that "so small a sum will not suffice to acquire a full cargo of spices."[29] Part of the constraint arose from the fact that Magellan's was not the only expedition for which costs were being incurred. As the Portuguese agent Sebastião Alvares reported to his king, three ships bound ultimately for China were being prepared to go to the west coast of central America, while Cristóbal de Haro was at work on a commercial venture with four ships to support or follow up Magellan's venture.

The figures on the fleet's expenditure are dubious at the margins because of inconsistencies in the accounts kept at the Casa de Contratación, but the initial outlay, including the cost of trade goods that, through negligence or as a result of overstocking, were never shipped approached 8.5 million maravedíes.[30] It was a frightening sum: the crown's annual income probably did not exceed 520 million. It did not include many costs that would mount up later: extra wages, pensions, arrears, accounting costs, inquiries, lawsuits (below, p. 273). Of the money that Magellan's expedition absorbed, nearly 6.5 million maravedíes came from the crown; Cristóbal de Haro supplied almost the entire balance: well over 1.5 million, in the form of trade goods to be sold for spices, or of cheap truck—including twenty thousand hawks' bells that Magellan's father-in-law purchased—to be handed out to indigenous people en route as payola for good behavior or in gratitude for support or supplies. Other items included 1,000 maravedíes' worth of combs, fifty dozen pairs of scissors, a thousand little mirrors, five hundred pounds of colored beads, and "four hundred knives of the lowest quality Germany can supply."[31] Haro spent a great deal of time and money, when the expedition was over, petitioning to have his investment returned to him. He got a payment on account in 1537, three years before his death. Officers and crew were allowed to take their own stocks of trade goods, the amounts in proportion to their ranks, to supplement their wages (though Magellan limited their freedom to make use of their allocations—see above, p. 94, and below, p. 256).

The balance of public against private investment was exceptional for the era: the crown met over 77 percent of the cost. The most revealing point of comparison is with other voyages in search of a route to the Moluccas, which continued sporadically until 1537, even after Spain's claims had been renounced in favor of Portugal. Although the unparalleled risks deterred most private investors, the crown's contribution never attained remotely the same level again and remained on average at a little over 50 percent.

For Magellan's voyage, the ships alone cost nearly 4 million maravedíes, to include a skiff and brig and all rigging and guns. The explorer's father-in-law, who busied himself as an agent of the Casa as well as a collaborator of Cristóbal de Haro's, procured nearly three tons of gunpowder in Bilbao and received over 84,000 maravedíes in reimbursement.[32] Further appropriations for spare parts, with planks and copper sheeting for repairs, added up to well over 1.25 million maravedíes.[33]

The sources are divided on how much tonnage this expenditure bought. Capacity was measured in terms of the number of barrels of various—and always approximate—sizes a ship could carry. According to Bishop Fonseca's report to the king, the fleet contained two ships of a burden of about 120 *toneladas,* two of 80 *toneladas,* and one of 60 *toneles.* There were no precise standards of measurement, and a slight difference between two barrels of the same class could add up to a significant one in the capacity of a ship. A *tonel* was merely a conspicuously small barrel. When standard measures were introduced in the nineteenth century, the *tonelada* was reckoned for convenience at 2.83 cubic meters, but barrels in Magellan's day were relatively small—perhaps a little over one and a half cubic meters on average.

What looks like a more accurate breakdown appears in another document, which records tonnage of 120, 110, 90, 85, and 75 *toneladas* for the *San Antonio, Trinidad* (which was to be the flagship), *Concepción,*

Victoria, and *Santiago* respectively. Small as the vessels seem, explorers typically used even smaller ships: in this case, however, the promoters of the voyage hoped it would be a trading venture as well as a reconnaissance. To suit the expected conditions, which would involve following winds at least as far as the further shore of the Atlantic, the ships were probably all round-hulled with two or three masts each and square sails. The records refer to them as "caravels" or "naos" indifferently, as was the practice of the time. Strictly speaking, a caravel was a peculiar type of ship with a relatively narrow hull and, typically, lateen rigging; but the term was used loosely; even mariners deployed the names interchangeably, and the landlubbers who typically drew up the documents can hardly be expected to know any better.

The vessel we know most about was the fourth in size, the *Victoria,* the only ship to complete the circumnavigation of the world: the rest were lost, abandoned, or deserted. According to a tradition that began with a note scrawled in the margin of a sixteenth-century manuscript, *Victoria* was built in Zarauz, Guipuzcoa, for Juan Sebastián Elcano— the expedition's last surviving commander, who piloted her home.[34] That sounds too good to be true, like a story of sundered and reunited lovers, but that she came from a yard on the Cantabrian coast is entirely believable. She cost 300,000 maravedíes, fully rigged with three masts, of which the central bore two sails.

Whoever owned her at the time of her sale to the crown in the Basque port of Ondarroa complained that the price was insufficient, but, per ton, her cost was huge: 38 percent above that of the rest of the ships on average. She must have displayed to the agents who bought her the quality that enabled her to survive the voyage. Her image, skimming the Pacific with sails billowing and pennants flying, on Abraham Ortelius's famous world map of 1589 has made her outline familiar, at least to historians. For, stamped in gold, she adorns almost every volume of the scholarly series of editions of the records of travel and exploration that the Hakluyt Society has published over the last 175 years. If she was renamed for Magellan's voyage, the Virgin of Victory at the friary in Triana, on the far bank of the Guadalquivir from Seville, was her namesake. Magellan, as we have seen, proclaimed his devotion to her. And as we shall see, he left alms for the Virgin in his will, to be paid from the proceeds of the voyage. It was before the altar of her church that he received the royal standard before embarking.[35]

Well over 1.5 million maravedíes went on shipboard provender.[36] The amounts provided show how long the voyage was expected to last: at an absolute maximum, 776 days at half rations, or 1,134 days at rations of one-third the normal allowance. The crown's promise to provide rations for two years was barely fulfilled. There were over two thousand quintals of ships' biscuit, the hard tack that was sailors' staple—edible only if crumbled or moistened. In Seville (measures were unstandardized and varied confusingly from place to place and kingdom to kingdom) a quintal was the equivalent of four *arrobas,* or 220 pounds in modern imperial measures. Large amounts of flour were carried so that ships' galleys could supply bread in the early stages of any voyage. Salt fish—as if in ironic tribute to the sea—was the main seasoning on this as for every voyage at the time: Magellan's fleet carried 150 barrels of anchovies and 245 dozen barrels of other dried fish. Each ship carried a barrel of fishing equipment and bait to supplement the diet and a share of the total of 10,500 fish-hooks, but fishing from on deck while a ship was in motion rarely produced results. Meat was a luxury: there were only 228 *arrobas* of bacon. The amounts of dried vegetable—beans, lentils, and chickpeas—were unspecified. The biggest of the many pots for the galley weighed fifty-five pounds.

Lubrication came from two hundred *arrobas* of vinegar (which, if the *arroba* is that of Seville, seems far too little), 475 of oil, and 415.5 barrels of wine, including eighty-two of sherry on the *Victoria,* and eighty-nine on the smallest ship, the *Concepción.* The equivalent amounts in modern measures are not easy to specify. An *arroba* was typically less for oil than for other liquids—perhaps about twelve liters as against about sixteen. Wine and vinegar were used to make water potable: especially in view of the stinted quantities of vinegar, the amount of wine required would be vast. For an expedition of the size and duration contemplated in Magellan's case, at least a thousand gallons of fresh water would be needed, and although every attempt to replenish stocks would be made en route, a quantity of that sort would have to be shipped, in the most nearly clean, watertight, and weatherproof casks the kingdom's coopers could supply. No water, however, could be protected from the putrescence that was an inescapable hazard for long-range explorers.

Even with all this expenditure, royal advisers exhibited anxiety over whether there would be enough food and drink to go round. The Casa had undertaken to find wages and provisions for 234 men. From early May 1519, officials began to issue orders that no more than 235 men be allowed to ship, ostensibly because of the limitations of available supplies (but also perhaps because of the costs of the wages bill, which, as we shall see, were daunting).[37]

For medical supplies the modest budget was of up to about 45,000 maravedíes. Most of it went on foods that would make up invalids' diets: 272 pounds of sugar, 112 *arrobas* of cheese and over fifty-four of honey, 250 clusters of garlic, unspecified amounts of figs, prunes, raisins, and almonds in their shells, a live cow on each ship for milk, and a single jar of capers (a trusted expectorant) kept aboard the flagship, where most of the medical kit was concentrated, including fifty-five tubs of quince jelly, of which other ships carried only three or four tubs each.[38] The sugar and preserved fruits had an obvious therapeutic value as sources of energy; they also supplied hope against hope as remedies or alleviants for scurvy—the disease that was to cripple some members of the crews on the Pacific stage of the voyage. The causes of scurvy were unknown, but preserves were among Spanish and Portuguese physicians' favored therapies.[39] The manifest also mentions "medicines, unguents, and distilled waters" in unspecified amounts.[40]

The expedition required a moderate budget—68,182 maravedíes—for maps, charts, and navigational technology. Work commissioned from Nuno Garcia de Toreno, formerly in Portuguese service (above, p. 84), included a dozen maps on parchment and another dozen on other skins, of which Faleiro had ordered seven sea charts and Magellan eleven. Faleiro had also purchased on the expedition's behalf six other charts, of which one was a presentation piece for the king and was not carried on board. Presumably to illustrate the adventurers' world-picture, the king also received a map of the world, which cost 4,500 maravedíes, in a leather folder, which added a further 340 to the cost, and a pair of gilt dividers worth 476 maravedíes.

The cartographers responsible for the expedition's maps included other Portuguese exiles, Reinel *père et fils,* Pedro and Jorge; they probably incorporated traditions absorbed from Javanese maps via the work of their fellow chartmaker, Francisco Rodrigues, Albuquerque's protégé, who had sailed with Francisco Serrão on the expedition bound for Maluco in 1512 and recorded places as far east as Amboina and Ceram.[41] Two planispheres by Pedro Reinel were among the effects Portuguese officials later seized from a captured ship of the fleet, from among the belongings of the cosmographer who replaced Faleiro, but these may have been private property not paid for out of the fleet's budget.[42] Sebastião Alvares reported that he had seen "a globe and map which Reinel's son made here and which his father completed, adding the Moluccas. They use this map as a model for all the maps, which are the work of Diogo Ribeiro, as are some quadrants and spheres."[43] There were also seven astrolabes, one of which Faleiro made, twenty-one quadrants, eighteen hourglasses, six pairs of dividers, some of which were gilded, with their cases, and thirty-six compasses, including four crafted by Faleiro: the repair of just one cost 136 maravedíes.[44]

The Casa officials had to provide for every other detail of life on board ship, including two suits of armor for Magellan's use, eighty-nine lanterns, fifty *arrobas* of tallow candles, sacred vessels for mass (which would not be said on board for fear of the effects of the rolling sea but would be an urgent duty at every place of landing), and a dentist's drill. Part of the specification for the flagship's royal standard has survived—eight yards of taffeta, all fringed ("de ocho varas de tafetán con sus franjas"), and painted by one Villegas, who seems to have been an official artist attached to the Casa de Contratación—but not the bill.[45] Eighty lesser flags were loaded. Stationery items for the accountants as well as the pilots included fifteen bound, blank ledgers.

The cost of the personnel has to be taken into account. Had all the crew survived, up to well over 3 million maravedíes would have been added to the expenditure; much of it would remain owing to heirs. In the event, something over 1.5 million had to be paid out in advances, arrears, and dues to widows. Magellan received 276,886 maravedíes in retainer and wages, and Faleiro, who never sailed, continued to receive his salary—originally fixed at 35,000 maravedíes a year and increased to 50,000 from August 10, 1519. It came to 636,540 maravedíes in all, including what went to his heirs. Most of the disbursements took the form of advances against salary. Juan de Cartagena, for instance, who was entitled to a lump sum as financial overseer of the fleet (70,000 maravedíes) and an annual salary as captain of a ship (40,000), received an advance of 10,000 maravedíes on June 30, 1519; no record of any further payments is known. Of the other financial officials, Antonio de Coca got a year's salary—50,000 maravedíes—in advance; Luis de Mendoza received an advance of half his annual salary of 60,000, plus a contribution of 10,000 maravedíes toward his costs. Gaspar de Quesada got half his promised annual 48,000 maravedíes in advance. Gómez de Espinosa was among those whose arrears were paid, but he had to wait until 1528 for them. He also received a pension of 300 ducats in 1529.

On July 31 all pilots got 7,500 maravedíes in costs, since they had to use their own equipment, and an advance of 30,000 maravedíes, corresponding to a year's salary—except for one who had a salary fixed at only 20,000. He was named as Juan López Carvalho but properly called João Lopes Carvalho, as he was yet another of the Portuguese exiles.

The sums involved represented a huge increase on the normal official pilot's pay of 10,000 maravedíes annually; the pay scale for Magellan's voyage reflected the difficulty of recruiting men of sufficient expertise. Navigators capable of coping in the latitudes Magellan proposed to explore, beyond sight of the Pole Star, were in a sellers' market and could command a premium. On June 30, four suitably experienced pilots successfully petitioned for a further raise of 600 maravedíes a month, which was eventually granted.[46]

There were disbursements to survivors' families, but probably not all are recorded; large numbers of petitions survive.[47] Before the fleet sailed, Magellan's wife received an assurance that she would garner his wages for the duration of the voyage. In 1523, the widow of the chief pilot, Andrés de San Martín (Faleiro's replacement), got one year of her husband's salary; in 1530, after her death, his daughter's guardian received a further 12,000 maravedíes. Juana de Durango, widow of Juan Rodríguez Serrano, who, as we shall see, played an important role in the expedition, was awarded a pension in 1534. As late as 1546, Ana de Oquintal, the aging mother of Martín de Magallanes, who died off the Guinea coast on the homeward voyage, petitioned from Lisbon for the pay he had never received. Diogo de Sousa, Magellan's brother, received a paltry 15,000 maravedíes to compound for arrears on October 25, 1524.[48] The Casa set a nominal budget of over 1 million maravedíes for the wages for other ranks, most of which was never paid.

It is hard in retrospect to make sense of the reasons for such a heavy investment in a vast and costly fleet. As we have seen, it was a double-purpose expedition, charged with exploration and commerce. Prestige dictated that Spain should make a lavish show in the Moluccas and at any place encountered along the way. Paradoxically, the longer and more hazardous the voyage, the bigger the fleet needed to be, as some attrition had to be factored into the calculations: properly so, in the case of Magellan's voyage, on which every ship but one deserted or was lost.

Above all, the makers of the arrangements had no hindsight and little relevant expectation to draw on. If the true size of the globe had been taken into account, Magellan would surely never have gotten sponsorship. But the whole basis on which the fleet was launched assumed a small world. If the weather had been anticipated, or the dissensions that racked and rent the expedition foreknown, or the insufficiency of the supplies foretold, or the hostility of native hosts predicted, or the threat of scurvy taken adequately into account, experts at the Spanish court would have been unwilling to send one ship, let alone five. None of those caveats was in place. In consequence, false expectations subverted the expedition at its inception. The voyage was compromised before it started.

. . .

Changes in personnel and organization made matters worse. On July 26, 1519, Faleiro received orders not to sail. No reason was given, other than ill health. His mental instability provided a pretext if one were needed, and adversaries already regarded him as dangerously irrational (above, pp. 89, 92), but no real evidence of a serious deterioration of his condition arose until after the fleet sailed. In the interim, the king proposed, perhaps disingenuously, to confide to him the preparation of a second expedition to follow Magellan's. Faleiro's wife responded to her husband's collapse by leaving him and returning to Portugal. There, when he followed, he was imprisoned, to be extricated with difficulty and restored to Castile, where he passed into now-impenetrable obscurity and ultimately into apparently incurable insanity. As we shall see, issues of confidence may have been in play in his dismissal, since other pilots and navigators rejected Faleiro's doctrines. In any case, whatever the reasons for the action, it brought into the open three sources of tension that would make the fleet unmanageable in the long run.

First, Faleiro needed to be replaced not only as the expedition's chief navigator but also as co-commander. Fonseca's protégé and nominee, Juan de Cartagena, who already exercised supervision of the financial affairs of the expedition, was the court's choice for the latter role, appointed as "conjoint person" (conjunta persona)—a style that allowed Magellan to think of Cartagena as his subordinate, and Cartagena to think of Magellan as his equal.[49] The explorer assented to the new arrangement with obvious reluctance, not to say naked resentment and jealousy, accepting Cartagena with weaselly evasions, only in order "to serve his Highness . . . as his Highness so commands" and "since his Highness orders in the provisos and instructions that the said Juan de Cartagena has from his Highness."[50] A power struggle for supreme authority in the fleet was under way, even before sails had been set.

The second of the revealed tensions was over what might be called the scientific side of the expedition and, in particular, the techniques the expedition would deploy to record the course and distance traversed, assign longitude, and work out the relationship of the Moluccas to the Tordesillas antimeridian. Magellan demanded that if Faleiro could not sail, his *Regimento,* the text of his "readings of longitude, East and

West, with all the rules accompanying them," should be on board and not retained in the Casa; should they prove unserviceable, Magellan suggested—presumably in confidence that his colleague's methods were impeccable—Faleiro should not get his due under the contract.[51] What Faleiro and his family thought of this presumptuous magnanimity is unrecorded.

The opinions of Faleiro's fellow pilots, however, are known; and they were unenthusiastic. The text in question, compiled over the years since 1516, does seem to have accompanied the fleet; at least, one of the participants made a copy of it, and the Portuguese chronicler Barros acquired it, perhaps among the papers confiscated from the expeditionaries whom Portuguese officials captured.[52] It focuses on how to read longitude by the methods mentioned above (p. 94) as characteristically Faleiro's, with emphasis on how to measure lunar distances and how to account for magnetic variation. As every schoolchild knows, or should know, no reliable method of recording longitude at sea was available until technicians perfected the marine chronometer in the 1760s. Faleiro, however, was in the vanguard of work on methods serviceable on shore. When Casa officials invited the pilots to evaluate Faleiro's methods, all but one of the respondents declared his work useless. Presumably they knew what they were supposed to say.

The exception was the pilot appointed to take Faleiro's place as chief navigator: Andrés de San Martín, to whom we shall have to return in some detail (below, p. 167), found the lunar distance method practicable. Although it demanded conditions calmer than any sea ever offered, he attempted to use it on shore (below, p. 168).[53] Unfortunately, of the five attempts that San Martín claimed to make, only one, the first, yielded a result that survives; a printer's error in the tables of longitude he used vitiated it.[54] Honest and open-minded readings would, of course, show that the Moluccas were in Portugal's sphere of navigation. In the event, that was what happened, although San Martín did not survive to render his accounts.

The final source of tension that Faleiro's removal revealed was between Portuguese or Portuguese-born members of the expedition and those of other nationalities, especially, of course, Castilians. Despite the permeability of the frontiers of the Iberian kingdoms, Castilian officials were not blind to the presence of foreigners. That the overseas dominions of the king of Spain should normally be reserved for Castilians to settle and exploit was a principle established between Castile and Aragon from the time of Columbus's second transatlantic voyage. In principle, exploration could be confided to people of one nationality, settlement to those of another. In practice, the distinction was hard to maintain.

An early sign of trouble occurred when Magellan demanded that Faleiro's brother, Francisco, be given command of a ship. Francisco, however, was detained in Castile, where he devoted himself subsequently to sorting out the problems Rui's collapse had brought on the family: negotiating his release from Portugal and the terms of his separation from his wife. Juan Rodríguez Serrano, who was promoted to take command of *Santiago*, the smallest ship in the fleet, apparently in preference to the formally well-qualified João Carvalho, was probably a Castilian. The Portuguese interrogators reported him as such.[55] He is not identifiable, despite frequent assumptions to the contrary, as a kinsman of Portuguese individuals with similar names, or as the "Serrão" who appears in a Portuguese agent's very imperfect list of fellow countrymen whom he believed to be with the expedition.[56] A shipmate is more likely to be right in identifying him as a native of Fregenal de la Sierra in Extremadura.[57] As we shall see, rivalry between Serrano and Carvalho persisted on the voyage, until the latter replaced the former—briefly before his own ouster—after abandoning him to his death in the hands of captors in the Philippines (below, p. 259). Meanwhile, the reshuffle took Juan de Cartagena to the quarterdeck of *San Antonio*. A veteran of several transatlantic voyages, Juan Rodríguez Mafra, a Castilian born in Palos, where Columbus had recruited his first transatlantic crew, was appointed pilot of the same ship, while a fellow Spaniard, Vasco Gallego, whose son, Vasquito, accompanied him as his cabin boy, became the pilot of *Victoria*.

Francisco Faleiro's exclusion seems therefore to have been part of a new policy of purging the expedition of personnel of Portuguese origin. On July 17, writing to the Casa, the king had voiced disquiet about the number of Portuguese on the roster and especially in Magellan's and Faleiro's retinues. They had, he understood,

> many Portuguese to take on board with them and because it seems that this would be inconvenient, I order you, in the best way as shall seem to you, to speak to the said master captains and tell them that each of them must not take more than, say, four or five attendants and that they must leave and dismiss all in excess of this number whom they may have proposed to take; and you shall take steps to see that no other outcome occurs; but this must all be done with as much discretion as possible.[58]

The reports Sebastião Alvares sent home to Portugal are not necessarily perfectly truthful and may have exaggerated Castilians' distaste for his fellow countrymen, but a dispatch he wrote on the following day strengthens impressions of increasing ill feeling. The Portuguese agent recorded growing dissension between Magellan and Casa officials, who found him insupportable; explicitly, Alvares wrote, "They cannot swallow him" (Como no poden tragar a Magalhaes), or, as we might say, "He sticks in their throats." According to the same dispatch, they issued other participants in the expedition with instructions contrary to Magellan's. As we have seen (above, p. 91) Alvares later taunted Magellan with this evidence of Castilian duplicity. In the document of July 18 he names, as recipients of conflicting orders, Juan de Cartagena, Cristóbal de Haro, and "Juan Esteban, treasurer"—perhaps an error for Luis Mendoza, who was properly the fleet's treasurer, or a garbled allusion to Juan Sebastian Elcano (or perhaps to a Genoese of a similar name), who had some financial expertise and may have been a subordinate to Mendoza, whom he later replaced.[59]

Alvares claims to have asked Magellan why the Portuguese João Carvalho, who was qualified for command, had only the rank of pilot and why so many of the pilots were Portuguese: at that point, indeed, they all were. Smilingly, Magellan answered that "once at sea he would be able to do as he liked" (respondió que haria en la Armada lo que quisiese sin dar les cuenta). Words ensued between Magellan and the royal officials—"so many and so bad" (pasaron tantas y tan malas palabras). In consequence, Alvares claimed (though no other evidence supports him), Portuguese participants' pay was suspended.[60]

On the day of Rui Faleiro's dismissal a further order from the king informed the officials of the Casa that

> I have learned that in the said fleet sixteen or seventeen have been recruited as gromets and at the time they were enlisted it was because of the need for recruits and that now that the lists are full for gromets and crew I command you to take other gromets . . . and no other foreign crewmen, beyond the numbers I have authorized who are sailing as attendants of the captains . . . because my intention is to adhere to what my lords the Catholic Monarchs, whom God receive in glory, have commanded.[61]

To requirements that he comply with these instructions, Magellan replied that he accepted Portuguese recruits "as he did many other foreigners—namely Venetians, Greeks, Bretons, Frenchmen, Germans, and Genoese, because at the time he took them natives of these kingdoms were unavailable." He was willing, he said, to take Castilian-born substitutes if the Casa supplied them.

As for supernumeraries, servants, and relatives, Magellan insisted that he would tolerate no orders in contradiction of his original contract.[62] Casa officials pointed out that everyone ought to obey the king without cavil and without raising objections that might delay the departure of the fleet. They repeated that the commander could retain his existing servants and gromets only if it should prove impossible to replace them and that the Portuguese members of his personal staff must be limited to five. Francisco Faleiro, in particular, having been denied a captaincy, could sail only in Magellan's entourage, and only within the specified limit.

The controversy had little practical effect. Other Portuguese recruits continued to receive favors. Estevão Gomes, whom we have already met as one of the lately arrived Portuguese exiles, was made an official pilot of the Casa de Contratación, authorized to take part in transatlantic expeditions, on September 10. At the same time, one of the pilots who had successfully petitioned for an increase of pay, the Portuguese João Lopes Caravalho (or, on this occasion, Johan López Caravallo in the rendering of the scribe), reappeared as pilot of the *Concepción*. In the end, at least thirty-one participants identifiable as Portuguese shipped on the voyage, along with at least 137 Spaniards and maybe between seventy and eighty of other nationalities. One pilot was Genoese, one Greek. Fifteen hands were listed as French, others as Italian, German, Flemish, Irish, and even in one case English.

On August 9, the day before the fleet left Seville, Magellan took witness statements from officers who had been involved in recruiting, including Elcano, Espinosa, and Mafra, as to why so many Portuguese and other foreigners figured in the expedition. He wanted to leave a record that made his position clear: that the problems of enlistment had led the agents far afield, to Málaga, for instance, Cádiz, and Huelva. In the circumstances, it had been impossible to exclude foreigners. The numbers he declared for the total complement were 62 aboard *Trinidad*, 57 on *San Antonio*, 44 on *Concepción*, 45 on *Victoria*, and 31 on *Santiago*, making 239 in all. The number accords with one of the surviving crew lists, but some names that occur in reliable accounts of the expedition do not figure in the lists, whereas some of those listed never boarded.[63] Last-minute changes, substitutions, recruitments, defections, illnesses, and deaths always made such lists unreliable.

. . .

Evidently, even before sailing, the expedition was split between contending factions: Fonseca's men against Magellan's, Spaniards against Portuguese. Fonseca's creatures constituted the apparently stronger party. Cartagena himself had the nominal position of Magellan's *conjunto*, however little notice his co-commander intended to accord him. The private instructions from the king, to which we have already alluded, strengthened Cartagena's hand in any potential conflict because they gave him explicit oversight of negligence or misconduct by other commanders—although the king urged him to exercise tact and prudence "to minimize damage" should he need to exercise his powers. His state was kingly. Ten servants accompanied him. His emoluments were large, albeit not (despite assertions to the contrary in most previous studies), greater than Magellan's. His private instructions granted him the captaincy of the first fort to be built or occupied in the Moluccas. And he was charged with particular responsibility for the settlement of Spanish personnel at the fleet's destination and the conversion of the natives to Christianity, "which," the king said, "is our principal desire."[64] Even had no quarrels broken out on board, these provisions would surely have guaranteed conflict with Magellan at the fleet's destination.

Cartagena captained the biggest ship, and his comrades included the captains of two other ships, Luis de Mendoza and Gaspar Quesada. They could, if they so wished, outgun and outnumber Magellan if fighting were to start. With Mendoza and Antonio de Coca, the faction controlled the finances of the expedition and could trim its records to prejudice Magellan's interests. They were able, in the event, to count on Juan Sebastián Elcano, master of the *Concepción*, who was recruited while in Seville on a pilgrimage of atonement for having lost a ship that he had mortgaged illegally.[65] He evinced no hostility to Magellan prior to departure but collaborated with Cartagena's faction on board and in subsequent inquiries into the disturbances that rent the expedition.

Except, as we shall see, for Estevão Gomes, who emerged as a rival, Magellan could count on the Portuguese, and, as events proved, on the sympathy of Gonzalo Gómez de Espinosa, despite the latter's previous associations with Fonseca's men; Haro's agent, Gonzalo Guerra, and Juan Serrano, who became captain of *Santiago* in the reshuffle that followed the extrusion of the Faleiro brothers, were also loyal to him when it mattered. His most reliable supporters were family members: Duarte Barbosa, Diego's nephew, and his mother's kinsman, Álvaro de Mesquita, both of whom were with him on the flagship, and presumably Martín de Magallanes, who shipped on the *Concepción*. Magellan's personal retinue included Cristóbal Rebelo, reputedly but improbably, on chronological grounds, his natural son, whom Magellan called "my page" and who was to die alongside him in battle on a beach in the Philippines (below, p. 258). Pointlessly, as it turned out, Magellan left him 30,000 maravedíes in his will "for services he has done and so that he pray God for my soul."[66] The imputed relationship rests on no contemporary or credible evidence. As we shall see in the next chapter, Magellan also took a personal chronicler with him. He did not, as Winston Churchill recommended, intend to write his story himself, but he relied on the next best expedient: to have it written for him.

. . .

Meanwhile instructions for keeping peace on board continued to tumble out of the royal chancery.[67] Their effect was to increase the chances of conflict.

On the one hand, the king's pronouncements emphasized Magellan's authority as commander. "All the seamen who sail in the said fleet," the king decreed, "shall be received under the supervision of our captain, Fernando de Magallanes, as he is the most experienced in such things." All participants "shall follow the opinion and order of the said Fernando de Magallanins [sic]" (sigais el parecer e determinación del dho fernando de magallanins).[68] Magellan had the right to appoint subordinate officers and ultimate responsibility for enforcing royal commands. Yet there was room to dispute the extent of his power. Before Faleiro's dismissal, authority to inflict punishments was vested in both co-commanders: "The captains general have power to devise and execute punishments against disobedient men." The oath all crewmen were obliged to take "to observe obedience and the king's service," including explicitly not to desert, was to be sworn before both leaders. As Faleiro's successor and explicit *conjunto,* Cartagena could be said to share the same joint responsibility with Magellan as had belonged to his predecessor. As we have seen, the private instructions the king gave him may have encouraged him to think so.

In any case, power to punish was qualified, implicitly by the obligation all the king's vassals shared to act in conformity with the law, and explicitly by the king's command to give offenders an opportunity to reform: disobedience had first to be met with a warning; if "it should be in our support and to our service" (sea nuestro servicio y provecho del armazón), punishment might follow persistence. The provision for reproof before punishment is an echo of the rules of monastic and religious orders from St. Benedict onwards. Ships were communities in which, as in monasteries, self-abnegation had, in the last resort, to be voluntary. Justice had to be tempered not necessarily by mercy but by prudence and deference to collective interests. Magellan's conduct on the voyage, as we shall see, would raise the question of how and where the practical limits lay between the proper exercise of command and its arbitrary or willful abuse.

One of the points on which his instructions were clearest—and which he scorned with breathtaking indifference—was his responsibility for what nowadays we might call "transparency": he was to share information about the route and destination of the voyage with his officers. The purpose of the provision was not to temper Magellan's responsibility for setting the course but to ensure that the fleet had the means to stay together and to reunite if dispersed by weather or other contingencies. As the instructions put it, "You will all make your way together, with good fortune, to the land which you shall specify to the other captains and pilots."[69] Yet Magellan would persistently refuse to comply, raising or reinforcing the presumption that he intended to ignore the route or destination his orders specified, or both. "Before," his instructions stated, "you leave the river of . . . Seville, or after you have left it, you shall call the captains, pilots, and masters, and you must give them the charts you have made in order to make the said voyage, and you must show them the first land for which you are bound in hope, so that they may know the route to it in order to seek it."[70] The language seems to give Magellan the option of disclosing the route bit by bit, as each successive intermediate destination goes by; but he did not even do that. The confirmation of the relevant instructions to the subordinate officers makes clear, in any case, that they were entitled to know the expedition's objective and to be consulted on how to find the best route to it:

> In the same way, you know that the said commanders-in-chief must declare to you the route that they intend to follow on the said voyage. I command you to receive notice of it in writing and that, in accordance with it, you and the said commanders-in-chief draw up a plan in which the said route is set out with all such readings of latitude and longitude as the commanders-in-chief have knowledge of, and that you show it to all the pilots that have to sail in the said fleet, and that you give to each of them a copy of the said plan.[71]

Navigation was so imprecise a science that pilots commonly differed widely about where a fleet might be on the surface of the world, amid indistinguishable seascapes, at any moment on any voyage. Conferences about the location of the fleet were therefore to be shared daily, if possible. Other methods of keeping the ships in touch with each other were standard: all must follow the flagship's lantern; should any ship be detached from the rest, Magellan was to wait at a prearranged rendezvous, further along the route, for a month; the fleet could then proceed, leaving a message under a pile of stones, topped with a cross to guide inquirers, should the missing ship turn up late. Clearly, these practices required that all captains and pilots know the route in advance, contrary to Magellan's defiance.

Orders were obsessively insistent on one of the king's greatest fears: that Magellan would infringe agreements with Portugal by ignoring the Treaty of Tordesillas or blatantly transgressing the antimeridian. "Our first charge and order to you," the king wrote on the last day of February 1519, addressing Magellan and Faleiro, "is to respect the line of demarcation and not to touch in any way, under heavy penalties, any regions of either lands or seas that were assigned to and belong to the most Serene King of Portugal."[72] On arrival in the Moluccas, the explorers must "erect a marker exhibiting our coat of arms"[73] and stating the difference in longitude and latitude between the kingdoms in question and the dominions of Portugal.[74] "Above all," the king insisted, "the domains and demarcations of the king of Portugal must be respected."[75]

Detailed instructions followed on what else was to happen on arrival. Magellan's discretion to make a treaty with the ruler or rulers was untrammeled:

> And when with good fortune you arrive at the lands and islands where the spiceries are, you shall make a treaty of peace and of trade with the king or lord of the land, as shall seem to you as best serves our interest and profit, and because in this matter I believe that you will do all that conforms to our service I do not impose any limitations on what you may do, because I well believe that you have the ability to do it, owing to the experience you have already had of such matters.[76]

Juan de Cartagena, with the help of Luis de Mendoza and Antonio Coca, was to be in charge of negotiating the best terms possible for the acquisition of spices and other products, keeping records of all transactions. When the ships returned home and the profits of the voyage had been realized, the gains would be distributed proportionately among sponsors and participants, with sums reserved, as was usual in such ventures that depended on God for success, for donations to the religious houses of Santiago in Seville and Nuestra Señora de la Victoria, across the river in Triana, and for the redemption of Christian captives held in ransom in the hands of infidels.

The expeditionaries might also expect to find previously unknown and unclaimed lands along the way, as they had to go further along the South American coast than prior explorers and cross the previously unexplored "South Sea." In such places they were to make records of "anything to our interest" and gather samples of all potentially useful products.[77] Suitable precautions had to precede the usual rites of the taking of possession—raising the royal banner and making a record of the event. Before landing, envoys had to be sent ashore to negotiate with the natives. Magellan was to risk his own person in negotiations only if necessary and only after taking hostages. Then, addressing the native leader,

> In token of peace and as a sign of the security thereof you shall tell them how we have it for a custom to order a marker displaying our coat of arms to be placed on the land, as a sign of security, and how it will be his responsibility and that of his people to keep the marker in good repair.[78]

Should Magellan find pagan merchants already operating, he should try to make suitable arrangements for trade with them. If they refused, he must warn them not to do business without leave from the king of Spain ("sin nuestra licencia"). Should they disobey, he was authorized to seize their persons, ships, and goods.[79] Muslim traders from outside the Spanish zone could be taken prisoner for ransom, exchange of captives, or sale as slaves, on the grounds that they were victims of "just war." The Spanish crown never made war without inquiring into the justice of it—which the court canonists and theologians almost always found grounds to affirm. But if Magellan were to capture enemy ships and find captives on board from lands outside the Spanish zone—including Muslims and pagans—he must treat them well in order to foster good commercial relations for the future: "It is good that you should treat them well, explaining to them the reason for which you are seizing the ships—which is that they belong to people with whom we do not wish to have peace or trade."[80]

If such folk proved willing to trade with Spain, they should be freed at their home ports, and any goods belonging to them should be returned "so that they may know that our will is not to do harm to those who wish to make treaties of peace and engage in the exchange of goods with us."[81] Operators of pirate vessels, on the other hand, "will be taken as in just war, and if it is necessary to use them with some degree of cruelty you may do so with moderation in order to provide an example and deterrent to others."[82] Magellan (jointly with Faleiro prior to the latter's dismissal) was entitled to seize choice items of booty: a jewel worth 500 ducats (about 300,000 maravedíes) if there was one, or up to the same value in any merchandise, plus 3 percent of any total in excess of 10,000 ducats. A tenth of the remainder of any loot or of the value of seized shipping was to be reserved, once again, for the friaries of Santiago and La Victoria.

The crown was, as usual, anxious that Spain's moral standing with God and the church should be unjeopardized by mistreatment of the native peoples the explorers might encounter. The king's instructions stress repeatedly that *indios,* as Spaniards usually called them wherever they were, must be well treated. The commonest cause of conflict—some defensive measure interpreted by Spaniards as insult or aggravation—should not be allowed as a pretext for violence: "And if by any means any of your people should be the victim of any unpleasantness, let the natives not therefore be maltreated by you."[83]

Women, the orders state, are to be respected. Goods or repair facilities must be paid for. The expeditionaries are to offer medical attention to natives free of charge, including for wounds sustained in battle against Spaniards. No discharge of firearms shall be allowed on any new land "because the *indios* fear this more than anything else." In common with a notable feature of Spanish colonization generally, the explorers should promote the learning of native languages by Spaniards: this was an old principle not only of good government of subject cultures but also of good missionary practice. If you want to command allegiance and make converts, talk the lingo. In the case of Magellan's enterprise, the royal command to learn native languages looks otiose, but it is evident from the instructions as a whole, as well as from the private commission to Juan de Cartagena, that some permanent establishment of Spaniards, at least in the Moluccas, was among the king's objectives. In sum,

> What we chiefly commend to you is that in any matter that you arrange with the indios you must maintain all truthfulness and that you break no agreement and that even if they have not responded peacefully you must work to establish concord; and you must not consent to any form of hurt or harm to them . . . and there is more to be gained by converting one hundred by these means than one thousand by any other.[84]

The king's instructions foresaw the likelihood of restiveness among the crew, especially as the expedition might be at sea for an unprecedentedly long time.

And because to some of the people who go in the said fleet it will seem that the time spent will have been long without finding anything, you shall notify them . . . that while they have sufficient supplies, no one should dare to say. . . . that in the said voyage or discovery the time lapsed shall be much or little, but rather that they must leave such matters to those entrusted with them.[85]

On the other hand, every crewman was to have total freedom to write home to Spain and to report any matter relevant to king's service, including complaints against Magellan. To encourage servants of the crown to inform against each other was a vital resource for a far-flung monarchy where most officials operated beyond the reach of recall or reproof. Excesses could be punished only, if ever, long after the event, either when the culprit returned home to face charges or, in the case of a colonial administrator, when a judge arrived from Spain to hear accumulated complaints. In either case, the prosecutors would compile a dossier out of the information laid by correspondents.

Other measures for ensuring the crew's contentment or passivity were standard: every man's name, salary, parentage, place of residence, and civil status had to be recorded, as had any property crew members might take on board. Any gold or objects of value had to be locked away with three keys, each in the hands of a different officer. Confession and communion before embarkation were compulsory for all. Temptations to indiscipline or to the vengeance of providence—including cards, dice, and blasphemy—were formally forbidden on board, in rules often honored in the breach.

When all the lading was complete, over 400,000 maravedíes' worth of material, described in the Casa accounts as trade goods and munitions, remained on shore, perhaps because there was no room or time or because payments remained outstanding. Magellan and the agents of the Casa were under orders from the king to take care not to overload the vessels—surely an unnecessary word to men wise in chandlery, but it may have had a cautionary effect. The munitions probably consisted of gunpowder, of which Magellan had already certified that he had more than enough.

. . .

After the ceremonies and mass in the friary church of La Victoria, the fleet left the quays for the port of Sanlúcar, beyond the river mouth, on the edge of the Atlantic. A few debts still kept the fleet in suspense, including bills for rice rations and materials—mainly lead and steel—

and labor for repairs to *San Antonio;* the costs of fitting Andrés de San Martín's cabin, presumably because the required modifications to the space designed for Rui Faleiro were still unmet; fees were due for messengers, including one who took official news of the departure of the fleet to Lisbon. In response to the king's plea, Cristóbal de Haro paid the amounts in question—263,345 maravedíes in total.[86]

The fleet was ready to sail—as ready as it ever would be—nearly six months after Magellan had assured the king that all preparations were complete, except for loading merchants' goods, and nearly five months after the deadline the king had stipulated.

During the delay Magellan dashed back to Seville to see his wife and son and to make his will.[87] He dictated its terms on August 24, 1519. "If I die in this city of Seville," he asked, "may my body be buried in the friary of Santa María de la Victoria, Triana,"[88] where he had taken custody of the royal banner on the eve of the fleet's departure from the city. Should he die at sea, he wanted to be laid to rest in the nearest church dedicated to Our Lady. The aspiration seems odd, as burial at sea was normal; if he had kept a realistic notion of the scale of his voyage in mind, he can hardly have expected to die near land, let alone land endowed with Catholic churches (unless he imagined building some himself). He envisaged a tenth of his estate going to charitable bequests: a third of it to the same friary in Triana, the rest to be divided between the monastery of Montserrat, which he may have visited when in Barcelona; the Franciscan friary of Aranda de Duero, which he might have seen on his way to his momentous meeting with the grand chancellor in Valladolid; and the nunnery of Santo Domingo de las Dueñas in his home town of Oporto, with conventional additional bequests for the building of Seville Cathedral, the redemption of captives from Muslim foes, the clothing of the poor, and the maintenance of hospitals.[89]

He was ready to leave. But where was he going? The great historian of seaborne exploration, Samuel Eliot Morison, who combined unique talents—insuperable scholarship, dispensed from his chair at Harvard; experience as an admiral of the American navy, uttered from the bridge; and swashbuckling prose—insisted that Magellan planned to circumnavigate the globe.[90] Today, Morison's statue stares imperiously along Boston's broadest boulevard. His peremptory imperatives, in his clipped quarterdeck lingo, still command attention but not assent. No evidence supports them. It would, moreover, have been contrary to orders to attempt a circumnavigation, which would infringe Portugal's exclusive right to the waters beyond the antimeridian.

The king of Spain thought he knew that Magellan was bound for the Moluccas, although his last words on the point seem to betray some havering: "First and ahead of any other destination whatsoever you shall go to the said islands of Maluco . . . and after you have done so it will be possible to seek whatever else may be suitable in conformity with the orders you carry."[91] It is hard to know what other destination the allusion might imply: perhaps Magellan had commended the Philippines privately to the king.

The man who, apart from Magellan himself, knew the explorer's intention best was Sebastião Alvares, on the basis of conversations we have already reconstructed (above, p. 90) and the intelligence his spies gathered. He expected Magellan to make for Cabo Frio, on the underside of South America's bulge, where the Brazilian coast trends westward. From there, "leaving Brazil to starboard," he would proceed beyond the Tordesillas line until he could turn west and, setting his course west and northwest, would proceed "straight to the Moluccas" (diretos a Maluco). According to Alvares, "From that Cabo Frio as far as the isles of Maluco by that route, the maps that the expedition carries show no inhabited land. May it please almighty God that this voyage unfolds like all the others in that direction"—all of which had failed, and perhaps the Portuguese agent meant to include, prospectively, the voyages supposedly under way or imminent from the Spanish colony in Central America. Then "Your Highness can relax."[92]

FIGURE 5. The turbulence of the South Atlantic is apparent in Levinus Hulsius's depiction of what he called "Oceanus Australis" or the Southern Ocean. Whales, flying fish, and ivory-snouted narwhal evoke the exoticism and commercial exploitability of the environment. Neptune and the Triton, who uses a nautilus for a horn, have weed-bedecked coiffures that evoke Native American plumage. A "fleet of Portugal on its way to Calicut" sails at top left. Magellan's name designates the strait and the vast Antarctic continent that geographers imagined to its south. Neptune holds the arms of Portugal framed in those of Castile because by the date of the engraving, dynastic alliances had united the Iberian crowns. Hulsius includes *legenda* in Latin, Spanish, and Dutch, disclosing part of the range of competition for Atlantic mastery. From Levinus Hulsius, *Kurtze, warhafftige Relation vnd Beschreibung der wunderbarsten vier Schifffarten, so jemals verricht worden* (Brief and true narrative and account of the four most wonderful sea-voyages ever accomplished) (Frankfurt, 1626), vi. (The other voyages covered were those of Drake, Tomas Cavendish, and Olivier van Noort.) Courtesy of the Library of Congress.

CHAPTER 5

The Cruel Sea

The Atlantic,
September 1519 to February 1520

If it had not been the Lord who was on our side, when men
rose up against us, then they had swallowed us up quick,
when their wrath was kindled against us. Then the waters
had overwhelmed us, the stream had gone over our soul.
Then the proud waters had gone over our soul.

—Psalm 123/4:1–5

The first religious of St. John were a shabby, shaggy lot, worthy of their
namesake, the hairy Baptist cloaked in camel's pelt. An early visitor to
their hospital in Jerusalem deplored the unkempt beards. The second
master of the order, Blessed Raymond du Puy, enjoined them, although
they were all of noble lineage and prosperous backgrounds, to share the
privations of the poor and sick they tended.[1] But the conditions of medieval Palestine, where crusaders struggled against resurgent enemies,
obliged them to take up arms, don armor, become knights, and continue
the fight at sea after expulsion from the Holy Land. Knights of Rhodes,
as they were called in Magellan's day after their island headquarters,
became adepts of naval conflict, as familiar with their ships as with their
mounts.

Antonio de Pigafetta, a nobleman of Vicenza, received a
commandership—a grant, in effect, of a livelihood from rents in the
order's lands—on October 3, 1524 (as we know from a copy of the
nominal roll made by an antiquarian in 1738).[2] At his reception at the
doge's court in Venice, where, in November 1523, he presented a digest
of his experiences on Magellan's voyage, a witness referred to him as a
chavalier hierosolomitano, a religious knight, of the order.[3] He signed

his name with the same title at least from February 1524 onwards.[4] References in Pigafetta's writings show that he was already familiar with the physical environment of the order and some of its literary and devotional traditions before he shipped with Magellan. They also reveal him as a consummate courtier, steeped in the literature of chivalry and the latest fashions in etiquette.[5] Baldassare Castiglione, the *arbiter elegantiarum* of the day—who wrote *The Book of the Courtier* and contributed to "the civilising process" of the Renaissance by prescribing rules of formal behavior for supposedly elevated company—knew him well and recommended him to publishers and patrons.[6] Pigafetta was respectably nurtured, with a humanistic education, insufficient to equip him with direct knowledge of Greek and Latin classics but good enough to make him exceptionally interested, by the standards of explorers, in the cultures and languages of the people he met beyond the oceans.

What possessed a well-born, well-mannered, well-schooled, and well-connected gentleman to take part in Magellan's adventure? Pigafetta expressed two motives for joining the expedition: to "experience and to go to see. . . . the great and marvelous things of the Ocean Sea" and "to gain some renown with posterity."[7] Both motives are believable. Curiosity was a Renaissance characteristic; in the 1450s a Venetian patrician, Alvise da Mosto, took service with the Infante Dom Henrique to extend—so he said—his knowledge and experience of the world.[8] Books Pigafetta read, he tells us, and conversations about them piqued his interest in the New World: indeed, he alludes to much of the literature available to him, including early editions and retellings of Columbus's first published report, writings by or attributed to Vespucci, and the first two decades, at least, of Peter Martyr's work in progress.[9] "Marvels"—which might include exotic products, monstrous deformations, scientific novelties, customs classifiable as savage or barbaric, and the comfortable horrors of distant dangers for readers back home—were staple ingredients of travelers' tales. And fame was one of the highest values of the time.

Pigafetta must have met Magellan when they coincided in Barcelona in February 1519. He was in the entourage of the papal nuncio, Francesco Chierichiati, who was there to canvas King Carlos for a projected crusade against the Turks, and whose continuing friendship and admiration are confirmed in a letter of recommendation he wrote on his fellow Vicentino's behalf when Magellan's voyage was over. The opportunity to join the voyage evidently appealed to Pigafetta as a chivalric deed, an instructive exercise, and a framework for the investigation of peoples and marvels. For Magellan, Pigafetta was a valuable acquisition: a literate supernumerary who could celebrate the voyage, eulogize the leader, disseminate success, vindicate failure, and deflect critics.

Pigafetta became the chronicler of the voyage par excellence. He had a draft, at least, ready on his return to present to King Carlos—among gifts, he said, not, as befitted a poor knight, of "gold nor silver but things worthy of being highly esteemed by such a sovereign."[10] The author then went on what, in today's book trade, would be called a lecture tour at the courts of Lisbon, Mantua, Venice, and Rome, scattering manuscript copies of his work as he went.[11] His account has eclipsed the influence of all other eyewitnesses and early recorders.

Four agendas wrench at the reliability of Pigafetta's narration: his partiality for Magellan, which inaugurated a tradition of hero worship, sustained in most histories; his taste for *mirabilia*—tales of marvels that readers of travel literature demanded; his snobbery, which packed the book with as many references as possible to great acquaintances and grand patrons; and the effects of the writer's narrow escape from death. He saw his deliverance as the hallowing effect of divine grace. It occurred while he was engaged in his favorite pastime—fishing over the side of his ship—when the fleet had reached the Philippines. His feet

> slipped, for it was rainy, and consequently I fell into the sea, and no one saw me. When I was all but under, my left hand happened to catch hold of the clew-garnet of the mainsail, which was dangling in the water. I held on tightly and began to cry out so lustily that I was rescued by the small boat. I was aided, not, I believe, indeed, through my own merits, but through that font of mercy.[12]

Pigafetta's sense of being spared by grace was enhanced by the fact that he survived the expedition—an apparently miraculous outcome, which, among those who endured to the end, he shared with only seventeen companions. His book has a kind of consequent spirituality, which sometimes makes him exalt apparently mundane events to the level of heavenly revelations. Combined with his deeply felt obligation to extol Magellan, the effect was to make the author skip or sideline the conflicts that pitted expeditionaries against each other, and to overlook or suppress much of the bloodshed in which Magellan took part. At Pigafetta's first mention of his hero, he praises the wide-ranging journeys for which Magellan "had acquired great praise" and mentions membership of the Order of Santiago—alluding to the bond of sympathy that chivalric status forged.[13]

. . .

No other surviving eyewitness narration rivals Pigafetta's for scope in coverage, vividness in detail, and ease in reading. Some of those written at the time are lost. Andrés de San Martín, for instance, wrote an account, or at least kept a log, which a prominent Portuguese humanist, Duarte de Resende, copied for Barros to use in writing his history of Portuguese Asia. It then disappeared. Of the other pilots' logs, three survive at least in part. One is convincingly austere, largely limited to sailing directions that are remarkably accurate, although missing the first two months of the voyage, and are only rarely elaborated with extra detail or commentary.[14] The man who brought the log back to Spain, known as Francisco Albo, was a seaman from Chios or perhaps from Rhodes who either had navigational training or took over a dead pilot's log at a late stage of the voyage.[15] The document varies a great deal in style and content. At times, dates are omitted for weeks in succession. There are phases, generally brief, when the writer takes an interest in his surroundings and makes occasional references to events, without any detectable difference between those included and those omitted. Some of the data involves cross-references, which must have been inserted at different times—some of them perhaps not until the end of the voyage. The likelihood is that the log passed from hand to hand, kept by different pilots at different times, or was a collation of various texts.

The second surviving log was by an author known to tradition as the Genoese Pilot. Long debate over his identity is now over: he was Leone Pancaldo of Savona, who served on the *Trinidad* and remained, with sixteen other survivors aboard that ship, as a prisoner in the Moluccas when Portuguese authorities seized it. Terrible privations followed, and a series of hair-raising escapes. Imprisoned in Cochin with the other survivors of the *Trinidad*, he broke out of jail, accompanied by a fellow Ligurian who had served as master of the ship, and stowed away on a homebound Portuguese vessel. Discovered in Mozambique, he was jailed again, and wrote plaintively to Bishop Fonseca and King Carlos for relief. Whether his pleas were heard or whether the local authorities wearied of his keep, he got back to Lisbon in or around July 1526, by which time his fellow stowaway was dead.[16] His return roughly coincided with that of the last living members of the crew of the *Trinidad*, released from Cochin in honor of another in the series of Spanish-Portuguese royal marriages. Pancaldo's narration is more elliptical than Albo's but includes occasional digressions, animadversions, and references to events on shore.

Finally, there is what looks like an abstract from a third log, described by Giovanni Battista Ramusio, the great compiler of voyages, who published it in 1554 as "Account by a Portuguese Companion of Duarte Barbosa" (Narratione di un portoghese, Compagno di Odoardo Barbosa).[17] Ramusio's designation suggests that the author was in Magellan's entourage aboard the *Trinidad*. The text covers the whole voyage sketchily without exceeding the proper sphere of a pilot or adding to material in other sources, save on two points: it tells us that when the fleet arrived at the strait now called Magellan's, "we named it the strait of Victoria because the ship *Victoria* was the first to sight it. Others called it the Strait of Magellan because our commander was called Fernando de Magallanes."[18] The text is unlikely to be a complete confection: Ramusio could have dressed it up with more appealing features if he had so wished; but the reference to Magellan as the strait's namesake, which occurs in no other sources before 1527, suggests a late interpolation.

Apart from Pigafetta, only one eyewitness authored a surviving account designed for publication. Ginés de Mafra was an ordinary seaman who shipped aboard *Trinidad* for pay of 1,200 maravedíes a month, of which he received four months in advance.[19] He was another of the captives the Portuguese held, first in Ternate, then in Cochin. When the royal marriage of 1526 prompted the release of the last Spanish prisoners there, he was one of only three who still lived. When he at last got back to Spain, he was the victim of a Martin Guerre moment: his wife, Catalina Martínez del Mercado, had arranged for him to be declared officially dead, before remarrying and selling his property. Accusing her of adultery and malversation, he took to sea again.

His manuscript, written "in his own hand," ended up in the hands of an editor, who describes the author as "an old man of few and true words" who "gave his relation of all that he witnessed on Magellan's voyage to the present author, knowing that I wished to write a book on all this."[20]) That book, as far as we know, was never written; the materials mainly concerned voyages that followed Magellan's. Mafra's work languished unremarked until a Spanish scholar published it in 1920.[21] The language exudes an authentic whiff of tar—which may of course itself be a rhetorical device, intended to enhance credibility; it is remarkably free of detectable special pleading, although, as we shall see, the author was strongly loyal to Magellan; if readers make allowances for lapses of memory and for pardonable exaggeration in the writer's claim to have "been everywhere and seen everything," it is a valuable check on other sources.

Another eyewitness narrative known only through an intermediate editor is the "Leiden manuscript," so called after the university library where it now reposes. It was among the materials assembled by the Portuguese court chronicler João de Barros for the history of Portuguese Asia that we have already quoted extensively, and that is a kind of palimpsest of lost sources. Fernão de Oliveira, tutor to one of Barros's sons, transcribed it in the 1560s.[22] Efforts to identify the writer have proved unavailing, and the text could be a confection compiled from other sources. Barros also possessed a now-lost narrative by Gómez de Espinosa, but it is hard to detect in the Leiden manuscript any resemblance to the materials that have survived in Gómez's hand—the report he wrote for the king of Spain when his own protracted adventures were at an end.[23] His account details his efforts to recross the Pacific from the Moluccas in command of the *Trinidad*, his failure in the face of adverse weather, his capture by the Portuguese, and the long incarceration that ended only with his repatriation in 1526, along with Ginés de Mafra, when the authorities in Cochin, prompted by the Spanish-Portuguese marriage, relented and packed their three surviving jailbirds home.[24]

Pancaldo and Elcano also submitted brief self-justifications to the royal court after returning home in solicitation of back pay, rewards, and favors—including, in Elcano's case, knighthood in the Order of Santiago in frank imitation of Magellan ("como lo dio a Fernando de Magallanes") and a share of any income that might accrue to the crown from trade with the Moluccas.[25] Elcano had evidently also written a more circumspect narrative of his own of at least part of the voyage. He mentioned it in passing while undergoing questioning from royal officials. "I did not dare," he said, "write anything while Magellan was alive. But after I was chosen as captain and treasurer I wrote down what happened."[26]

Some of the participants who left written accounts also reported orally to official investigations, inquiries, or interrogations into the conduct and results of the expedition. Elcano and Albo were interrogated jointly with another survivor, the former shipboard surgeon-barber, Hernando de Bustamante, in October 1522 after returning to Spain with the *Victoria*, the only ship to survive the voyage.[27] The investigators were more concerned with possible peculation—discrepancies in the records of cargo (below, p. 251), or uncertainties about the fate of prizes and prisoners taken in the course of the expedition—than with reconstructing the events of the voyage; but they garnered some useful details. Meanwhile, deserters who preceded them back to Spain had submitted to arduous questioning about the reasons for their early return (below, p. 163), and those who had fallen into Portuguese hands with the capture of the *Trinidad* had to give accounts of their deeds to hostile interrogators (below, p. 201).[28] A cabin boy, Martín de Ayamonte, whom the Portuguese had picked up separately—after he deserted or, according to his own testimony, was abandoned in Timor for unexplained reasons—was the subject of a further interrogation and gave testimony broadly favorable to his former commander.[29]

Chroniclers alert for copy interviewed survivors too. Writing at speed, from his post as official historian at the court of Carlos I, in order to get a narrative to the pope as quickly as possible, Peter Martyr of Anghiera (the Italian town now known as Angera) completed his hurried research within seven weeks of the return of the *Victoria*. He relied mainly on his conversations with a Genoese seaman known as Martinus de Iudicibus (perhaps properly dei Giudici), who had originally shipped on *Concepción,* and who was among the exhausted, emaciated survivors who made a barefoot procession of thanks, in threadbare shirts with candles in hand, when *Victoria* landed at Seville. The evidence Peter Martyr heard in the course of the interview made him suspicious of Magellan's probity (below, p. 257) or at least of his obedience to orders and fair conduct to his crew.

The account Peter Martyr produced is brief and general,[30] but a rival in the quick turnaround of material was even faster and immeasurably more detailed: Maximilianus Transylvanus's history of the voyage offers a corrective to Pigafetta's because, whereas the latter was Magellan's mouthpiece, Transylvanus based his account largely on interviews with the commander's shipboard enemies.[31] Francesco Chierichati, Pigafetta's patron, received copies of both works; he passed Transylvanus's on to a colleague in the bureaucracy of the Holy See.

The writer's name has deceived readers into supposing that Transylvanus came from Hungary, whereas—befitting a humanist who, like Peter Martyr, wrote in Latin—it was a Latinization he adopted for unknown reasons. He was a native of Brussels, son of a goldsmith, Lucas van Zevenbergen: why he did not call himself "de Septem Montibus" or something of that sort is a mystery.[32] He said he was "of Brussels" in a lightheartedly erotic poem he wrote in 1507. In that city he married Francisca, Diego de Haro's daughter and Cristóbal de Haro's niece. That would not necessarily prejudice him against Magellan, but it helps to account for the profound nature of his interest in the voyage.

Transylvanus's first job was as amanuensis to the learned archbishop of Salzburg. As a result of connections he made in the Netherlands, he moved to the household of Carlos I as one of the royal secretaries and was present in 1520 at Molins del Rey at the reception of an embassy that arrived from Germany to acclaim the king as Holy Roman Emperor. He completed his book on Magellan's voyage in Valladolid by October 24, 1522—less than a week after Elcano and his companions made their reports to the king. Pigafetta had, at best, only just arrived, as he was a few days behind his fellow travelers; so Transylvanus would have been unable to incorporate his positive view of Magellan, even had he so wished. Such are the penalties of journalistically tight deadlines. The book, which the author dedicated to his first employer, the archbishop of Salzburg, "my only lord," was a publishing coup, appearing the following year and going through three editions in a twelvemonth.[33]

Explorers lied. Like graduates writing "personal statements" in pursuit of employment today, they found modesty an encumbrance and accuracy a superfluity. The object of writing a petition for rewards—a document, in which many explorers' narrations originated, in solicitation of graces from the crown for services rendered—was not to tell the truth but to make a favorable impression. Those who wrote for publication had to please a public that thirsted for "marvels" and, then as now, preferred fiction to fact.

Still, we are lucky to have so many sources from which to try to reconstruct Magellan's great voyage. *Contraria sunt complementa*, as Niels Bohr said: contradictions complement each other. Writers partial to Magellan—notably Pigafetta and Mafra—balance the testimony of witnesses who hated or resented him and make up for the lack of an account from his own hand. Had he lived, like Vespucci and Columbus, he would no doubt have foisted apologiae and counter-recriminations on the world. As it is, we have to make do with those who were willing to speak for him, or against him.

. . .

As we have seen, Magellan intended, once he was at sea, to take control of the expedition and reverse the leaching of his authority to royal nominees. Sixteenth-century partings were far more wrenching than their modern counterparts. When a ship left the shore, she became a world unto herself, where communication with home rapidly dwindled and disappeared. The last letter Magellan received, as far as we know, arrived by pinnace on Thursday, October 3, the day after the fleet left

Tenerife, where the ships took on wood and water—the outermost Spanish port the ships would sight until their mission was accomplished or abandoned. We do not know what the letter said, but the need to pay the messenger appeared as a note in the records of the commissariat.[34] Thereafter, Magellan was free of interference from on shore. He knew he had at least two years in command—nominal command, which he could try to convert into unfettered power—before he had to account for his actions. As he had smilingly told Alvares (above, p. 117), "At sea he would be able to do as he liked." On the other hand, as we saw in the previous chapter, the existing balance of forces was against him. He acted with a nicely calculated mixture of resolve, ruthlessness, and prudence.

Tensions with his co-commander or *conjunta persona,* Juan de Cartagena, emerged almost at once, when Magellan refused to share details of the proposed route. Refusal was contrary to orders as well as to custom. We have seen that the king's instructions had been clear: "You will all make your way together with good fortune to the land which you shall name to the other captains and pilots," and Magellan "must give them the charts you have drawn to make the said voyage."[35] Elcano's testimony to the inquiry of October 1522 mentioned that Cartagena pointed out how "both commanders jointly were to attend to all things that might be necessary, while the said Fernando de Magallanes replied that in that case the orders were in error, nor did he understand them in that sense."[36] The words ring with truth: they express Cartagena's position perfectly. They capture Magellan's authentic tone: trenchancy underpinned by evasiveness; insubordination masked by quibbles; arrogance straining for utterance.

Pigafetta's defense of Magellan against the charge of breaking orders confirms the facts. The commander, says his apologist, was

> resolved to make so long a voyage through the Ocean Sea, where furious winds and great storms are always raging, but not desiring to make known to any of his men the voyage that he was about to make, fearing that they might be cast down at the thought of doing so great and extraordinary a deed.[37]

The text reeks of special pleading, which makes nonsense of the author's protestation that Magellan desired "to accomplish that which he promised under oath to the emperor, Don Carlo [*sic*], king of Spain."[38] The subordinate commanders knew the official destination of the fleet; unless Magellan proposed to change it, he could not keep it from them.

Pigafetta himself had already "heard" that the fleet was "going to discover the spicery in the islands of Molucca."[39] Every seaman had seen the scale of the provisions the fleet carried, even if rumor had not already acquainted everyone aboard; they could all calculate that they were bound for the far side of the world. Evidently, moreover, Pigafetta realized that the reason for declaring the route to all captains and pilots was to minimize the risk of scattering the fleet: he offers, as if in apology, Magellan's precautions for keeping the ships within sight of each other's lanterns. The king's orders, however, had made it clear that Magellan was to employ both methods: keeping a tight formation and confiding details of the route. Pigafetta was being disingenuous or deceptive.

Tension between Spaniards and Portuguese exacerbated the ill feeling Magellan's secrecy aroused. Pigafetta claimed to be able to think of no other reason than patriotic odium for the animosity of "the captains who accompanied him."[40] Equally improbably, Elcano represented the Portuguese presence, together with that of other foreigners, as a menace so formidable as to make the Spanish hands afraid.[41]

. . .

In early October, however, all the squabbles recorded between Magellan and other captains focused on what surely mattered far more than the expeditionaries' diversity of provenance: the issues of disclosure of the route and compliance with royal orders. The sources differ about the date of the first altercation, but perhaps on October 3, or at any rate not later than the fifth, Magellan announced a decision that precipitated serious discord: the fleet turned south, rather than following the usual transatlantic route to the southwest, and maintained the new course past Cape Verde and along the African coast beyond Sierra Leone.

The course made no navigational sense. On his second transatlantic voyage Columbus had discovered the best route across the ocean: southwest from the Canaries with the northeast trade winds. He had completed the crossing to the Antilles in twenty-four days. All travelers in his wake on their way toward what are now the Guyanas and Brazil followed his example. He had tried a more southerly route on his third voyage and got stuck in the doldrums: after that, most navigators knew better than to repeat the experiment. Magellan's course was even odder, because he ordered the ships to follow the coast beyond the outermost point of the West African bulge—in a direction opposite to the expected trajectory.

Cartagena approached the *Trinidad* "with fury and want of respect" (com fúria e desacatamento), according to the eyewitness whose reminiscences ended up in the Leiden manuscript,[42] to ask the reason for the change. He was within his rights, both as co-commander and, if Magellan refused to acknowledge the status of his *conjunta persona*, as a fellow officer, entitled to daily briefings on the choice of route. The pilot of the flagship leaned over the side to sing out Magellan's reply: Cartagena should desist from inquiries and follow the flag.[43] Presumably, the peremptory order was designed not only to put Cartagena in his place but also to alert him to his powerlessness. Short of responding with violence, he had no choice but to swallow his pride and do as he was told.

If Magellan had a reason for the new course—apart from a desire to provoke Cartagena—he never divulged it. He may have been trying to elude Portuguese attempts to waylay his fleet, but at the time Spanish-Portuguese relations were good, mollified by Manuel's recent marriage to Carlos's sister. No record is known of any Portuguese mission to intercept Magellan before he was expected or feared at the Moluccas.[44] In the Atlantic, a violent confrontation was unthinkable if Magellan stuck to orders. Nor would progress further into waters the Portuguese frequented be the best means to keep out of his fellow countrymen's way. He may have had some notion, similar to a chimera Columbus pursued on one occasion, that a southerly course would lead to richer lands on the far side of the ocean, or get him closer to the putative strait he was seeking.[45]

In any event, the result was disastrous. The new course led through "furious squalls of wind," as Pigafetta said, "and torrents of water struck us head on." To stay safely offshore, poles were bared, "and in this manner did we wander here and there on the sea, waiting for the tempest to cease, for it was very furious," while St. Elmo's fire flashed dazzlingly, disorientingly, around the mastheads.[46] The Atlantic was a cruel sea, where the waves churned restlessly or—between the zones of the trade winds—stilled into breathless immobility. A voyage across it was transmutative: liberating, despite the constraints of onboard life, because far from home and free of nagging families, censorious parishes, and officious constabulary. It could accentuate existing character traits. Or, like a masked entertainment or travesty, any long voyage could make a traveler behave unpredictably or exhibit traits repressed at home. Because sailors can move on from ports of call "with vows of returning yet never intending to visit them more," seafaring encourages irresponsibility. That, I suspect, is why explorers' books had so much sex in them.

The Atlantic did not change Magellan very much—the main personality changes came further on in the voyage (below, p. 285)—but it freed him to be himself: decisive, daring, vengeful, ruthless. Its caprices were crueler than the Mediterranean's, where tideless placidity was typical and disasters rarely found seafarers far from shore. No sooner were the adventurers through the storms than they stalled, becalmed in the doldrums, for at least a fortnight. In all, it took nearly two temper-trying, patience-fraying months, from the last halt in harbor, to reach the equator. The pilots recorded passing the line on November 20.

. . .

Pigafetta spent the becalmed days in the recourse he always used to beat boredom: fishing from deck. He hoisted sharks, but "They are not good to eat unless they are small, and even then they are not very good."[47] Others occupied their time less innocently. On October 30, Antonio Salomón, master of the *Victoria,* was arraigned for sodomy with a cabin boy. He was found guilty, but with a stay of execution, granted for an unknown reason, for nearly two months, until he became the expedition's first fatal casualty—garroted on shore at Rio de Janeiro.

For present purposes, the importance of the incident is that Salomón's condemnation to death required, or at least justified, a council of deliberation that brought all the fleet's leading officers together aboard the flagship. The occasion was unprecedented: the opposed factions were face to face, and Magellan's opponents were in his power. Cartagena, nevertheless, felt secure enough to take the opportunity, after the legal proceedings were over, to challenge Magellan again on the questions of why he had changed course, what route he was proposing to follow, and why he was flouting royal instructions to share information with his colleagues.

Magellan evaded the questions but replied with a counteraccusation. He pointed out that for days past Cartagena had been guilty of breaching protocol during the daily ritual, at the hour of the evening Angelus, of exchanging salutes with the flagship. Instead of hailing Magellan as captain-general, Cartagena had saluted him merely as "captain" in the standard formula, crying, "God save you, Captain, sir, and Master and Good Company!" (Salve, señor capitán y maestre y buena compañía).

Cartagena's first offense of this sort had occurred four evenings previously. Magellan had asked to be called by his proper title. Cartagena, according to some testimony, responded contemptuously. He said that in future he would send a mere cabin boy, instead of an officer, to tender his compliments. He then went further: he suspended the custom altogether. The outrage transcended matters of etiquette: if Cartagena was Magellan's inferior in rank, it was an act of insubordination. From

Cartagena's point of view, however, he was Magellan's *conjunto*, deprived of his due share of authority. By treating him as a subordinate Magellan was committing another transgression against the king's order. The exchange of insults touched both officers' honor, and on a point of honor sixteenth-century gentlemen would be as pernickety as Hotspur. Beyond honor, real power was at stake.

Magellan responded by staging a coup de théatre, which he had evidently prepared in advance. Taking advantage of the security of his own cabin, guarded by his own men, he seized Cartagena by the throat, according to the recollection of Juan Sebastián Elcano, and sent him to the hold. The tactics of shock and awe succeeded. Magellan threatened his prisoner, if Elcano's testimony can be believed, with marooning on the coast of Brazil, but "at the other captains' plea, he did not cast him ashore at that time."

Magellan appointed his kinsman Álvaro de Mesquita to replace Cartagena in command of the *San Antonio*. The largest ship was therefore now in hands Magellan could regard as safe. Perhaps that was why he felt secure enough to indulge in a magnanimous gesture. He released Cartagena into the custody of a member of the captive's own faction, Gaspar de Quesada, "under pledge and promise that he would hold him prisoner."[48] Magellan was fulfilling his plan to "do as he pleased" once he was at sea.

. . .

The long delay had served its purpose—if that purpose was to await a pretext to eject Cartagena from command of the biggest ship—but at serious cost. By the time the fleet sighted the Brazilian coast at its easternmost point, on November 29, three months' rations had been consumed. The direct route would have got them to the same place in as little as a month or six or seven weeks, at most, from Tenerife. Brazil represented an opportunity to acquire fresh food. Pigafetta's reaction to the prospect—relief and delight—seems representative. He had suppressed all mention of the crimes and violence at sea. He had devoted his time before landfall to observations of marginally monstrous birdlife, including one species "that had no anus," another without feet, and a third that fed on guano dropped by other birds and seized in flight. He extolled the "plentiful refreshment" the expedition found ashore, including native pineapples, samples of sugarcane that had only lately been introduced, and flesh of a local quadruped—perhaps a tapir— "that resembles beef."[49]

At Cabo Santo Agostinho the fleet had reached a region where the visits of Portuguese and French loggers, who acquired shiploads of the dyewood that gave Brazil its name, had accustomed the native Tupi to Europeans. Albo's log described the local amenities with characteristic concision: "There are many good people, and they go naked and they trade food for fishhooks, mirrors, and hawks' bells. And there is a lot of brazilwood."[50] The Tupi were fairly recent arrivals themselves—present for perhaps a couple of generations, after migrations in search of their mythical "Land without Evil," celebrated in tales early ethnographers recorded. Some of their customs—cannibalism, internecine warfare, polygamy, even nomadism, which implied the instability of savagery—were offensive to missionaries. Secular Europeans, however, found a modus vivendi easy to establish.

Here, as almost everywhere European imperialism was successful, newcomers benefited from a propensity to receive strangers with exceptional honor. I call this phenomenon the stranger effect.[51] In modern Western societies the propensity is rare. We mistrust strangers. We reject them. In the US, where I live, some of us call them "illegal aliens." We impose on them bureaucratic or fiscal burdens. If we admit them, we often make them unwelcome and typically assign them low-status and demanding or demeaning work. In other times, however, and in other parts of the world, people have behaved differently. Sacred rules of hospitality oblige people in some cultures to greet strangers with their best gifts and goods and women and even actual deference. When Spaniards found themselves treated in this way in parts of the Americas it made them feel godlike: hence, I suggest, the myths of Spaniards as "returning gods" or "men from heaven." As the anthropologist Mary W. Helms has pointed out, the stranger often brings an aura of "the divine horizon," which makes him or her numinous.[52] A possibly useful analogy is with the value added to exotic goods and long-range imports in modern Western commerce. So it is, in many cultures, with people from afar. In Christendom in the past, as still in Islam, pilgrims have profited from a similar effect, acquiring prestige with neighbors on returning home, in rough proportion to the distance traveled or the sanctity of the furthest point.

To defer to the stranger—given an appropriate cultural context—is often a highly commendable, rationally defensible approach. The stranger is useful as an arbiter or judge because he or she is uninvolved in existing factional and dynastic conflicts and can bring an objective eye to matters in dispute. The early colonial archives of Spanish America are full of cases in which native elites confided in Spanish arbitration, gradually thereby shifting power into Spanish hands.[53] For similar reasons, strangers, untainted by prior associations with local rivalries, make first-class allies, bodyguards, or close counselors for rulers, and marriage partners for elite families. Of course, the stranger effect is always a waning asset: imperialists commonly outstay their welcome or forfeit goodwill, like Captain Cook, by returning uninvited. Arrogant or exploitative behavior by European visitors could dissipate the advantages; but they were still operating at Cabo Santo Agostinho when Magellan arrived.

Mutual economic advantage confirmed good relations between natives and newcomers. Pigafetta provides details: for a fishhook or for one of those cheapskate German knives the ships carried (above, p. 108) a native would give "as many fish as would be sufficient for ten men; for a bell or a strip of lace, "one basketful of" what Pigafetta calls potatoes but were really cassava "as long as turnips." Further down the coast, "for a king of diamonds, which is a playing card," Pigafetta reported, "they gave me six fowls and thought they had even cheated me."[54] The fabulous rates of exchange European goods commanded in Native America were an inexhaustible topos in what was, after all, in part promotional literature addressed to potential European traders, colonists, or conquerors, and were reported to portray natives as simple-minded and exploitable. From an indigenous point of view, however, it made sense to give surplus food, which was effectively valueless to those dispensing it, for the tawdry bric-a-brac that Europeans despised but that commanded rarity value in a place like Brazil.

. . .

The fleet enjoyed the amenities of Cabo Santo Agostinho for a few days. For the next leg of the journey, João Lopes Carvalho, who had been on a previous expedition in the region, took over the role of pilot on the flagship. Pigafetta said that Carvalho had spent four years in Brazil, perhaps as a captive of the Tupi. The returnee took the opportunity to retrieve his son by a native woman or "negra," as Martín de Ayamonte called her.[55] The young man was lost to headhunters in the vicinity of Brunei later in the voyage. Carvalho does seem to have fancied himself as a ladies' man. "Sleeping with black slave women," according to Ayamonte, was one of the reasons for his ultimate removal from command: they must have been the three captive women Carvalho "kept for himself" from among the spoils of a junk he captured off Brunei.[56]

Estevão Gomes yielded his place to make way for Carvalho and transferred to the *San Antonio,* giving himself a possible reason for resentment against Magellan. Yet transfer to the biggest ship in the fleet was hardly derogation; it suggests rather Magellan's abiding confidence in his fellow countryman, whose presence on *San Antonio* might have been expected to strengthen Álvaro de Mesquita's hand in his new and controversial captaincy. If the gesture was one of reliance on Gomes, it was, as we shall see (below, p. 203), sadly miscalculated.

It took a surprisingly long time—until December 13—to reach the next recorded anchorage, at a bay that Albo's description identifies: a low island at the mouth of a river, large with many inlets and a muddy bottom, shallow on the left as you enter. It was Guanabra on the estuary of Rio de Janeiro, which the explorers called after St. Lucy, whose feast it was.[57] Modern Rio is just to the south. Either the fleet had dawdled at Santo Agostinho for the entire thirteen days that Pigafetta counted as spent in Brazil, or they undertook a minute examination of the intervening coast. They stayed in the vicinity of Rio until December 27.[58] There were onshore supplies to be had and a good deal of business to transact.

The deferred sentence of death on Antonio de Salomón was carried out on December 20. According to a poorly attested but in any case curious story, Magellan pronounced a further penal sentence: he clapped his wife's cousin, Duarte Barbosa, in irons to prevent him from staying in Brazil.[59] It seems unbelievable, as applied to Barbosa; but Brazil did exercise allure, at least temporarily, for some Europeans at the time. Seafaring is not a profession for the contented and unadventurous. For one un-nostalgic for the comforts of home, Brazil offered some characteristics of the "Land without Evil" the Tupi sought. Pigafetta summed up the attractions: reputed longevity, abundant food, generous sexual hospitality, and women with insatiable sexual appetites.

"One or two of their young daughters" could be bought from Tupi men for "one hatchet or one large knife. . . . One day a beautiful young woman came to the flagship . . . for no other purpose than to seek what chance might offer."[60] Pigafetta's language of "chance" (as his best translator, Theodore Cachey, pointed out) evokes sexual predation. "While there and waiting, she cast eyes upon the master's room and saw a nail longer than one's finger. Picking it up most gracefully and gallantly, she thrust it through the lips of her vagina, and bending down low immediately departed."[61] The place of concealment she selected was presumably for security—in awareness that she had stolen an item that her people, at least, would consider valuable; but the episode suited the sexual utopia European observers often saw when they contemplated or desired distant lands.

. . .

Prior reading and fabulous expectations colored Pigafetta's observations. His interest in ethnography, however, was genuine, and his data on the people he saw in Brazil are too important and interesting to dismiss with the insouciance some readers have shown. He did not have as long in the country as in Patagonia or the Philippines; some of the material he recorded, therefore, relied not on personal witness but on occasionally misleading inferences, extrapolations, or analogies from what he had read, or on what João Carvalho told him.

In the former category belongs almost everything he claimed to know about the natives at the expedition's first Brazilian port of call: the vast dimensions of the country they inhabited; the longhouses in which they supposedly lived; the nakedness—perhaps innocent, perhaps bestial—in which they defied shame; the hammocks in which they slept, warmed by fires lit below them; their canoes operated with paddles "like baker's peels," such as Columbus had described; the inhabitants' surprising failure to conform to the images of monsters with which medieval maps enlivened the recesses of the Earth; their reliance on stone technology; their indifference to law except the law of nature; their apparent conviction that Europeans "came from the sky"—evidence not of credulity but of the fact that the stranger effect is often associated with the touch of the divine horizon; and their want of any recognizable religion.[62]

All those details could represent Caribbean or peri-Caribbean peoples as well as or better than Tupi and could have been cribbed from earlier writers. A vivid image of how "black, naked, and shaven, they resemble, when paddling, the inhabitants of the Stygian marsh" (asimigliano, quando vogano, a quelli de la Stigia palude) sounds like a classical allusion worthy of Peter Martyr. In fact, however, the reference derives, as the Argentinian scholar, Antonio Gerbi spotted, from the naked, beslimed, and angry swamp-dwellers of Dante's Inferno.[63] Dante was ingrained in every Italian gentleman's education. Vespucci had quoted him too.[64]

The Stygian image hardly accords with the rest of the material Pigafetta took from books, most of which depicts people redeemable for God: fully rational, free of monstrous deformities, ripe for conversion. Some of the genuine observations Pigafetta made for himself accorded with this positive view or added to the favorable impression. The nakedness, which was a literary topos variously for innocence, dependence on God, madness, or savagery, looked less naked in the light of real observation, verifiable from other evidence, of the bulbous dresses of colored feathers for special occasions. The "natural" way of life, to which Pigafetta alluded, not with a romantic or ecological eye but with at best equivocal approval for wild ways, seems in some measure civilized when Pigafetta's references to cultivation are taken into account—of domesticated peccaries and farmed cassava and sago. Sexual depravity is tempered, in some of his remarks, by evidence of uxorial fidelity. "Those people," moreover, "are not entirely black, but olive-skinned" (non sono del tuto negri, ma olivastri). It was not necessarily bad to be black in Renaissance Europe: most of the widely known images of black people were depictions of one the kings of the Epiphany, or portraits of favored, cossetted, or professionally specialized slaves; on the other hand, blackness evoked plenty of the negative connotations suggested in Dante's Stygian scene, and on the whole to be "olive-skinned" was better.[65] Above all, the natives seemed to Pigafetta to show a kind of predisposition to piety—as if God had confided to them a partial revelation in preparation for evangelization: during mass, "Those people remained on their knees with so great contrition and raising clasped hands aloft that it was an exceeding great pleasure to behold them. . . . Those people could easily be converted to the faith of Jesus Christ."[66]

. . .

On the other hand, the information Pigafetta obtained from João Carvalho related to a morally equivocal custom: cannibalism. The expeditionaries were heirs to an unresolved medieval debate about the order of creation. That order was hierarchical: no one doubted that images of the "ladder" of creation and the "chain of being" fitted the reality of the world. Nor was any serious challenge mounted to humankind's place near the top, under the angels, with the divinely confided stewardship of all the rest. In the stained glass of the nave of León Cathedral the whole scheme sparkles vividly. Under saints in glory and mundane man, beasts trample earth from which plants spring. But how many rungs did the ladder have, or how many links the chain?

The big debate concerned creatures known only by report, from classical texts or popular legends or biblical allusions, especially the *similitudines hominis*—the monsters whose deformations from normality of physique might be so deviant as to exclude them from the category of the rational and redeemable, as St. Albertus Magnus thought, or might merely be variants of beauty discernible to God but hidden to imperfect human scrutineers, as St. Augustine opined. Should such monsters be classed with men, or with beasts, or in some intermediate category? "For it is reported," said Augustine, summarizing the most commonly depicted monsters,

> that some have one eye in the middle of the forehead; some, feet turned backwards from the heel; some, a double sex, the right breast like a man, the left like a woman ... and ... others are said to have no mouth, and to breathe only through the nostrils; others are but a cubit high, and are therefore called by the Greeks "Pigmie." ... So, too, they tell of a race who have two feet but only one leg, and are of marvelous swiftness, though they do not bend the knee: they are called Skiopodes, because in the hot weather they lie down on their backs and shade themselves with their feet. Others are said to have no head, and their eyes in their shoulders. What shall I say of the Cynocephali, whose doglike head and barking proclaim them beasts rather than men?

Augustine points out that humans are no less human for looking odd or emerging from the womb with physical peculiarities. "Accordingly," he concludes, "it ought not to seem absurd to us, that as in individual races there are monstrous births, so in the whole race there are monstrous races."[67] Many readers agreed, including, among the chroniclers of Magellan's voyage, the humanist Maximilianus Transylvanus, who thought that one of the scientific achievements of the expedition was to help undermine belief in the reality of mythical beasts.[68] The twelfth-century portion of the monastery Church of Vézelay—known throughout Christendom because of the pilgrims who beheld it—shows a vital Last Judgment,[69] with Christ in majesty, flashing thunderbolts from his fingertips: among the people coming to judgment are the halt and the lame, the maimed and misshapen, and representatives of the monsters whose descriptions Augustine culled from the classics. In Eastern Orthodox images, a cynocephalus could be a saint: St. Christopher often appears with a dog's head. The message is clear: a normal physique is not required for salvation.

Moral deformities were not so easily accommodated if they amounted to offenses against nature. Standards of what is natural change. Strictly speaking, every vice and evil we commit is natural to us, as it is impossible for us to act other than in conformity with our natures. Few of Magellan's contemporaries, however, would have been inclined to agree. Sodomy was unnatural: Magellan put two members of his crew to death for it. Incest incurred similar revulsion. Blasphemy, many lawyers and theologians claimed, was an offense against natural law because humans' nature is to praise God. Human sacrifice, despite many prevalent exceptions to the inviolability of life at the time, qualified as an unnatural way of disposing of people. Cannibalism was peculiarly abhorrent, perhaps because observation showed that it is rare in nature, at least among mammals. All such offenses deprived the guilty of the protection of natural law: they could therefore be enslaved or put to death.[70]

The Tupi "do not eat the bodies all at once," as Carvalho explained, "but every one cuts off a piece, and carries it to his own house, where he smokes it. Then every eight days, he cuts off a small bit, which he eats smoked with his other food."[71] According to European legal traditions, cannibalism, as an offense against natural law, deprived practitioners of any right to protect their own sovereignty or freedom from encroachment by conquerors or slavers. So the image Pigafetta transmitted was politically and economically charged. Apart from the authority of Carvalho, who had lived among the Tupi, many other observers' and visitors' narratives of the sixteenth century confirmed it.

Pigafetta, however, was the first not only to provide a vindication for cannibalism but also, I think, to understand its true nature. Las Casas's apologia of 1550 is well known: he pointed out that cannibalism is a universal human custom, discernible in the historical records of every people, including the ancestors of Romans and Spaniards, if one goes back far enough, or—if you like—that under the stones of every civilization lie the bones of cannibal feasts. Gonzalo Fernández de Oviedo was aware of the stratigraphy, although he did not make the same moral inferences in favor of the cannibals.[72] Even more famous is Montaigne's essay "On Cannibals," published in 1580, which accuses Europeans of abusing cannibalism as a distraction from their own customary and even more reprehensible transgressions of natural law. Jean de Léry's flattering comparison of Tupi cannibals with their more reprehensible Huguenot counterparts at the Siege of Sancerre in 1573 is less well known but has received just attention.[73] Yet Pigafetta deserves credit for an at least equally remarkable insight. "They eat the human flesh of their enemies," he explained of the Tupi, "not because it is good, but because it is a certain established custom."[74] He goes on to give two reasons for it: revenge, which he illustrates with a story of an old woman who, confronted with one of her only son's slayers in battle, "remembering her son, . . . ran upon him like an infuriated bitch, and bit him on the shoulder"; and commemoration—"to remind" the Tupi "of their enemies."[75]

Pigafetta seems to have realized a fact that long eluded many students of cannibalism: that normally, where it is normal, cannibalism is not practiced for nutrition or gustation but as a moral act, designed in some cases to dispose reverently of the dead and in others in order to express moral sentiments—such as love or vengeance or piety—or to appropriate moral qualities of the victim, and nearly always for self-transformation or the ritualization of the eater's relationship with the eaten. It is socially or physiologically functional for the Gimi of Papua New Guinea or the Huari of Amazonia, whose women eat their menfolk out of respect, or the Hua of New Guinea who consume corpses out of a need to conserve *nu,* the vital fluid they believe to be nonrenewable in nature. Among the Aztecs, eating selected bits of a war captive was a way of acquiring his prowess. In Fiji human meat was gods' food, comestible under "a mythical charter of society."[76] Of course, there are cases when other motives supervene, such as survival, as in "the custom of the sea" or in face of starvation, or irrational rage, as in the notorious case of the citizens of Hautefaye, who ate a well-known, well-liked local gentleman on the supposition that he was a Prussian infiltrator; or in excesses of ghoulish psychopathology, such as those of Jeffrey Dahmer, whose fridge in Milwaukee in 1991 was full of human body parts.[77]

. . .

Pigafetta's final contribution to Tupi ethnography was linguistic. Humanist ethnography was always hungry for word lists in exotic languages, as a contribution to comparative philology and, in particular, to the search for the original language of humankind—a priority for scholars at least from the fourteenth century to the seventeenth. As we shall see in future chapters, Pigafetta was an assiduous collector of lexical trivia. Among the Tupi, perhaps because of the brief duration of the fleet's stay in Brazil, his gleanings were sketchy. His words for knife—*tacse*—and scissors—*pirame*—do look like genuine Tupi terms, similar to those recorded in a vocabulary by a French trader on the same coast, Jehan Lamy, in the 1540s.[78] These were manufactures for which the market in Brazil remained insatiable for decades.[79] On the other hand, *maiz,* which Pigafetta also lists, or anything like it cannot have been a native Tupi term. It derives from a word or group of words in Arawak languages, recorded by Columbus and Latinized as *maizium* by Peter Martyr before Pigafetta used it. Unless the Tupi had absorbed it from Arawak or European traders or visitors, it must belong with other words of Caribbean origins that Pigafetta culled from readings and misattributed to Brazil.

His versions of local terms for houses, hammocks, canoes, and chieftains, none of which figure in his word list but all of which are part of his description of natives of Brazil, are all Arawak in origin and accessible in works of Columbus and Peter Martyr.[80]

. . .

The ships spent a fortnight in Rio de Janeiro, taking advantage of fresh provender. When they left, native canoes saw them off, as in regretful farewell. They were now on a coast where a strait might have been overlooked on previous voyages. So the pilots' logs show them proceeding cautiously, checking each bay, as they quitted Paranaguá on the last day of 1519, and, pushing against the Falkland current that imposed ever harder going, they reached, on January 10, 1520, an estuary with a "hat-shaped hill."[81] A squall momentarily concealed the land. It cleared, revealing to Carvalho a shoreline recognizable as the place where Juan Díaz de Solís had met his death more than three years before. They were at the limits of the world any of them knew from firsthand experience or even reliable reports. The Portuguese called it Cabo de Santa Maria, and knew it as the furthest point to which their own explorations had attained.[82]

The fresh water that poured out of the River Plate might have convinced Magellan that there was no point in lingering in order to explore the estuary. The shallow bottom was unpromising. Conceivably, however, the river or rivers that produced the fresh water debouched into a strait, obscuring the possibility of an outlet to open sea further to the west. The vastness of the river mouth, 220 kilometers wide, nourished the illusion. Pigafetta, indeed, seems to have expected to find the strait at this point: at least, prior to the exploration Magellan undertook, "It was thought that one passed from there to the South Sea."[83]

So the exploration began. On January 12, Magellan sent *Santiago* to reconnoiter. While other ships took on wood and water—cautiously, in remembrance of what had befallen Solís—boats probed the inlets on the far side of the estuary. Juan Serrano took his vessel far upriver until the banks, closing in, and the soundings, persistently shallow, made the theory of a strait untenable. Meanwhile, bad omens accumulated. A cabin boy fell overboard and drowned on January 25. A few days later a seaman died in a fistfight.[84] The fleet set sail again on February 3.

Southward progress was slow from this point on. The current, which had borne the ships along the Brazilian coast, was now against them. Storms were increasingly frequent. On February 12 and 24, for instance, ferocious winds forced the fleet out to sea for three days at a time. A

tempest then sent the ships splaying in all directions. The cold was starting to bite. On the twenty-seventh, a shore party, having scrambled
onto an islet in search of fresh water, was cut off by a storm. Rescuers
found the survivors shivering and soaked the next morning, Six days of
storms more terrible than before now sent the fleet scurrying out of
sight of land again, eluding the sandbanks, until, according to Pigafetta,
apparitions of St. Elmo, St. Nicholas, and St. Clare—specialists in spar-
ing seamen—bade the waves cease.[85] Francisco Pascasio Moreno sailed
along this coast on his way to explore the Santa Cruz River in 1876; he
confessed the fear the high seas caused him—the groaning of the masts,
the waves that grasped at the crow's nest, the sense of danger from the
impossibility of taking soundings in a turbulent sea.[86]

Even had the weather been kinder, Magellan had to grope for a tentative course, probing every inlet in case it should prove to conceal the
longed-for strait. On the twenty-fourth, for instance, the Gulf of San
Matías (or of San Martín, as Magellan called it) delayed and disappointed him. The inhospitable coast offered no refuge from the falling
temperature and worsening winds, but on the twenty-seventh they
found a bay where penguins and seals were plentiful: literally innumer-
able, Pigafetta said, and "In one hour we loaded the five ships. . . . They
were so fat that it was necessary to skin them rather than pluck them."[87]

The progress of the fleet amid these banks and bays, islands and
inlets is hard to follow because of the pilots' conflicting calculations of
latitude. Practiced navigators could tell by the "feel" of their location
roughly how close it was in latitude to a known place with a similar
feel. By squinting at the sun or glancing at the Pole Star (while it
remained visible), one could make a rough but reasonably reliable guess
with sufficient experience to back it. Such methods yielded figures too
approximate to reconstruct the ships' whereabouts on a slow voyage
and ragged coast such as Magellan followed in the fading Southern
Hemisphere summer of 1520, where each cape or bay was too close to
the last to make guesswork serviceable.

All accurate methods depended on fair weather. Columbus likened
navigation to "divine prophecy," and Vespucci represented himself as a
magus, conjuring the heavens. But that was partly because they wanted
the kudos that mystification brings and partly because both were concealing their inexpertise. They used the simplest, most amateurish method
available: calculating the hours of daylight by timing, with an hourglass,
in northerly latitudes, the passage of the guard stars around Polaris and
subtracting from twenty-four; they could then read the corresponding

latitude from printed tables.[88] Even for so elementary a procedure, the stars had to be visible, the tables accurate, and the memory of the cabin boy who turned the glass reliable.

Only a brief gap in the clouds was necessary for the technically more complex methods on which professionals relied. The angle of elevation of any celestial body above the horizon was a key to a calculation every trained navigator could make. The noon sun or, in the Northern Hemisphere, the Pole Star was the most favored object for the purpose, because both are conspicuous and were used by the compilers of printed tables that could be used to check readings of the heavens. The sun, moreover, of course monopolized the sky for much of every day. Albo, or whoever kept his log, relied entirely on sightings of the sun and used a four-year table identical with one that Martín Fernández de Enciso published in Seville (above, p. 79).[89] In the Southern Hemisphere, any star in the Southern Cross might serve, but printed tables were not available as cribs for the diffident. When the sky was clear, navigators typically learned how to read the sky with an astrolabe, but for shipboard use a quadrant or backstaff sufficed: all very well on a placid sea, but unmanageable in adverse weather. For almost the whole of March, from the second to the thirty-first, for instance, Albo was unable to see the sun. Some of the readings that survive in logs kept on Magellan's voyage were made on land—but even then could happen only under a clear sky.

. . .

In other words, the expeditionaries literally did not know where they were, except that they were on a coast beyond the edge of the known world. Other circumstances conspired to make nerves grate: intensifying cold and life-threatening storms; a sense of summer's shortening lease; a permanently adverse current; repeated disillusionment as one inlet after another led nowhere; diminishing rations, replenished by increasingly blubbery and unpalatable animal carcasses; a cannibal coast on one hand and a raging sea on the other. Factional strife, natural antipathies, national rivalries, secrecy, and suspicion had made the fleet an unhappy little world almost from the moment it first set sail. Now tempers quickened as the fleet slowed.

When they were in the high forties, latitude south, as Elcano later recalled, an alliance of ships' officers renewed the effort to get Magellan to comply with orders. "The other captains, jointly with the said Cartagena"—who was still under arrest in Gapar Quesada's keeping— "required the said Magellan to take counsel with his officers, and that he tell them the route to where he wanted to go, and that he cease to wander lost, as he was."[90] Elcano must have been exaggerating in at least one respect: the "other captains" can hardly have been united, since Álvaro de Mesquita would not have taken part in a front against his kinsman, on whose goodwill he depended.

Elcano can be believed, however, when he reveals that new fears were afflicting the malcontents: that Magellan would make them spend the winter in inhospitable latitudes and that supplies would give out. "And let him not," the captains said of Magellan, according to Elcano's testimony, "make port where they might winter and consume the rations, and let them continue the voyage to where they may be able to bear the cold in order to continue ahead if they find a place to do so."[91] The meaning Elcano sought to convey in his information to the royal officials was clear: the "other captains" wanted to follow the king's orders by finding the strait without delay. Once they were through it, they could turn north into warmer climes. Elcano was trying to deceive his interrogators. Since the whereabouts of the strait, if any, remained occluded and its very existence doubtful, the demands he puts in the captains' mouths would have made no sense. The real points of contention, as we shall see in the next chapter, were whether to abandon the expedition or switch to an easterly route, via the Cape of Good Hope: an approach known to be viable but preempted by Portugal's exclusive right to navigation in that direction. To take the eastward route to the Moluccas would be an even more flagrant violation of royal commands than Magellan had already committed.

Hostile camps were preparing for conflict. A kind of class war seems to have been developing. The evidence Portuguese investigators later gathered from those of Magellan's men whom they took as prisoners suggested that "the lower sort, the majority" (la gente baja, la mayoría) of hands supported Magellan and that "the ordinary seamen were on his side" (los marineros estaban bien con Magallanes) or at least content with his leadership. That is believable: Magellan's offenses, so far, had been apparent only to the officer corps and injurious only to the faction that gathered around Cartagena. Castilian-Portuguese animosities, moreover, were still at work. Elcano complained that Magellan's kinsmen were "beating and maltreating Castilians" (maltrataban e daban palos a los castellanos).

Somewhere amid the storms and frustrations along the Patagonian coast, Cartagena and Quesada, who though cast as prisoner and jailer, were old comrades and now coconspirators, invited Elcano into the mutiny they were planning. Elcano represented it as a prospective loyal coup. They asked, he said, "that he give them his favor and help so that they could see to it that the king's orders were obeyed, as specified in his instructions."[92] He complied. "And this witness," he complained, "was obedient, and is under arrest for trying to follow the king's will and to make the said Fernando de Magallanes follow it."[93] The conspiracy Elcano joined played out, with terrible effect, in the events of the next chapter.

Primaleon.

Los tres libros del

muy esforçado cauallero Prima
leon et Polendos su hermano hijos del Emperado:
palmerin de Oliua.

FIGURE 6. The decisive evidence that Magellan read tales of chivalric romance during his education at the Portuguese court is in his recorded allusions to the *Book of Primaleon,* shown here in a Spanish edition (*Primaleon: Los tres libros del muy esforçado cavallero Primaleon et Polendos su hermano hijos del Emperador Palmerín de Oliva* (The three books of the very brave knight Primaleon and Polendos, his brother, sons of the Emperor Palmerin de Oliva) of 1535. Magellan named his native captive "Patagon" after a character in the book—a giant who exhibited features, physical and mental, that Magellan and his men detected in the natives. The region is named "Patagonia" accordingly. Encounters with giants occur frequently in the novel, as in the scene shown here, where Primaleon confronts the giant Gatarú, who is holding an enchanted maiden hostage on an island. In the background, the onion dome conveys exoticism, and the gallows and withered branch evoke danger. Magellan seems to have modeled his own ascent to knighthood on Primaleon's. Courtesy of the Biblioteca Nacional de España.

CHAPTER 6

The Gibbet at San Julián

Patagonia,
March to October 1520

And it may be that I will abide, yea, and winter with you,
that ye may bring me on my journey whithersoever I go.
—Corinthians 1:16.6

They saw, first, "rocks like towers." Then many islands in the bay appeared. They chose one as headquarters. In June 1578, when Francis Drake—on the most ambitious and spectacular of the pirate raids he launched against Spanish possessions—went ashore in what is now Argentina at San Julián, his men found the remains of a gibbet of spruce or fir. It stood "close by the sea upon the mainland, . . . fallen down, with men's bones underneath it." Of the part that was "sound and whole," as Drake's staunchly bigoted but creditable chaplain recorded, "our cooper made tankards or cans for such of the company as would drink in them, whereof for my own part I had no great liking."[1]

The wood that provided those grisly souvenirs was all that was left of the structure on which Magellan hanged some of his alleged mutineers, or displayed the quartered corpses of those he beheaded.

Drake was aware of the episode. Magellan was, in a sense, a spectral companion on what developed into the first English circumnavigation of the globe. Almost all the contemporary accounts of the voyage, including Drake's own, allude frequently to his predecessor, both as a model for imitation and as an adopted rival for emulation. For in his own estimation, Drake "outreacheth in many respects that noble mariner Magellanus, and by far surpasseth his crowned victory."[2]

The English crew recognized the gibbet they found for what it was. It is hard to resist the impression that Drake chose San Julián not only as a useful spot to take fresh water, gather firewood, trade with natives for victuals, and see out the southern winter, but also for its symbolic connotations as a place of execution. He intended to follow Magellan's example and put his own alleged traitors to death.

. . .

Magellan arrived on March 31, 1520, the eve of Palm Sunday. The solstice had just passed. The expedition had reached a little over forty-nine degrees south. Chances were receding of finding a strait and turning north into warmer latitudes before winter set in. The only route toward a possible passage led further south, into lengthening night and deepening cold. According to Transylvanus, whose narrative at this point differs markedly from all other surviving accounts, and yet is full of such vivid detail as to show that the author made good use of his eyewitness reports, the bay first attracted the explorers' attention because they hoped it might be the mouth of a strait ("sinus hic vastus videbatur et speciem freti referre").[3] In that respect it was like every other deceptive inlet they had probed, as the frustrating coast dragged them down into unknown dangers.

The weather had been horrible: cold, for the men's taste, squally, and hard on the ships' planking. The sheltering bay, though no strait, was at least a refuge. Its pinched, narrow entrance would be easy to guard against interlopers from outside and deserters from within. With rations wasted and his ships leaking, Magellan envisaged, or came to envisage, a long stay for repairs and refreshment in winter quarters, which he proposed to build onshore. Pigafetta, who can be relied on to speak for Magellan, said simply that "we found a good harbor, and as winter was approaching we deemed it convenient to spend the season there."[4] For the phlegmatic pilot Francisco Albo, whose priorities were always narrowly linked to his job, the scraping of the ships' bottoms was the main outstanding task and the chief motive for the halt. In his log of the voyage he does not even mention the events that dominated the stay: the crisis of command and the condemnation of Magellan's victims.[5]

It is hard, however, to resist the impression that on arrival at San Julián Magellan's mind was already focused on how to intimidate his opponents, exact vengeance on his enemies, and eradicate his rivals. He struck not only because it was propitious to do so: his need to assert himself, sometimes violently, arose from his character and values. López de Gómara, who was only a boy when news of Magellan's voyage first reached him, summed up what happened at San Julián with nicely judged irony: "As a man of spirit and honor, he bared his teeth" to his foes (Mostrándoles dientes, como hombre de ánimo y de honra).[6] The speed with which events unfolded suggests premeditation.

There is a discrepancy in the evidence: Transylvanus's account deferred the mutiny and murders, in which those events culminated, until near the end of the time in harbor. He worked at least in part, as he tells us (and his fellow littérateur, Gonzalo Fernández de Oviedo confirms) from an apologia the mutineer Juan Sebastián Elcano submitted when he led the survivors of the expedition to Spain.[7] The chronology Transylvanus imposed, however, was his own. In his account, the breakdown of order appears as the result of the long and comfortless winter, which revived resentment over "the old eternal hatred between the Portuguese and the Spaniards, and about Magellan's being a Portuguese."[8] When Magellan punished the murmurers, according to Transylvanus, he turned them into mutineers. The narrative makes dramatic sense. But it is wrong: here, as in other places in his book, Transylvanus conflated events in his rush to get his narrative finished and published ahead of potential competitors. Other sources concur in dating the fatal events at the start of the stay—"almost as soon as we were in harbor," said Pigafetta.[9]

Before there was any evidence, let alone any occurrence, of potential mutiny by malcontents in the fleet, Magellan prepared a preemptive strike. If later recollections by Ginés de Mafra can be believed, the captain-general's first act on arrival at San Julián was to summon the crew of his own ship and tell them "with sweet words and big promises" (con palabras dulces y grandes promesas) that rival captains were plotting to kill him at mass the following week, on Easter Day.[10] Mafra's memory was not always reliable, but as a fervent supporter of Magellan's he had no reason to invent the "sweet words" or the bitter prediction. In any case, Magellan exposed and punished the alleged plotters within a few days, with the first execution and condemnations taking place on Holy Saturday, perhaps in order to heighten the horror for waverers who questioned Magellan's authority or doubted his resolve. After the conflicts of the Atlantic crossing, a showdown was predictable. When a mongoose confronts a cobra, it is pointless to complain about fair play. The winner will be he who strikes first. The tensions that had accumulated in the fleet since it set sail from Spain demanded resolution.

Magellan needed an opportunity to crush the clique of officers whom Fonseca had introduced into the fleet and whose opposition had solidified around Juan de Cartagena. How he managed is hard to elicit from the sources. Some narrators misremembered events when they recalled them long after the event: Ginés de Mafra's relation, for instance, is studded with insights but in places contradicts the testimony he gave to the earliest official inquiries. Under interrogation, witnesses conflated events in muddled memories. Almost all historians have made nonsense of the story of San Julián by following Antonio de Herrera, who, writing eight decades after the events, telescoped episodes that cannot have unfolded in the order he utters and that probably occurred months apart.[11] If, however, one resists the temptation to cram in every assertion, however self-interested and unlikely, from every surviving document, and if one excludes recourse, for want of more systematic data, to the fantasies of early historians who were trying to tell a good tale or serve a patron's agenda, it is possible to construct a credible narrative.

. . .

It starts on Palm Sunday, April 1, when Magellan invited the leading men of the fleet to dine after mass. Dinner with the Borgias? Or with Titus Andronicus? Or the Godfather? The summons to a deadly meal has been a topos of art from Absalom and Amnon to Agatha Christie and the Mob.[12] A seat at dinner is a convenient place for an assassination: the victim is pinioned behind the table, disarmed save for unmurderous cutlery, easily approachable to a cutthroat from the rear, and vulnerable to poison in what may be set before him or her. It is perhaps unsurprising that of Magellan's invitees only Alvaro de Mesquita, his kinsman, turned up for the entertainment. Luis de Mendoza and Antonio de Coca, who, as Fonseca's appointees and Cartagena's allies, were surely on Magellan's hit list, risked the mass but excused themselves from dinner. Gaspar de Quesada and his nominal prisoner, Juan de Cartagena, were among those conspicuous by absence.[13]

Though Ginés de Mafra muddled Palm Sunday with Easter Day when he recalled the proceedings years later, his account of the conversation at mass is engaging. "Speciously and disingenuously" (con disimulado semblante), Magellan and Mendoza exchanged civilities.

"Why did the other officers not come to mass?" (Cómo no venían los demás capitanes a misa?), Magellan asked.

Mendoza replied that he did not know. Perhaps, he ventured, they were unwell.[14]

The invitation served, at least, to flush out Magellan's foes. Whether or not he intended to abuse his own hospitality by arresting or killing his guests, the invitees exhibited understandable signs of alarm. They rebuffed the invitation, like *grandes dames* snubbing a parvenue hostess. Having done so, they had to act. The aborted dinner party ignited mutiny. Cartagena and Quesada went into action—if the witnesses who testified can be believed—the same night.

The earliest testimony we have gives a vivid account of what happened as darkness fell. The document is highly partisan: a petition addressed to Magellan by Álvaro de Mesquita, demanding justice against the conspirators, with the record of the inquiry that followed his complaints. The petition bore the date of April 9—more than a week after the event. Enough time had elapsed for Magellan to conspire with his friends and allies to concoct a version of events favorable to himself. He officially acknowledged receipt of the petition on the fifteenth and assented to Mesquita's request on the seventeenth—a day of ill omen, when a total eclipse blotted the sun. More time for collusion.

The inquiry took place on Thursday, the nineteenth, aboard *San Antonio,* under Magellan's eye. So there is no question of impartiality. The concurrence among the witnesses is suspiciously exact. It is surely not, therefore, an accurate account, but it is perfect evidence of what Magellan wanted the world to believe. And it is the only version recorded at the time, or, rather, within three weeks of the alleged mutiny. What clinches its general reliability is the remarkable fact that the self-defense of mutineers who later escaped to Spain largely confirms it, with some elaborate special pleading along the way. The surviving transcript was made in Seville in May 1521, from the original record of the scribes, Martín Mendes and Sancho de Heredia, who also acted as interrogators of the witnesses; so—with allowance for minor errors in transcription—the authenticity of the contents is assured.[15]

Mesquita's testimony resounds with righteous indignation and takes us straight to the moment when rivalry between Magellan's and Cartagena's factions became violent aboard *San Antonio*. "I make known," Mesquita began in the correct legal style of the time,

> that on the night of the said Palm Sunday, April 1 of this year of 1520, when I was in my cabin in the said ship, and all the crew were at rest, there came Gaspar de Quesada, captain of the ship *Concepción,* and Juan de Cartagena, in arms, with about thirty men, all armed, and they broke into my chamber with swords drawn, and they seized me, with the points of their weapons against my breast.

After claiming the ship for their own mutinous purposes, they took their prisoner below decks and they put me in the cabin of Gerónimo Guerra, scribe of the said ship, and they shackled me, and it was not enough for them to put the said shackles on me, but they also locked and padlocked the door of the said cabin, and what is more they set a man at the door to keep it under guard.[16]

Other witnesses took up the story. Pedro de Valderrama, priest and ship's chaplain of the flagship, claimed to have confronted Quesada, calling him to return to his duty in the words of Psalm 17: "Cum sancto sanctus eris et cum perversis perverteris" (With the upright thou wilt be upright and in the company of the perverse thou shalt be perverted).

"Who says so?" retorted Quesada, who did not need a translation.

"The prophet David."

"We don't know any prophet David here, Father."[17]

Valderrama's presence aboard the *San Antonio* is unexplained. The exchange, however, is believable: the priest implicitly exonerating the penitent whose reformation he tried to secure; the hell-bent malefactor postponing his reckoning with God.

Juan de Elorriaga, master of the *San Antonio,* intervened. "I require you," he told Quesada, with apparently becoming formality, "in the name of God and of King Don Carlos, to return to your ship, because this is no time to be wandering through the fleet with armed men, and further I require that you release our captain."[18]

Quesada refused. The master then told his immediate subordinate to summon the crew to arms "and let us demand our captain's release" (e demandemos nuestro capitán).

Quesada responded by stabbing the master four times in the arm[19]— or, by other accounts he hit him six times, leaving him, according to the most trenchant witnesses, "for dead."[20] The victim took two hours to come round. The mutineers removed him to the *Concepción* and by their later claims tried to heal his wounds. But, after a long agony, he died—on July 15, according to the "Declaración de fallecidos," a sort of muster of the dead, a tally, compiled in Seville in September 1522, of the deaths of members of the expedition.[21] With help from Antonio de Coca, Quesada disarmed the crew, whom the fate of Elorriaga seems to have quietened. He had all weapons locked away, imprisoned all dissenters, and brought in Juan Sebastián Elcano, who was serving as the ordnance master on Quesada's own ship, to take command of the guns.

Witnesses at the inquiry depicted ensuing disorder aboard *San Antonio,* with men lowering and raising anchors, muddling cables. and rushing around in confusion. The tellers of the tale knew how to scandalize the bureaucrats who would eventually review the evidence back in

Spain: with denunciations not so much of mutiny as of anarchy, to the prejudice of the king's interest and investment. Hence, perhaps, the witnesses' extraordinary emphasis on "very great damage to the stores, without heed to weight or measure, but with all left open to the depredations of anyone who wanted to help himself."[22] The following day, Monday, April 2, the mutineers "handed out food without observing regulations" (daban pan a la gente sin regla) for their supporters.[23] The abuse of the stores got more attention in the inquiry than the culprits' alleged acts of violence against their victims. What could be worse than wasting the king's patrimony? It seemed less lamentable that in dealing with Magellan's loyalists Quesada allegedly "near-killed men with blows and would not give them their rations."[24]

Testimony later collected in Spain from surviving mutineers does not cast doubt on the evidence from Mesquita's inquiry. Under interrogation, they admitted the intrusion that led to the petitioner's captivity, in terms similar to his own, albeit without the tone of indignation. They offered, however, the assurance that their initial intention was not to coerce or kidnap him but to try to persuade him to take their part and join them in demanding that Magellan comply with the king's orders.

As in previous shipboard disputes, they saw themselves as the king's men, their patron Fonseca as the king's representative, and their leader Cartagena as the king's nominee. Their role was not to usurp command but to exercise their due share in it, in pursuance of royal wishes and orders. They wanted Mesquita as a spokesman, they averred, because they feared the captain-general's wrath. Only when Mesquita refused to comply did they bind him. As for Elorriaga, his claim to have demanded Mesquita's release was confirmed, as was the wound he got for his pains, though the mutineers claimed that they then healed the gash for him.[25] If so, evidently, their ministrations were good only for a few more weeks of life.

. . .

With Cartagena in command of *Concepción,* Quesada in control of *San Antonio,* and the *Victoria,* under Luis de Mendoza, already in the hands of the Fonseca faction, the conspirators felt strong enough to dictate terms on the morning of April 2. Their proposals emerge in the records of another, equally unobjective interrogation, conducted in May 1521 among mutineers who escaped to Spain.[26] Confirmation is available, in every significant respect, in the record of the interrogation in Malacca on June 1, 1521, of Martín de Ayamonte, cabin boy of the *Victoria,* who had deserted and fallen into Portuguese hands.[27]

On the face of it, the mutineers' proposals seem eminently reasonable; but readers should bear in mind that they were framed with potential charges of mutiny in mind, which specious reasonableness might head off or counteract. And once the mutineers got Magellan into their clutches there was no guarantee that they would have observed the courtesies they promised him.

Their demands were three, all seemingly innocuous and conciliatory. First, Magellan must respect the orders the emperor had laid down for the voyage: differences over the interpretation of those orders had, of course, caused conflict throughout the voyage. Second, he must abort the plot of which his opponents suspected him. Their suspicions were left vague but obviously related to the fears excited by the menacing dinner invitation. Third, he must come to an agreement with the proposers "about what would best be for His Majesty's service": implicitly, in other words, he would have to submit to their agenda for the rest of the voyage. Should he accept the terms, the proposers promised that they would cease to call Magellan by the—to him—derogatory style of "Your Grace" (Vuestra Merced) and use instead "Your Lordship" (Vuestra Señoría). It may seem a trivial matter to modern readers, oblivious of the distinctions of rank that meant so much to Magellan and his contemporaries. In any case, it was as much as the mutineers were willing to concede.

Magellan responded by inviting the leaders of the mutiny to discuss it with him. They declined on the grounds that "they would not go in case he mistreated them." The allusion to his violent exploitation of the last chance that brought the officers together (above, p. 140) is unmistakable. Their counteroffer was of a meeting on one of the ships they commanded, where "they would all work together in accordance with the king's commands."[28] It looked like a standoff, the mutineers presumably relying on their superior tonnage and ordnance to prevail, by violence if not by intimidation. A meeting was impossible: there was no neutral ground on which to hold it, where each side could feel secure from the other's potential treachery.

Magellan's next move, later on the same day, April 2, was to send a letter by boat to Luis Mendoza on the *Victoria*. But the letter was a blind: the six boatmen and the Captain-general's messenger had their orders. When Mendoza, allegedly smiling or sneering with a conviction of security, opened the letter to read its contents, Gonzalo Gómez de Espinosa drew a concealed dagger and plunged it into the reader's neck. Another stab to the victim's head finished him off.

"Necessity hath no law." The assassination was a desperate measure: a crime that perhaps helps to explain why the crown never honored Magellan's memory or compensated his heirs. No sooner had the act been perpetrated than a second boat from the flagship arrived, with (allowing for conflicting testimony) at least fifteen armed men under the captain-general's relation by marriage, Duarte Barbosa. They hoisted the commander's flag without encountering resistance. The recaptured ship and the *Santiago* closed ranks with the flagship. The balance of power between Magellan's partisans and enemies had been reversed. The odds were now on the commander's side.

The next day, Tuesday, April 3, with the two ships left to them, Cartagena and Quesada seem to have decided that their best hope now lay in escape. The flagship, however, was blocking the narrow neck of the bay. After apparently trying in vain to get Mesquita, who was still their prisoner, to intercede for them, they resorted to using him as a hostage: from the prow of the *San Antonio* he would call to Magellan as they approached and beseech him not to fire. The plan never had a chance to be put into practice. The same night, after dark, while the leading conspirators slept, some of the crew, dragging the anchor, steered the *San Antonio* closer to the flagship in an apparent attempt to defect to the captain-general's side: that, at least, was the story they told when they got back to Spain. Volleys of shot met them.[29]

A counter-boarding party appeared on *San Antonio's* deck, demanding, "What side are you on?" (¿Por quién estáis?).

From all who knew what was good for them, the reply came back, "For our lord the king and for you."

Quesada, Coca, and the other leading men who were aboard *San Antonio* were clapped in irons. On *Concepción* resistance subsided and Cartagena was soon a prisoner too. Mafra's recollection was that he surrendered on realizing that the crew would no longer support him. Magellan sent to ask "for whom the ship stood. 'For the king,' Cartagena replied, 'and for Magellan in his name'" (though Mafra spoiled the effect of this obviously dramatized scene by repeating his habitual mistake: misremembering Magellan's name and calling him Sebastian).[30] What was in Cartagena's mind? Were his words craven or courtly? He had tossed and lost; the dice were in Magellan's fist. There was no point in bravado but maybe some virtue in dignity.

. . .

On the fourth, Mendoza's body was dragged ashore, quartered in accordance with the convention for traitors and spitted on a gibbet *pour encourager les autres*. The seventh was Holy Saturday, when Christ is in his tomb and the voice of prayer is silent. It was a good day for a judicial murder. To spare Quesada the hanging to which Magellan condemned him, a servant severed the victim's head. The body was quartered and strung up alongside Mendoza's rotting limbs.

On the same day Magellan sentenced Juan de Cartagena and Pedro Sánchez de Reina, a priest accused of inciting mutiny, to be marooned.[31] According to the mutineers, the priest's real offense had been to refuse to disclose secrets of the confessional. Was the commutation of death a gesture of mercy or the result of a loss of nerve? Magellan spared forty other mutineers, including Juan Sebastián Elcano, who, according to their accusers, deserved to die: Mafra said that they were condemned to hang but pardoned them at the pleas of fellow seamen.[32] Like Cartagena and Fr. Sánchez, they were kept in chains and put to hard labor while the fleet remained in port. According to reminiscences gathered by Gaspar de Correa in Goa, probably in the 1540s, Magellan made some of the prisoners reconnoiter ahead on foot. They "went for more than forty leagues and returned without news."[33] Necessity, rather than clemency, saved their lives. Manpower was valuable, and large-scale slaughter would have weakened the expedition. Cartagena and Sánchez were spared the gibbet for reasons of policy. To execute a priest was to invite arraignment for sacrilege. To put the king's *veedor* to death, especially without what we should now call due process, was a form of lèse-majesté.[34] Marooning, by a legal fiction, was a way of deferring the ultimate sentence to God. Had the victims any chance of survival, Magellan would surely not have left open the possibility that they might live to tell tales against him.

Magellan kept his distinguished prisoners chained until the sentence was executed—leaving the victims stranded "with only a little biscuit and one sword each" (cum biscocti panis pera, singulisque ensibus).[35] According to the official register of deaths, it was August 11, less than a fortnight before the ships left San Julián. Literature is full of tales of castaways who survived—tales variously of heroism or redemption. For Cartagena and Sánchez, who died in obscurity, how long did it take, as they watched the ships sail away, before they turned to appraise each other? In popular pseudohistories, taking their cue from an assumption of Herrera's, Sánchez has been identified with another priest, called Bernaldo or Bernardo Calmeta or Calmetas in surviving ships' lists, probably of Gascon origin, who was enrolled to sail on *San Antonio* as a member of Antonio de Coca's entourage.[36] But there is no reason to conflate two apparently distinct persons or to invent a backstory of subterfuge or concealed identity on Sánchez's part. Although other sources do not mention him, except for the register of deaths, in which his was recorded, he was a real person. According to some depositions, he threatened Magellan with hellfire for the injustice of his proceedings. There is no evidence in any of the witness's testimonies that he was a conspirator or that he had been close to Cartagena's faction. On the face of it, the principled confessor and the chief mutineer made an odd couple. Fleeing mutineers later had a chance to return to San Julián to rescue them: they discussed the possibility among themselves. But the fugitives high-tailed it for Spain, leaving the castaways to the mercy of the wilderness. Cartagena and his companion evidently did not have the gift of attracting sympathy.

. . .

According to some deponents, another chaplain of the fleet suffered torture—suspended by ropes, with weights attached to his feet—apparently because he had complained of the insufficiency of the ship's provisions, but perhaps in an attempt to make him disgorge sacred secrets and reveal more conspirators' names. A seaman called Hernando de Morales, whose name was in the crew list of the *San Antonio* of August 8, 1519, when the ship was about to leave Seville, got the same treatment and ultimately died of it.[37] According to mutineers' later denunciations back in Spain, Andrés de San Martín was tortured with similar brutality because he had aroused Magellan's suspicion, allegedly by throwing overboard potentially incriminating papers said, in one source, to have been a record of the route.[38] If the event really occurred, it seems unlikely that it was part of the proceedings that disfigured that Easter season. The evidence that on April 17 San Martín was in good health and making astronomical observations at Magellan's request is incontrovertible. On that day, an eclipse of the sun gave the astronomer a chance to make a reading of longitude.

To understand what might have happened we have to make a brief excursion into the obscure and imperfect record of San Martín's astronomical work during the voyage. Magellan needed as good a record of longitude as his pilots could contrive, in order, if they ever got to the Moluccas, to vindicate Spain's case for appropriating the islands. That was why Faleiro, with his convincing mumbo jumbo about how to find longitude, had been such a loss to the expedition when he withdrew to a madhouse in Seville. The same reasoning led to San Martín's ascription: he knew better than most how to handle astronomical instruments. Prior to arrival at San Julián, however, his efforts at reading longitude had failed. One version of his attempts we owe to Barros, who had San Martín's papers to hand when he wrote his history of Portuguese Asia in the 1540s, and quoted convincingly from them.[39]

The astronomer's first effort, at Rio de Janeiro on December 17, 1519 (above, p. 115), was doomed. The task demanded great precision: to measure the angle suspended between the positions of the moon and Jupiter. The result was wildly, ridiculously wrong—twenty-three degrees, forty-five minutes west of Seville, an underestimate of nearly fourteen degrees. San Martín dismissed it, on the grounds that printing errors in his astronomical tables were responsible. Barros, who treated the matter contemptuously—presumably because he thought a Spaniard's readings prejudicial to Portugal's interests—gives no figures, but the tables in question, do, indeed, contain printers' errors that must have vitiated San Martín's attempts. The figures transcribed by Herrera confirm the fact.[40] San Martín tried twice at other points on the South American coast in February 1520, when there was a conjunction of the moon and Venus on the first of the month; but his readings have not survived. Transylvanus had information—we do not know from where—that at San Julián a reading of longitude placed the expedition "fifty-six degrees west of the Canaries": about six degrees out, according to the early twentieth-century historian Denucé (see below, p. 282), who assumed it was San Martín's reading; but the latter's papers, as reported by the Portuguese officials into whose hands they fell, mention nothing of the sort.[41]

The reading of the solar eclipse on April 17, on the other hand, was uncannily accurate. San Martín calculated that San Julián was sixty-one degrees west of Seville—an error of only thirty-seven minutes. No one took a better reading before the invention of the marine chronometer.[42] An eclipse is an ideal opportunity for computing longitude, because the difference in time between observations of the event at different points corresponds exactly to the difference in longitude between them. If you are at San Julián, and know the time at which the eclipse would be visible at some place in Europe, you can make the calculation with ease. But in any case it seems an extraordinary achievement for someone who had just been tortured. And here is a mystery: no known printed European tables predicted a time for the eclipse San Martín observed. Indeed, it was invisible in Europe.[43] The explanation seems obvious: San Martín witnessed the eclipse but did not use it as the basis for his calculation, which must have been made according to the lunar distance method.

According to Herrera, who had data at his disposal apparently unavailable to other historians, San Martín made a further attempt reading longitude at San Julián on July 21.[44] But the author made an obvious error: the reading was of latitude and was almost exactly correct, at forty-nine degrees and forty-seven minutes. Herrera also ascribed to San Martín a second observation of a solar eclipse on October 11, when the fleet had resumed its search for a strait. But the eclipse on that day, though Herrera may have had European sources about it, was not visible that far south.[45]

A persistent claim in Portuguese sources maintains that Magellan and his pilots conspired to falsify the record of their readings in order to make the distance they traversed seem less than in reality. The tradition originated with Fernão Lopes de Castanheda, an old India hand, born and raised in Goa, where he collected materials on the history of Portuguese India before returning to Portugal in 1538 to write up his results. His account of a meeting at which Magellan's pilots, including San Martín specifically, rejected, as either commonplace or fantastic, Faleiro's methods, which Magellan still treasured and recommended, is plausible but unconfirmed by other sources. At the same meeting, Castanheda ventured, all present agreed to distort the reading.[46] It sounds like scuttlebutt or sour grapes or a combination of the two, in a writer always keen to belittle Portugal's enemies and rivals, especially in view of the scrupulous accuracy of the reading—that of April 17—for which we can be reasonably sure that we possess San Martín's own record. But there may be some merit in the tradition that a meeting of some sort took place and that San Martín had a part in it.

So was San Martín tortured or not? If he was, it must have been in the course of a later phase of the altercations at San Julián. And whose bones were rotting under the gallows when Drake arrived? Maybe just whatever the vultures had left of the quartered corpses of Quesada and Mendoza. The victims of the Holy Week mutiny, however, were not the only malcontents to die at San Julián. Magellan, according to his detractors, went on "dealing dark judgments and death."[47] On April 27, for instance, Antonio, the Genoese cabin boy previously convicted of sodomy (unless there had been a further, unrecorded case), was thrown overboard from the *Victoria*. His bloated, sopping body floated back into view on May 20, according to the relation of the dead compiled in

Seville, like a revenant wraith. No other victims of Magellan's vengeance or justice appear in the register for the period at San Julián, but the
record was very incomplete.

. . .

That trouble and rumors of mutiny continued seems certain. Until now,
historians have assumed that Magellan's decisiveness quelled opposition.
The reverse is true. He had scotched mutiny but had not killed it. In some
ways, his display of ruthlessness stoked disloyalty. "Hatred," Transylvanus observed, "settled more deeply in the hearts of the Spaniards" (altius
itaque hoc odium pectore Castellanorum insedit).[48] Magellan had obviously acted beyond the bounds of his authority. The king had enjoined
obedience to him and empowered him to administer punishments (above,
p. 105)—but only within the limits of the law. Mendoza's death was
extrajudicial assassination, Quesada's—at best—judicial murder. The
marooning of the castaways was an illicit death sentence disguised as
clemency. An oblique proof of the unlawfulness of the proceedings is
that Pigafetta suppressed all mention of the fate of the mutineers, just as
he always avoided reference to anything that was ever to Magellan's
discredit. Having already—in the recollection of the survivors—"made
up his mind either to die or to complete his enterprise." Magellan was
growing reckless.[49] His policy of dealing out death, torture, and isolation
to his enemies was less a calculated strategy than a defiant reflex. This
was not the way to quell dissent but rather to justify it.

The evidence of continued unrest has remained occluded because
Herrera, writing his compendium more than eighty years after the
event, caused confusion among successor historians. He transferred to
the start of the stay at San Julián an episode that arose well into April,
if we follow the chronology of Transylvanus. The humanist, who wrote
on the basis of interviews with survivors and, as we have seen (above,
p. 136), completed his narrative within a few weeks of their arrival
home, was more likely to be right than Herrera. Unhappily, however, he
compounded the problem by a chronological error of his own, trans-
posing the events of Holy Week to a time near the end of winter. The
obvious way to reconcile the data is to excise the error and adhere to the
rest of Transylvanus's outline.

The truth is then revealed: after the Easter tragedy, Magellan's prudent parsimony with the ships' stores provoked a further convulsion of
unrest. He had shortened rations before. At San Julián he did it again.
In the hinterland of the harbor and on nearby shores there were pen-

guins, seals, and guanaco to hunt and salt for the future. But supplies of hard tack were dwindling. Wine was irreplaceable and precious. Fresh vegetables were virtually unobtainable on the shores of Patagonia, except perhaps for the herbs that natives taught the men to chew. Winter quarters would deplete, rather than replenish, vital provisions. On short rations, in increasing cold, crews' enthusiasm always wanes. A winter of discontent beckoned. "With the month of May close upon them," a south wind promised malcontents a way home, if they could induce their leader to take it.

Transylvanus captured the tenor of the men's reprimands and of Magellan's response.[50] The men's opening complaint was that they feared "the severity of the winter and the barrenness of the country." Even Magellan, they ventured, could see "that the land stretched interminably to the south and that no hope remained of its coming to an end, or of the discovery of a strait through it." The commander, however, was the last person to admit any such thing. He had his precious map. He had staked all on the expectation of finding a strait. "A severe winter," the complainants asserted, "was imminent." But Magellan had decreed winter quarters for that very reason. Many shipmates already, the malcontents declared, "were dead of starvation and hardships." Magellan, however, was bound to judge such remonstrations as unjustified. San Julián is an inclement place, but not far enough south to justify complaints of extreme cold; and the products of hunting and fishing eked out the rations. Pigafetta, as usual, spoke for his hero in commending the resources of the bay. His stories sound promotional, of incense trees and of shells that compensated for their inedible contents by yielding pearls, parodying the complaints with even more incredible counterassertions.[51]

The further particulars the complainants added—"that they had gone farther than either the boldness or rashness of mortals had ever dared to go as yet"—were neither strictly accurate nor likely to impress their leader, whose notions of adventure were expansive and who liked to reserve self-congratulation as his own prerogative. To invoke the authority of the king of Spain was routine in confrontations between commanders and crew, but the malcontents' assertion "that Caesar had never intended that they should too obstinately attempt what nature itself and other obstacles opposed" was a claim calculated to arouse Magellan's anger: it challenged his sagacity, his courage, and his authority. The murmurers' real agenda, along with their most outrageous threat, came in the midst of all this unpalatable rhetoric: they told Magellan that "they could not bear the rule that he had made about the allowance of provisions, and begged that he would increase their rations and think about going home." The language of the original petition was probably more peremptory than Transylvanus's paraphrase: the humanist's informants were naturally circumspect in confessing mutinous sentiments.

Though Transylvanus puts it in *oratio obliqua*, Magellan's speech in reply is so evocative of his own voice that the writer seems to have had a copy to hand; but like the well-trained classical humanist he was, he probably followed an ancient model and put into his protagonist's mouth the words he deemed best. They sound uncannily convincing.[52]

First, Magellan disclaimed power to gratify the complainers' demands. His course had been laid down for him by Caesar himself. He neither could nor would depart from it in any degree. "He would sail till he found either the end of the land or some strait." This display of bravado might have been provoking, but Magellan followed it with something like cajolery, backed by implicit reproach. The voyage would be easy if resumed in summer, and meantime, "there were means, if only they would try them, by which they might avoid famine and the rigor of winter," including abundant wood, fish, fowl, and prey for hunting. Biscuit and wine would last well enough "if they would only bear that they should be served out as needed or for health's sake and not for pleasure or luxury." Were animadversions on the men's lack of industry and austerity likely to touch consciences? Magellan took the reproaches further: the expedition, he said, had done nothing yet worthy of admiration.

The captain-general then made a risky move, taunting crewmen of other nations with Portuguese superiority. It is hard to imagine a surer way of fueling xenophobic tension. Moreover, he got his figures, if the text is correctly transcribed, dangerously wrong: his fellow countrymen, he pointed out, frequently reached twelve degrees below the Tropic of Capricorn. His own comrades had only got to four degrees below and "would be thought very little worthy of praise." For his part, Magellan would not suffer such ignominy "and trusted that all his comrades, or at least those in whom the noble Spanish spirit was not yet dead, would be of the same mind." Yet surely everyone knew that they were well south of either mark—some twenty-six degrees beyond the tropic. If the error was not the fault of a scribe, maybe Transylvanus was at fault, or transposed something Magellan said on the Brazilian coast, at Santa Catarina or thereabouts, to San Julián.

The captain-general concluded, in accordance with all the best rules of rhetoric, as of song writing, on the "money note." He promised his men rewards, "the more abundant the more difficulties and dangers they had endured in opening to Caesar a new world, rich in spices and gold."[53]

"Within a few days" of this unpromising exchange, according to Transylvanus, murmuring resumed on the usual topics: first, "the old eternal hatred between the Portuguese and the Spaniards, and Magellan's being a Portuguese"; next, the miseries of "snows and ice and perpetual storms"; and finally the hopelessness of Magellan's mission: "Nor was it credible that he should even wish to discover the Moluccas, even if he were able" (Neque credendum etiam si posset Moluccas insulas reperire, cupere).[54] Barros, with records in his hands of depositions that expeditionaries later made to Portuguese captors, was able to amplify: the ordinary crew members, who were not privy to the officers' debates, suspected their leader of wanting to return to Portuguese allegiance.[55] His purpose in embarking on his venture, they suspected, was to serve Portugal by distracting Castile from more profitable ventures. Magellan, "very much enraged by these sayings, punished the men, but rather more harshly than was proper for a foreigner."

Barros adds details of an exchange with officers that might have occurred at any time but that seems to belong to a period at an advanced stage of winter, because the subject was an analogy Magellan broached between the South and North Atlantic. "Since," he supposedly said, "the seas along the coasts of Norway and Iceland, which were at a higher degree of latitude, were as easy to navigate in summer as those of Spain, the same would apply" in the Southern Hemisphere. That, replied the malcontents, was because the seamen concerned were native to the boreal region or from very nearby and could reach their remotest destinations in the space of about a fortnight. A voyage of six or seven months across every kind of climate was not comparable for cost, danger, or loss.[56]

. . .

Amid the exchange of recriminations, toward the end of April, Magellan sent the *Santiago* ahead to explore for the strait. On or about May 1, the ship reached an inviting estuary, over a mile wide, which, on May 3, the captain, Juan Serrano, named Río Santa Cruz, after the feast (as it was in the calendar in force at the time) of the Invention of the Holy Cross. They killed game: a huge sea lion and plenty of seals, "clubbing them on the snout."[57] Next day, about three leagues further on, a storm tore the sails, drove the ship onto rocks, and smashed it. All the crew escaped, except for Serrano's black slave, whose death the Seville register confirms.

.

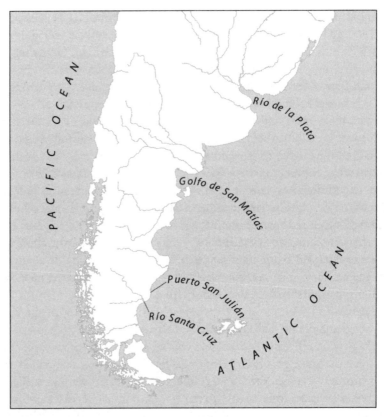

MAP 3. Patagonia: Places named in the text.

The survivors' reminiscences conflict, but the likely course of events began despondently, as the men picked the tough mollusks called *lapas*—disproportionately prized in South American gastronomy—from the rocks, perhaps for as much as eight days, awaiting a rescue that never came, before resigning themselves to attempting a return to San Julián. They were on an uncomfortable coast where vivid plant life studded the stony soil. A nineteenth-century voyager described it—the oppressive lie of the land, where the ground rises abruptly to the vast flat uplands, like the reveal of a gaunt, gray steppe:

The soil is sandy, but dense, under an almost uninterrupted crust of loose stones. Numerous shrubs—aromatics, spiny barberries, and the like)—with leaves of different colors enliven the scene, and great silvery clumps of half-withered grass add a bright metallic gleam to the earth. Endless little guanaco tracks score the surface, making it less difficult for humans to walk where cacti, spiky brush, and the incredible number of rats' burrows tire and torment the hiker.[58]

The castaways dragged planks to the river mouth. "And because they were very weak," according to reports that presumably grew like sprigs planted by Munchausen, "they took four days" to reach it (i por estar muy flacos, tardaron cuatro dias en llegar). There were thirty-six of them, but they could only build a raft big enough for two. In any case, a return by sea was impracticable in the face of the storms.

After rafting over the estuary the two appointed messengers walked for two days, without finding good green stuff to eat. They suffered cold and battled snowstorms. They deviated from the straight route north to get shellfish from shore but had to return inland because the shoreline was impassable. "After eleven days they arrived so faint that they were unrecognizable."[59] Magellan was dismayed rather at the loss of provisions from the wreck of the *Santiago* than at the predicament of the wrecked men. He sent a party of twenty with wine and biscuit to bring the survivors back "because the sea was so wild that it was impossible to go by sail."[60] Their drink on the journey was melted snow. When they arrived with the biscuit "the men of the shipwrecked vessel said that they had had none to eat for thirty-five days."[61] If they were right, it must have been June 9 or thereabouts. Some survivors recalled time spent salvaging useful material from the shattered *Santiago*. When the crew's rescue was complete, Magellan divided them among the remaining ships and made Serrano captain of the *Concepción*. How did the disaster affect morale? Mafra and the Genoese pilot seemed able to shrug it off. But there was now less food to go round. And the survivors were only about halfway through the winter.

Still, *Santiago's* death-berth looked attractive from one point of view: there were plenty of sea mammals there to kill and salt and stave off starvation. On August 24 Magellan took advantage of a lull in the weather and, after two days under sail, transferred his base to Río Santa Cruz. Or was it an accident that took the fleet there, fleeing from heaving seas and flailing wind when "the whole squadron was on the point of shipwreck," as Pigafetta declared? "But God and the denizens of heaven succored and saved us."[62]

For once, all the surviving narratives agree about the time spent in the haven, taking in water and wood, catching "many fish" of a scaly variety "about two feet long," and salvaging goods left from the wreck of the *Santiago*.[63] It all took until St. Luke's Day, October 18. Meanwhile at least six more crewmen had died "of sickness," according to the register. Summer was at hand and it was time to try Magellan's promise that navigation would now be easy. Everyone confessed and communicated before leaving.[64]

. . .

Meanwhile, back at San Julián—two months after arrival, so around the beginning of June, with snow on the ground and the crew of the *Santiago* still languishing on their rocks—encounters with the native inhabitants had begun.[65] Pigafetta, who traveled, as we have seen, to satisfy his curiosity, was always the most ethnographically inclined of the narrators of the voyage and the most sharp-eyed and sympathetic observer of alterity. At San Julián, he saw much that seemed to distinguish the natives he met from those encountered previously. The first thing he noticed was their size. "One day, when we least expected it, a man of gigantic stature appeared. . . . We scarcely came up to his waist."[66] The language suggests someone bigger than the cannibal, "in stature almost a giant" (de la statura casi como uno gigante) with "a voice like a bull" (una voce simille a uno toro) who had paid a brief visit to the flagship at Río de a Plata.[67] Pigafetta kept up the use of the term *giant* with every further reference. Native men measured, in different onlookers' assessments, from nine or ten spans (perhaps ninety inches), according to Transylvanus and the Genoese pilot, to fifteen spans, according to Correia's informants. The women, though shorter, had dugs a foot long.[68]

Almost every visitor over the next two centuries confirmed the inhabitants' gigantic stature. A survivor of an expedition of 1525 recalled women nine feet tall and others so big that he only came only up to the level of their pudenda.[69] Drake absolved them of being "so monstrous, or giantlike as they were reported; . . . but peradventure, the Spaniards did not thinke, that ever any English man would come thither, to reprove them; and thereupon might presume the more boldly to lie." Special pleading apart, he admitted that "Magellane was not altogether deceived, in naming them Giants; for they generally differ from the common sort of men, both in stature, bigness, and strength of body" and put them at typically seven and a half feet in height "if not somewhat more." Drake's chaplain confirmed the "huge stature" of the natives at San Julián.[70] Some giant finders can be dismissed as doctrinaire sensationalists, like Anthony Knivet, a shipwrecked English pirate, who, when he got home in the 1580s, set up as a purveyor of literally tall tales. He measured, at least in his own mind, footprints four times normal size and a corpse of fourteen spans.[71] Captain John Narbrough (below, p. 209) had no opportunity to measure people at San Julián when he passed through in 1670. The normal specimens he did measure were in the vicinity of the Strait of Magellan. A commercially successful account of Commodore John Byron's voyage of 1741 described a native chief "of a gigantic stature," who "seemed to realize the tales of monsters in a human shape. . . . If I may judge of his height by the proportion of his stature to my own, it could not be much less than seven feet." A fellow officer, "though six feet two inches high," seemed "a pigmy among giants; for these people may indeed more properly be called giants than tall men."[72]

No one as tall as that inhabits the region today. But part of the discipline of history is treating with respect the evidence of the dead. When we look at the unfamiliar, with no slot in our minds in which to fix it, we do not see what is really there but, rather, what we expect. Our struggle to make sense of it scans what is in our experience until we come up with a good match. Vicarious experience, from what we have read or heard, is the default setting for our efforts. As we have seen (above, p. 146) monstrous deformations peopled the margins of maps in the late Middle Ages and the texts the Renaissance inherited from antiquity. So explorers expected them. Giants were monsters, even at the modest near-eight foot-high of the onlookers' more sober assessments. People who saw them had the satisfaction of confirming received wisdom, fulfilling their own expectations, and, paradoxically but gratifyingly, being able to report a marvel to a marvel-hungry world back home.[73] The eagerness with which Magellan sought to capture a sample giant to take home to Spain for exhibition to the curious—a project he did not extend to routine specimens of encountered people, such as the cannibals he had met up the coast—suggests the giants of San Julián made a peculiar impression.

When people could be classified as monsters, there was usually a further, underlying agenda, derived from the centuries-old debate about the rationality of the abnormal, the redeemability of the naturally deformed, and the gulf between animal and rational souls (above, p. 147). In the debate, Pigafetta and the Genoese pilot were emphatically with Transylvanus, who, as we have seen, was on the side of St. Augustine. They stressed the normality—in proportion to their size—of the giants' physiques. They were "well formed," said Pigafetta, "proportioned as we" (Sono disposti huomini et femine como noi) and, according to the pilot, "very well made."[74] Their frequently displayed nakedness was equivocal:

perhaps a relic of the preclassical Golden Age or of the innocence of Eden, or, as Columbus suspected in the case of the people he encountered in the Caribbean, a sign of dependence on God. Even in nakedness, they had what Christians of the time classed as a proper sense of shame: their women, "though far from lovely," covered their pudenda.

In any case, the giants could make and don clothes of the skins of guanacos—monstrous creatures in themselves, combining, in Pigafetta's description, mules' heads, camels' bodies, deer's legs, and horses' tails. The Genoese pilot insisted that the inhabitants wore clothes and shoes, except "when they do not wish to be clothed from the waist upwards"—making nakedness, which was anyway only partial, into a rational option. Pigafetta detected traces of protoscientific learning: "a certain kind of medical knowledge"; they used arrows, for instance, to introduce vomiting and practiced bloodletting—a therapy with the best classical Western authority to back it—for headache in conformity "with their theory, which one of them, whom we captured, explained to us: pain is the result of blood that does not want to be trapped in a particular part of the body." Most impressively of all, they showed some inkling of divine revelation in practicing something like marriage and were "jealous of their wives." One man, exceptionally tall and elegant, who learned to repeat the name of Jesus, qualified for baptism: a big boon for the explorers to concede, as it confirmed unquestionably human status, with rationality and redeemability, and conferred exemption from enslavement. The Genoese pilot's apologetics went to unbelievable extremes, relocating the giants in a pastoral idyll that existed only in the romantic fiction of the time. "They do no harm," he averred, "and thus they follow their flocks."[75]

Most of Magellan's men do not seem to have noticed these supposedly redeeming features. The natives' behavior seemed rather to bear the mark of the beast or the barbarous. They were nomads "like gypsies," transporting their huts, and thereby eluding capture. They ate raw meat—a literally crude sign of the uncivilized—which they consumed with ravening appetites. They were scared of their own images in a glass—a test still used in experiments to evaluate nonhuman cognition. Worthless truck attracted them—hawks' bells, combs, beads, "and other bagatelles." They made a show of "barbarous pomp, which they thought a royal one." They left their women to carry Spaniards' truck or gifts, "as if they were beasts of burden." Their religion was "apparently limited to devil worship," under the supreme demon, Setebos. Shakespeare borrowed the name for Caliban's mother's god. Their

devils were "painted like the natives" but also equipped with parapher-
nalia suspiciously reminiscent of a Christian imagination, including
horns and flames.[76]

In other words, whether traduced or flattered, the giants appear in
our sources not as they were but as characters from existing literary
traditions, iconic models, and juridical categories. Only a few real
observations are retrievable. The first giant the explorers encountered
greeted them in song and dance. Dance initiated subsequent meetings.
Arrow swallowing—surely no mere party trick but a variation on diplomatic hatchet-burying—was part of some performances. At the first
meeting, the dancer dusted his head; on a subsequent occasion a native
initiated contact by touching his head—presumably in an abbreviated
version of the same gesture. Pigafetta seems to have realized that he was
witnessing a ritual but did not of course know its meaning. Dusting
one's head, in every culture in which the gesture is recorded, indicates
deference, and dancing and singing were actions consistent with an invitation to the newcomers to make themselves known. The natives were
keen to establish contact. When Magellan sent a man ashore to communicate by imitating the denizen's actions, he felt that the native
understood him well: that can hardly have been so but is evidence of an
attempt to establish a friendly relationship, confirmed when the giant
accompanied his interlocutor back on board. Natives unhesitatingly
shared food and to some extent maintained the habit of hospitality even
when their guests brought robbery and violence to the huts.[77]

Spaniards attributed a frequent gesture in the dance—a finger pointed
upwards—as evidence "that they considered us as beings who had
descended from above." That was a typical piece of self-flattery. Columbus's men congratulated themselves as being mistaken as divine. Conquistadores routinely thought their native victims so cowed or stupid as
to mistake them for gods. Myths of "the god from the sea" do occur in
some cultures. But at San Julián we seem to confront rather the phenomenon from which the explorers had already benefited: the stranger
effect (above, p. 142). The giants' culture was hospitable to strangers.
What the Spaniards mistook for gestures of worship were probably
steps unrecoverably meaningful in native dances; at most they may have
been acknowledgments of that "touch of the divine horizon" that jour-
neyers bring from afar (above, p. 142).

The stranger effect is an expendable asset when intruders abuse it.
The giants were not quite the bucolic, idyllic characters of the Genoese
pilot's imagination. They had weapons—flint-tipped arrows and bows

strung with guanacos' guts. Flint axes "for working with wood" were to hand when wanted. But the giants usually left their weapons behind when consorting with Spaniards or carried only their bows, which were unthreatening at close quarters. Although peace, therefore, was the dominant mode, at least in early encounters, Magellan became determined, as Pigafetta put it, "to catch some one of these giants to take to Caesar on account of their novelty," while his men tended to see the natives as convenient sources of plunder. Mafra's recollections were all of violent incidents: raids by Spaniards on the natives' guanacos; sorties to seize giant specimens; Magellan's frustration and anger at repeated failures to secure captives to exhibit.

The earliest native visits to the ships occurred before Magellan developed his longing for captives. Transylvanus described an expedition to get them. "Our men," he wrote, invited a group of natives them to return with them to the ships, "with the whole family. When the Indians had refused for a considerable time and our men had insisted rather imperiously," the men withdrew, as if to consult with their wives; but they emerged in war gear, "covered, from the soles of their feet to the crowns of their heads, with different horrible skins, and with their faces painted in different colors." The explorers opened fire. Scared into submission by gunshots, three natives joined the party for the trek back to Magellan's camp. Two escaped because their long strides made them uncatchable. The third died soon after arrival "according to the habit of the Indians, through homesickness."[78]

Eventually, Magellan trapped some captives in manacles, which he represented as gift bracelets. "As soon as they realized the trick they became furious, sighing, howling, and invoking Setebos" (avedendose poi de l'ingano, sbufavano como tori, quiamando fortemente Setebos). An expedition to seize women ended in failure, with one Spaniard killed: according to the register of deaths he was called Diego de Barrasa and died on July 29. Mafra recalled the dispatch of a party to bury the victim, and the rage and revulsion the captain-general felt when a task force could find no one on whom to wreak vengeance. The tribe had fled, and Magellan's avengers burned their village.[79]

Pigafetta omitted to mention the more violent episodes, except for efforts to collect what to him were scientific specimens, one of whom Magellan confided to his care. He "looked after him as best I could" aboard the flagship. Pursuing his vocation as an amateur humanist (above, p. 130), Pigafetta set about making another of his vocabularies in a native language, asking "the giant . . . by means of a sort of

pantomime the names of various objects" and actions. Both participants grew accustomed to the game. Pigafetta had only to pick up his
pen and paper for his informer to supply words in the guttural fashion
Pigafetta noted. Of eighty-nine or ninety items in the lists printed across
all early editions of the text, forty or forty-one related to body parts; so
it is easy to picture the giant and the humanist, ticking off bits of their
bodies—like Kate and her nurse in *Henry V*, albeit with the addition of
penis (*isse*) and testicles (*scancos*). The maritime environment supplied
the next-biggest category. Pigafetta thought he got the word for a storm
(*ohone*); so they can be pictured trading more expansive gestures. They
did a bit of coming, going, looking, and running too. The Patagonian
demonstrated fire lighting (*ghialeme*) by friction.

In different surviving versions, the terms Pigafetta collected coincide
fairly closely, with allowances for differences of transcription in a few
cases, but without enough consistency or overlap to justify an assess-
ment of his accuracy as a Patagonian lexicographer. It is hard to detect
any resemblances with available glossaries of Tehuelche, the native language that barely survives in the region today, with only one mother-
tongue speaker currently reported.[80] The giant shied from kissing the
cross Pigafetta whittled for him, crying, "'Setebos!' and made signs to
me that if I again made the cross it would enter my stomach and die."
But later, when the captive knew he was dying (below, p. 218), he
embraced the cross, imitating Pigafetta, who understood him to want
baptism. He died bearing the only name we know him by: Paul.[81]

· · ·

What did Magellan think of the giants? And why was he so anxious to
exhibit one at home? As we have seen (above, p. 24), sample gathering
was an implicit part of explorers' job at the time, enhancing royal
menageries and the *Wunderkammern* of collectors of curiosities and
admirers of the panoply of creation. The spurt in Western science that
we conventionally call "the Scientific Revolution" depended on the
availability of materials for study. Magellan's interest, however, was not
primarily scientific, nor was it merely an exoticist itch for unveiling marvels and monsters. A single line of Pigafetta's gives us a clue: "Our captain," he tells us, "gave this people the name of Patagones" (Il capitano
generale nominò questi populi Patagoni),[82] from which Patagonia, as the
region's name, derives. For a rough-hewn, imperfectly educated reader
such as Magellan, the classical lore on monsters came, not directly from
Greek and Roman originals or from scientific compilations in Latin, but

through the medium of allusions in popular literature, much as a barely literate teenager today might absorb some references to Homer or Virgil from Harry Potter.[83] I tell my students that you become a historian when you "get" an old joke—for humor is the hardest register to detect across chasms of time and culture. When he applied the name of Patagones to his giants, Magellan made one of his rare jokes, for the name literally means "big paws." But the text he had in mind was a romance of chivalry: the *Primaleon*—one of the books we have already encountered in his mental baggage (above, p. 41). In the relevant episode, Primaleon is on his travels, in a faraway land, where

> there lives a people very removed from all others; for they live there like beasts and are very wild and isolated, and they eat raw meat from prey they hunt across their mountains; and they are just like savages, who do not wear clothes save for the skins of the animals that they kill; and they are so unlike other creatures that it is a wonderful thing to see. But nothing compares with a man who is there now among them, whose name is Patagon, and this Patagon, they say, was engendered by a beast that there is in those mountains, who is the most monstrous there is in all the world: save that he has great understanding and is a great friend of women. They say that he had to do with one of those Patagona women, for thus we call them who are so savage, and that that animal engendered in her that son.[84]

A cascade of topoi tumbles through the lines: raw food as evidence of beastliness; the antisocial connotations of savagery; the allusion to nakedness; the bestializing effect of animal skins; the evocation of singularity and wonder; the theory that monsters are the half-breeds of humans; the alien threat to the purity of womanhood. Yet, on the other hand, the "great understanding" that betokens reason appears, just as in the case of the giants of San Julián. Next comes a particular description of Patagon as a cynocephalus—one of the "dog-headed men" whose existence Pliny validated, on whom debate about the rationality and redeemability of monsters focused in the Middle Ages (above, p. 147).

> And this they hold for very certain, wherefore he emerged very monstrous, who has a face like a dog's and big ears that reach to his shoulders, and teeth very sharp and big that stick out of his mouth all twisted, and feet in the manner of a deer; and he runs so lightly that there is no one who can catch him.

His uncatchability is another link with Magellan's giants, whose stride outran pursuers. Further in the text of *Primaleon*, we learn that the cynocephalus hunts lions with a bow, summons other Patagones with his horn, and is both invincible and deadly. But Primaleon defeats him, along with—for the bargain—two lions who accompanied him.

Then comes the revelation of Magellan's motives: the explorer wanted a giant of his own in imitation of his hero, who, "when he saw how monstrous" Patagon was, "and what a strange thing he was to behold, conceived it in his will to take him prisoner; and thought that if he could carry him off in his fleet it would do him very great honor if his lady, Gridonia, could behold him."[85] The similarities between Magellan's Patagon and his fictional namesake deepen, for in the book his captors chain him, at which "he gave out mighty roars" (le dava grandes bramidos). "Let us take Patagon," they exclaim, "to where everyone may see him!" (Llevemos a Patagon vivo porque todos lo vean!).

Later in the fable, Patagon, sulking in captivity, refuses a cure for wounds. Seluida, the daughter of the knight who is lord of the magic island where Primaleon has his base, comes to see the monster. At first, she is frightened. But the story of *Beauty and the Beast* is reprised. Seluida's beauty pacifies Patagon. Her kind soul responds. "And may you know that he was of that same condition as the monster who engendered him in that he was very pleasing to women."[86] Seluida overcomes her fear, tends Patagon's wounds, and "with sweet words" induces him to take medicine and nourishment, even though they cannot understand each other's language. Here was another topos: late medieval stories and images are full of examples of wild men—the "green men" of inn signs, shaggy or leafy, drawn from their native woods—whom ladies domesticate, often by teaching them chess or peaceful or amatory arts.[87] Moral: beauty is stronger than brute force.

. . .

Before leaving San Julián, according to Pigafetta, a detachment planted a cross on a rise they named Monte Cristo, a few leagues inland, "and took possession of this land in the name of the king of Spain."[88] Drake's men noticed only the gibbet, a more fitting memorial of Magellan's winter of discontent. They replayed the events of 1520 with uncanny fidelity. As in Magellan's case, factional divisions, reproducing those at court, split Drake's expedition. His own clique, radically Protestant and belligerent, opposed that of the queen's confidant, Sir Christopher Hatton, who was an enthusiast for overseas imperialism and American colonization but who, as a moderate in religion, wanted to avoid a state of active war with Spain. Some of Drake's charges against Hatton's representative, Thomas Doughty, echoed Magellan's complaints against Cartagena.

Doughty was "too peremptory and exceeded his authority, taking upon him too great a command." Other accusations veered between vagueness and fantasy. Doughty was "a conjurer and a witch," or "a very bad and lewd fellow," with a brother who was a witch and a poisoner. "I cannot tell from whence he came," Drake said, "but from the Devil, I think." Most of the testimony against the accused was tittle-tattle and scuttle-butt that hardly bore on any serious issue, unless to show that Doughty was tactlessly ready with criticism of Drake. Complaints that he was partial to the Portuguese demonstrate what was really at issue: whether the voyage should preserve peace or provoke war. Impartial opinion reckoned him God-fearing, truthful, erudite, of excellent qualities, an "approved soldier," and "a pregnant philosopher." After a show trial he was buried, with his severed head, on Magellan's execution ground, or, in Drake's chaplain's words, "near the sepulcher of those who went before him, upon whose graves I set up a stone."[89]

FIGURE 7. The picture of Magellan, made in Frankfurt by Theodore de Bry in 1594, and shown here in a nineteenth-century reprint, was copied, in reverse, from a Florentine original of 1589, engraved by Johannes Stradanus. Magellan is both aristocrat and artisan: impassive in knightly armor, aligned with his ship's mast, holding dividers amid other nautical impedimenta. The fallen mast evokes the storminess of the voyage. The banner bears the eagle of Carlos I's dynasty. Neptune presides (top left), while wind strains to blow the ship back. A roc clutches an elephant, perhaps in allusion to Pigafetta's record of a similar myth (v. infra, p. 272) and to hint at the Orient beyond the strait. Patches of fire signify Tierra del Fuego. A Patagonian giant swallows an arrow; other natives and sea creatures cavort. Apollo, blessing the afflicted ship and bearing his lyre, seems to promise that the narrative of the voyage will be poetic in quality, that the sun will disperse the lowering clouds, and that the voyage will imitate the sun in circling the Earth. From F. H. H. Guillemard, *The Life of Ferdinand Magellan* (London: Philip, 1891).

CHAPTER 7

The Gates of Fame

The Strait of Magellan,
October to December 1520

Strait is the gate, and narrow is the way, which leadeth unto
life, and few there be that find it.

—Matthew 7:13

"You are afraid to speak."[1] Magellan was taunting his officers with the
truth, as he invited—or incited—them to challenge his policy. The scene
in the cabin is imaginable: the dangerously exalted captain-general,
exhibiting paranoia, uttering threats; the comfortless subordinates,
with pursed lips and averted eyes. Outside: relentless sea and the glow-
ering mountains that frame the strait. "The Strait of All the Saints" was
the name that stuck among the sailors. The world came to call it "Magellan's Strait." It was what they had been seeking: the supposed gateway
to Asia—though they did not yet know how far away their destination
still lay.

The leader expected expressions of joy from his followers. Instead,
fear choked them. Their goal was near: that, at least, was what Magellan felt, or proclaimed, as the flotilla twisted and squirmed through the
disconcerting labyrinth of gulfs, narrows, and channels between the
American mainland and the islands that fringed it to the south. Finding
this strait had cost a year's suffering. It was a kind of climax. Even the
captain-general's most skeptical critics might have been caught between
expectation and hope: expectation of more travail and disillusionment,
hope of a happy issue. It should have been a moment of triumph—rare
enough in a voyage that, so far, disasters had disfigured and almost
doomed. No doubt some of the men felt a brief spell of satisfaction. But
the salvos fired to celebrate success echoed meaninglessly from the bleak

cliff faces. There was no one to respond, except the indifferent penguins and the incurious seals.

. . .

The voyage was already an irremediable failure. The strait was too far from Spain. The way was too long and arduous, the weather too cold, the available food too scarce and unnourishing, the winds too adverse, the coasts too hazardous. Magellan's route, even if it eventually led to the Spicery, would never be able to compete with the faster passage the Portuguese already followed.

The narrative of disasters is all too easy to reconstruct, but the course and chronology of the voyage through the strait are elusive, partly, as usual, because of the mutual contradictions of the sources, and partly because of the deficiency of data at crucial moments. It helps to have an idea of the setting—the trend of sea and shore, the enormities of the environment, the monstrosity of the task.

Pigafetta, Magellan's avowed apologist, praised the beauty of the strait. It has, indeed, a terrible beauty, under dazzling mountains from which glaciers seep into the sea. To the south, Tierra del Fuego looks as if it has been wrenched and clawed from the heel of South America, strewing the strait with islands like scraps of flesh or blood clots from a Procrustean bed. The result is a maze of confusing channels and bewilderingly variable soundings, with rocks and shoals threatening to rip or strand passing vessels. Though there are mild meadows toward the eastern end, for much of the way forbidding cliffs rise precipitously from the shoreline. There are nearly 350 miles of hard going to traverse. The weather lurches unpredictably. Fogs descend and vanish like traps flung and flicked by an evil retiarius. In any case, beauty, as most love stories attest, does not "live with kindness" or necessarily induce calm or placate rival appetites. Even though it was high summer when Magellan sailed here, the weather was almost unbearable; for the strait is a kind of wind tunnel, where the howling westerlies of the Southern Ocean drive back vessels that try to force their way along it—the very experience *San Antonio* endured, the very winds that made *Trinidad* and *Victoria* scurry for shelter. It is not uncommon for sailing ships to be ejected like pellets from a popgun, flung back almost as far as the Falklands.

Nowadays, loungers at the rails of cruise ships can go to the strait to escape the northern winter at their ease, as they contemplate with equanimity features that struck horror into the hearts of sailors in the age of sail. Sailing directions, before the nineteenth century, make tense reading: summons to constant vigilance, untouched by romantic sensibilities or rhapsodical reactions to the sublimities of nature. Where, as they approach the strait, voyagers now see the gentle slope of the land toward Cabo de las Vírgenes and the white cliffs of Tierra del Fuego to the south, navigators in Magellan's wake saw barren shores, with no wood or fresh water, and a long barrier of treacherous sands barring the route. In today's comfortably powered and balanced vessels, passengers hardly notice the racing tides and frighteningly inconsistent soundings. Rocks that now look picturesque seemed to be waiting to rip ships apart. The penguins that nest winsomely in sand burrows along the shore were, in the hungry eyes of early mariners, hard-won prey. The well-charted bays and side channels were like Sirens' hair, streaming in the wind, luring ships astray through deceptive mists. Where photo opportunities now catch the notice of passersby, ships' officers formerly scanned the shores desperately for life-saving rivulets of water or stands of timber.

The strait, in crude outline, forms a chevron, tapering toward the south, at the cuff of South America. Cabo Froward, the southernmost point of the continent, is at the apex. The eastern arm curls a little, as if the cuff were slightly rucked. A broad channel, usually known as Paso Ancho, pokes north and slightly west from Cabo Froward before curving northwest toward the Atlantic in the form of three roughly round bays, separated by tightly pinched narrows, like bubbles in a toy balloon. A strong tidal race runs from west to east, first into an irregular space, blistered with surrounding bays and speckled with four little islands, two of which are no more than uprearing rocks; the passage to the next "bubble," usually called "Bahía Victoria" nowadays, is about two miles long and four wide. Wide banks of weeds cling around the islands. The gaunt, dark mountains on the southern shore rise to over a thousand feet and display a snowy rim even in summer. A further twenty miles to the east, beyond another broad stretch, the shores draw in to form narrows only two miles wide and some twelve miles long. Magellan, arriving from the east, approached from a cape a little over fifty degrees south and a mere 135 feet high. He named it "de las Vírgenes" in honor of the feast day—celebrated on October 21, the day he entered the strait—of St. Ursula and her eleven thousand companion-refugees, ultimately martyrs, from Hunnish invaders of Cologne. Magellan's pilots got their latitude almost exactly right and reckoned the longitude—the reading again being in San Martín's hands—at 44.5 degrees west of São Antão in the Cape Verde Islands: an error, passable for the conditions and means available, of just over 2 degrees.[2]

To penetrate what his pilot called "an opening like a bay," Magellan had to pick his route around a long but luckily visible spit of sand and among vast dunes, forcing his way in against the current and a fierce wind, contending also with what Pigafetta claimed was a universal conviction among the crew that, like all the bays and river mouths they had explored so far, the opening would lead nowhere. "Had it not been for the captain-general, we would not have found that strait, for we all thought and said that it was closed on all sides."[3] Two of the strait's main hazards immediately became apparent: the confusing soundings, which must have been a constant strain on the expeditionaries' morale, especially in the narrow stretches of the passage; and the ferocious and almost relentless wind, which kept driving the ships back on their course, consigning the crews to despair of their hopes and lives as it trapped them in apparent culs-de-sac. At times, storms separated them, severed by narrows. Dispersals forced them to sail back and forth in search of each other as they fired forlorn cannon shots as signals. Emotional periodic reunions ensued, with the shouts of joy, blasts of Lombard shot and occasional tears, all of which Pigafetta describes, echoing or heightening the emotional charge. Obviously, almost everyone (except, perhaps, the ever-unflappable Pigafetta) seems to have been on edge. The enmities that had riven the crews in the Atlantic grew sharper beyond it. Mutinous moods, which had brewed and simmered since San Julián, bubbled at the rim of violence. As safe anchorages were few and uncertain, and storms wrenched at anchor cables, the combination of perils from winds and shoals was nerve-racking. When the wind failed, fog often succeeded it.

. . .

However much faith Magellan put in his vaunted map, the fact was that no one knew where they were going. The succession of bays and channels marked manic mood swings from optimism to pessimism: with every widening, reveries in the hope of riches; with every narrowing, orisons in the face of death. "Now he was happy, now sad," was Mafra's summary of his commander's mood.[4] The passage of the east arm of the chevron was taxing, among mainly barren and waterless shores, and seems to have taken some ten or twelve days. (The pilots' reckoning faltered; so we cannot be sure.) Not for nothing did later navigators give the name of "Port Famine" to Punta Santa Ana, toward the end of the run. The closed channels, bays, and creeks along the edge of the strait grew increasingly deceptive, especially toward the southern end of the Paso Ancho, where the tragically named "Bahía Inútil"— "useless bay"—and Admiralty Sound beckon eastwards, broad and cruel, like the Sirens' wide-mouthed screams. These and other wickedly alluring dead ends had to be explored.

When the most advanced ships—probably *San Antonio* and *Concepción,* which Magellan favored for reconnaissance—approached Cabo Froward from Paso Ancho, through an entrance of between five and eight miles wide, they found themselves in a triangular sea-space. The western shore of Isla Dawson, which lies, shaped like a Phrygian cap, with its crown pointing northwards into the Paso Ancho, formed one side of the triangle. Another was the south coast of the Brunswick peninsula, which seems to dangle like a butcher's hook from the rest of the South American landmass. To the south lay the first of the many ragged-fringed, narrow-coved, moss-spattered islands that lead westwards toward the Pacific alongside the remainder of the strait. More, narrower siren-channels lead south and east. At this point the strait veers abruptly to the west by north. Reconnaissance beyond the bend, on or around November 1, probably swung opinion in the fleet in favor of Magellan's view that they were in a strait that would lead to the western ocean. Henceforward, All Saints' Day, the feast celebrated on November 1, began to share its name with the strait.

It took Magellan some twenty-six days to ascend the western arm of the "chevron." It was amazing that the flotilla was able to make it through in relatively good time: the captains were lucky that the winds abated for a while and let them make headway. The expedition completed the passage in about thirty-six days, albeit with enough trouble to demonstrate the intractability of the route and the improbability of ever making it serve as a viable highway of commerce.

A good deal of that time was spent in waiting for reconnaissance expeditions to report, while Pigafetta dawdled, fishing for what he called sardines,[5] or in searching for *Concepción* and *San Antonio,* which went astray—the latter never to return, as we shall see. But in any case, the western half of the strait is harder to navigate than the eastern: consistently narrow, rarely more than eight miles wide until it splays open at the western end of the Península Córdoba, squeezing the wind against the ships' course, with scanty anchorages and rocky, craggy shores for most of the way. If the passage to Cabo Froward failed to convince the crews that they were on a commercially unviable route, the completion of the course to the ocean left no doubt that the strait was better understood as an obstacle than a gateway.

. . .

Most of the crew surely knew that, even before they emerged into the ocean. We can still read the avowals of failure in and between the lines of the two officers whose opinions have survived. Both were experienced "pilots," or navigators, as we might rather say today. They knew what they were talking about when they called for the mission to be aborted.

When Magellan demanded their counsel, as he did at least twice during the passage of the strait, he was baiting a trap. If they endorsed his decision to continue, they would fail in their duty and forgo their right to offer independent advice. They would have to share the blame that their commander was rapidly and recklessly accruing. If, on the other hand, they dissented, they would be going against the royal instructions that the expedition had received at its outset: to reach the fabled Spicery. They would therefore be liable to accusations of insubordination, and at Magellan's mercy.

So far, he had shown little of that quality. Dissenters would risk his wrath and incur vindictiveness, such as they had already seen gape from the wounds and glare from the eyes of earlier dissidents, flayed, garroted, or marooned, by the captain-general's orders, on the Atlantic shores of South America. Magellan flourished the fates of previous victims in the survivors' faces. "It is because of what happened," he wrote, "in the harbor of San Julián, when Luis de Mendoza and Gaspar de Quejada died, and Juan de Cartagena and the priest Pedro Sánchez de Reina were cast away ashore, that you now fear to speak or to say all that seems good to you for the service of his Majesty."[6]

It is hardly surprising that he should have recalled the horrors of the expedition's time at San Julián. They were on his mind because he was still in the same kind of danger from the ill-concealed malevolence around him. Opponents still outnumbered friends. Few officers shared their commander's vision of the mission; the gap was still wide and deep between his hopes for himself—of a permanent domain and command in a new world at the edge of Asia—and the ambitions for quick returns or conventional honors that animated almost everyone else on the expedition. The singularity of his responsibilities oppressed Magellan more and more as his fleet sailed further into the unknown. The chivalric and romantic self-image that he formed in his youth was under terrible strain from the squalid, stinking, rat-ridden realities of shipboard life; his grand notions of conflict, sprung from crusading and jousting, faltered in ignoble squabbles with murmurers and mutineers. He felt lonely, and the sense of loneliness showed up in vexation, frustration, anxiety, and defiance. He was right to say his officers were afraid. Did he acknowledge, inwardly, that he was, too? His dialogues with dissenters, which we can explore over the next few pages, reveal his state of mind.

. . .

Magellan issued the Order of the Day in which he made his demands for advice (and his threats at the consequences) on November 21, 1520, when the fleet was very close to the exit, already scouted, from the western end of the strait.

The reality of the passage can no longer have aroused any doubt. When the fleet had first turned into the strait, a month previously, "we all," Pigafetta reported, doubted that the opening would lead to a way out on the far side.[7] How long did the skepticism last? As we have seen, the name "Strait of All the Saints" suggests that a decisive breakthrough must have been made on November 1. In any case, by the twenty-first, reconnaissance by land and sea had doubled and redoubled assurance that the western ocean was near. On the other hand, the hazards, hardships, and unreliability of the route were equally obvious. It was navigable—just—but not viable as a long-term proposition. Magellan had already made a show of consulting his officers on at least one occasion during the struggle through the strait. His purpose, perhaps, was to assert his own conviction that success was still possible; or perhaps he wanted to ensnare subordinates he detested or suspected. He may also have aimed to indemnify himself against more mutinies or to appease his own obsessions. Maybe he wanted merely to exult over the doubters who had never believed in the existence of the vaunted strait or the accessibility of the Spicery via the Atlantic. Might he have genuinely wanted his subordinates' advice? That would contradict everything we have seen so far in evidence of his character. Self-reliance was part of what was great—and a great part of what was tragic—in him.

The first officer to call for abandonment of the mission was Estevão Gomes. He, like Magellan, was a Portuguese renegade, who, for unknown reasons, had fled his native country to serve Spain. His recorded Spanish phase dates from about the same time as Magellan's and invites the presumption that they may have cut and run together.

His prior service is unrecorded or unidentified, but his record was good enough to secure immediate appointment to the post of pilot in the Casa de Contratación in Seville (which licensed all Spain's Atlantic ventures), and his adscription to Magellan's little armada. A story by Magellan's pet chronicler, Antonio de Pigafetta, has suckered historians: that hatred for the captain-general consumed Gomes, who had hoped for command of the expedition for himself: "He hated the captain-general exceedingly, because before the fleet was fitted out, he had gone to the emperor to request some caravels to go and explore, but his Majesty did not give them to him, because of the coming of the captain-general."[8] The story served the interests of Pigafetta's patron by making Gomes's views seem unobjective and by explaining the latter's opposition without reference to its merits. No other evidence supports the tale, and it is unlikely to be true because navigators were professionals rarely appointed outside their proper roles, whereas captaincies demanded further specific qualities and generally fell to noble or charismatic amateurs. In any case, Gomes was documented as a volunteer member of the crew: he received exemption from liability to billeting as a reward.[9] He obtained no independent command until 1524. When he did, his competence in navigation and mapping became evident. His counsel may have been self-interested, but it was soundly based.

Why did Gomes dare to be forthright in the advice he gave his chief? Common betrayal of Portugal and joint ascent in Spain gave the two émigrés a basis for fellow feeling—the security that comes with shared experience, the sort of honor that binds thieves. Gomes may have felt able to contest Magellan's views in the expectation that his chief was doomed, in any case, by an errant record, unpromising prospects, and a growing tally of enemies in the expedition. Gomes was soon to turn mutineer and may already have been planning his escape from the fleet. Any of these considerations might suffice; but it is hard to resist a further, decisive impression: that Gomes knew that he was right, and Magellan wrong.

Two witnesses recorded the pilot's opinion: the ships should return at once to Spain. As they had already found the strait "for reaching the Moluccas," Gomes said, the expeditionaries had accomplished their primary objective. If further days of calm or more storms should delay or threaten them, they might lose all they had accomplished. "A great gulf" had still to be traversed, and they needed better ships and more supplies for the job.[10] As ever, we have to retrieve the image of the world current at the time. The New World's relationship to Asia was still undetermined. It might be the last, longest, easternmost of Asia's peninsulas and promontories; it might be a separate landmass. In geographers' argot the "Great Gulf" meant what we now call the Pacific Ocean: the "South Sea," as they sometimes said, the arc of water that separated the Americas from the East, or curled between regions of Southeast Asia that Portuguese navigators had already visited—the Malay peninsula, the islands to its south and east, and the as yet uncharted, almost unvisited western shore of the New World. All attempts at world mapping at the time show such a gulf. None displays awareness of how wide it was (above, p. 81). But Magellan had banked his reputation and founded his plans on the assumption that it was very narrow. As we have seen (above, p. 80), he followed Columbus and Vespucci in their gross underestimates of the girth of the globe.

Gomes, evidently, had a more realistic view or at least a more prudent suspicion. He knew that the fleet's depleted resources of food and water would not be sufficient for so long a voyage. In order to understand his point of view, it is also helpful to glance at his future career, which was largely dedicated to the search for a "Northwest Passage" from the Atlantic to the Orient around or across North America, at latitudes that would guarantee a relatively short passage: the further you are from the equator, the shorter the lines of latitude. Gomes realized from the expedition's experience so far, without even going beyond the strait, that the route Magellan pioneered was impracticably long.

Magellan's reaction survives in his own words, addressed to the assembled officers—in reported speech in the sources, but easily retrieved: "Even if we have to learn to eat the cowhides that wrap the masts," he said, "we must press on, and discover what you have promised for the emperor."[11] The text is reliable and endorsed by various eyewitnesses; the implied reproach to anyone who forgot or despised a promise is characteristic. But the words have a tincture of poetic irony, like a cinematographer's shot of a life belt on the *Titanic*: Magellan's figure of speech proved eerily prophetic, as before the voyage was over the starving crew would be reduced to chewing on those very hides, with jaws painfully rotted by scurvy.

After hearing Gomes, the captain-general ordered that, on pain of death, no one be permitted to make any adverse reference to the route the fleet was following or to mention the state of the stores: a tacit admission that the route was unserviceable and the stores dangerously depleted. The effect can hardly have been to encourage candor among those whose counsel he continued to demand.

Nonetheless, at least one respondent stepped up to offer an opinion, on November 22, on the threshold of the ocean, near the western mouth of the strait. The ships were still probing to find the best route through, but the prospect of breaking out into the ocean was imminent. Andrés de San Martín, who now sat at his desk to compose a memorandum for Magellan, was more than a practical navigator. He was admirable for expertise in geography and mastery of the latest technology for mapping and navigation. At San Julián he had used a solar eclipse (above, p. 168) to determine longitude, succeeding to within less than half of one degree of accuracy—an achievement unexcelled at the time. He had been an official pilot of the Casa since 1512 and a candidate for the headship of the institution's school of navigation. His salary in the Casa had been raised to the highest level in 1518. He had replaced Rui Faleiro, on the latter's withdrawal or extrusion, as the resident cosmographer of Magellan's expedition (above, p. 115). His estimate of the size of the globe is unknown, but there is no reason to suppose that it was distorted by the same considerations of prejudice and self-interest as affected those of Columbus, Vespucci, and Magellan. San Martín can hardly have been willing to take part in the voyage unless he thought there was a chance of locating spice islands on Spain's side of the line of demarcation; but that would not necessarily imply that he expected the "Great Gulf" to be easily crossed.

San Martín was in a dangerous dilemma. I imagine ink drying on his quill as he strives—brow furrowed, perhaps, eyes narrowed—to find words candid enough to fulfill his duty and diplomatic enough to save his skin. His duty to declare his opinion was clear. On the other hand, he was fully aware of the solemnity of Magellan's threats. In the aftermath of the mutiny at San Julián, he had faced the prospect, if not the reality, of ferocious torture, at Magellan's command, on mere suspicion, in an attempt to extract a confession (above, p. 169): slung by the wrists from a pulley, with feet weighted, while torturers jerked the straps to increase the pain. San Martín's only real offense, even then, as far as we can tell, had probably been to disagree with Magellan, saying that the expedition had not made as much way as the captain-general reckoned. He survived—spared the torture, as argued above, or released before it wrenched his limbs from his body—and resumed his office. Still, it required courage to defy Magellan again. Perhaps in consequence, the memoir San Martín wrote to his chief was temporizing, but its import was clear: the mission was over and the expedition should head home.

In the early part of the memorandum, diplomacy prevailed. Irrespective of the best way through the strait, San Martín averred, there was a genuine prospect of eventually reaching the Moluccas. While good weather persisted—"inasmuch as we are now in the heart of summer"—there was no reason to turn back: "It seems that your Grace must therefore proceed for the time being." By mid-January, however, "let your Grace find good cause to turn back toward Spain, for from then on the days suddenly shorten, and because of the tempests they are bound to be more wearying than at present." The writer went on to remind Magellan of the brevity of last winter's days in far southern latitudes. He discounted the possibility of adopting his leader's declared fallback plans of seeking a better passage to the south or turning east to follow the Portuguese route via the Cape of Good Hope,

> both because it will be winter when we get there, as your Grace will know full well, and because our crews are weakened and devoid of strength, and, although we have at present enough provisions to keep us going, they are not sufficient nor of sufficient quality to renew our strength nor to fit us for very much further hardship, without the men feeling the effects in their spirits and their bodies. And furthermore I see that those who fall ill take a long time to recover. And although your Grace, thank God, has good ships, well rigged, some of the rigging is missing, especially in my ship, the *Victoria;* and besides I repeat that the crew is wasted and undone, and the supplies are insufficient for a voyage to the Moluccas, thence to return to Spain.[12]

He warned of the danger of damage to the ships on unfamiliar coasts, especially if Magellan were to continue his policy of spending the nights under sail. "I have spoken as I feel and as I may in order to fulfill my obligation to God and to your Grace, and to what seems to me the service of his Majesty and the good of the expedition." With a critic's usual cop-out, San Martín concluded, "Your Grace may do as he thinks fit." Despite the caution, the message is unmistakable.

Yet in defiance of the best available professional advice, the fleet sailed on, oblivious to failure, indifferent to risk, and bound for disaster.

. . .

To understand what was happening, two strategies help: on the one hand we can recall Magellan's habitual mindset, which always responded to setbacks with obduracy, like a compulsive gambler on a losing streak. In every strait of life, chivalric and romantic notions inspired him to revel in quixotic odds and embrace unrealistic, even fantastic, ambitions. When fortune failed, the grip of his obsessions tightened. To return when his contract with the crown remained unfulfilled and his record of command left him so exposed to the reproaches of enemies would, for his own purposes, have been unthinkable. As for why most of his men continued to go along with him: what his subordinates' discipline and inclination to obedience could not supply, his own charisma and force of character perhaps compelled. Ships were machines for sailing in, and crews were like springs or cogs, with little more freedom of movement than the give or take of the great, creaking mechanism demanded. As in any machine, friction could cause breakdown. In Magellan's fleet, parts of the system had already failed; but the mutinies, so far, as we are about to see, had detached only some of the men and one of the ships.

It may also help to reconstruct, with imagination restrained by the sources, the extraordinary circumstances that surrounded the expedition. On the embattled little ships the mariners constituted, in a sense, a world of their own, where the vastness of the distance they had traversed and the strangeness of the environment they had penetrated warped standards of reality. They had gone further south than any explorer, as far as they knew, had ever been. Even Vespucci, with his talent for exaggeration, had not claimed to venture beyond fifty degrees south. The passage through the strait led Magellan and his men more than three degrees further. On the way they had encountered giants, as if in fulfillment of ancient myths of monstrous beings at the edges of the world (above, p. 147). They had experienced cold most of them had never sampled before. Those who still lived had survived mutinies, tempests, shipwrecks, and the threat of starvation.

The facts of their lives had already exceeded fiction. They were beyond hope of help—literally thousands of miles from the closest friendly settlements. The nearest Europeans were hostile Portuguese or French loggers single-mindedly felling dyewood around the bulge of Brazil. Internecine hatreds, clashing opinions, ethnic rivalries, and mutual suspicions divided Magellan and his companions—but they shared the extraordinary shipboard life. Under sail at sea, in their frail craft, with hulls condemned to rot, ropes to fray, water to sour, and food to perish, survival depended on collaboration. The sense of comradeship, which cramped conditions induce, was enhanced by the vastness of the sea, the menace of the shore, and the hostility of the elements. In each ship, these conditions intensified. Each crew remained able to act on its own—as would happen during the passage of the strait, when *San Antonio* broke ranks and deserted. But the paradox of adversity is that it often enhances resolve, like strain stimulating adrenalin.

Magellan's own mental state is readable, opaquely, in the document in which he commanded his officers to declare their opinions on the priority of continuing the mission.[13] It is, by any standards, a psychologically disarming, disturbing text. There is no reason to doubt its authenticity. Andrés de San Martín carried his copy across the Pacific, where it fell into the hands of his Portuguese captors, and thence to an archive in Lisbon, where, some twenty years or so later, the generally reliable official chronicler João de Barros copied it, long before the great earthquake of 1755 engulfed it. "I have sensed," Magellan admits, addressing his officers, "that you all esteem it a grave matter that I am determined to go forward, because you think that time is short to accomplish this voyage on which we are embarked." A barefaced lie or stunning self-deception follows: "I am a man who never discarded any opinion or counsel, but rather all my proceedings are undertaken and communicated generally to all, without affront by me to any person." The threat implied in the precedents Magellan goes on to cite—the fatal sentences executed upon the mutineers at San Julián—has deterred, he acknowledges, frank avowals of contrary opinions.

The tone of menace then resumes:

> You err in the service of the emperor, our king and lord, and you go against the oath and pledge of homage that you have made to me. Wherefore I command you in the name of our said lord, and for my part I request and recommend, that all that which you feel and which may be convenient for our voyage, whether to press onward, or to turn back, you confide to me in writing, each man for himself, declaring the substance and reasons for which we must go forward or turn back, without heed to any cause for which you might fail to declare the truth. With the which reasons and opinions, I shall pronounce my own and my determination of how to come to a conclusion concerning what we must do.

The language mingles sly use of cajolery and threat with paranoiac delusions and a lack of self-awareness so blatant as perhaps to be disingenuous. The internal self-contradictions of the document are frightening. The trenchancy with which the writer declares at the outset that he has already made up his mind belies the assurances, with which he closes, that he is willing to be advised. A reader today is left unsure whether to fear more for the captain-general's state of mind or for the safety of any subordinate who dared to disagree with him. For Magellan the Strait of All the Saints was the object of his hopes and the fuel of his ambitions. But it was not, in today's lingo, "a good place."

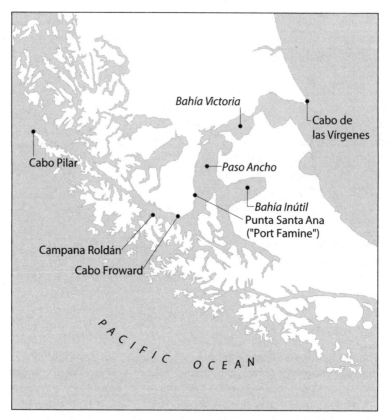

MAP 4. The Strait of Magellan: Places named in the text.

• • •

The remaining, fateful events of the passage of the strait can be pieced together with confidence about the way they unfolded, though not with chronological certitude. Every historian who has made the attempt has come up with a peculiar version, inclining now to one, now to another of the mutually contradictory sources. The conflicting agendas of witnesses fly in incompatible directions, like ships sundered by storms. The mutineers' account, elicited in an official, judicial inquiry when they got back to Spain, is condensed and evasive. Antonio Pigafetta's, as ever, is dedicated to vindicating Magellan and at crucial moments is frustratingly elliptical. The pilots focus, as usual, on stating the changes of course without adverting to matters of human interest. But one body of evidence is both objective and revealing: the accounts that the survivors of the *Trinidad* rendered when they fell into Portuguese hands and were interrogated in Ternate, nearly two years after emerging from the strait. The chief interrogator, the local Portuguese commander, António de Brito, was in no mood to treat the interlopers gently, but they had no reason to suppress the truth or to invite torture by refusing to cooperate. The tale they told is credible. Because Pigafetta's version shows signs of being modified after the event, the *Trinidad* testimonies also constitute the earliest surviving eyewitness narration.

Brito summarized it for his king.[14] On October 20, 1520, or thereabouts, Magellan's ships entered "a strait, and they did not know what it was." By other participants' reckonings, as we have seen, they did so on October 21, feast of St. Ursula and her mind-boggling tally of virgins. The discrepancy is negligible, as there was no standard way of reckoning the start and close of a day. At first, the mariners related, they thought they were in a big bay from which there was no exit to the west, but events proved otherwise.

In the interim, the narrative continued, Magellan had attempted to get a view of the prospects by sending a reconnaissance party under João Lopes Carvalho, a Portuguese pilot, to survey the strait from a hilltop. Carvalho saw no way out, but when, on one or two further attempts at reconnaissance, *San Antonio* and *Concepción* explored thirty (or according to some witnesses fifty) leagues ahead—that is, up to about halfway through the strait or nearly so—"they returned to tell Magellan that the waterway continued further, but they did not know how far it might reach. In view of this, he set sail with all three ships and navigated the strait as far as they had already penetrated," ordering, on the way, the *San Antonio,* "of which Álvaro de Mesquita, his relative, was captain, with Estevão Gomes as navigator, to go and discover what lay beyond an opening in the strait toward the south."

Pigafetta recalled the events with more fervor than facts. He heightened the drama by describing three days of terrifying storms that tore away the anchors and made the flagship take refuge near the mouth of the strait "at the mercy of waves and wind." *San Antonio* and *Concepción,* meanwhile, were stoppered in a further bay. Driven westwards toward what they supposed was an implacable and uninterrupted shore, and thinking that they were lost, they saw a small opening, which did not appear to be a channel but only a cove. Like desperate men they hauled into it, and thus they discovered the strait by chance, as they pressed on through two further channels. "Very joyful, they immediately turned back to report to the captain-general" (Molto alegri subito voltorno indrieto per dirlo al capitano generale).[15] Pigafetta's sequence of channels and bays seems baffling, especially if one takes literally the assertions about their relative sizes, but his further descriptions make it clear that his "third bay" was where the strait broadens suddenly beyond Cabo San Isidro into the triangular space before Cabo Froward. "It was from there," says the account of the survivors of the *Trinidad*, "that they emerged" into the Pacific.

The conviction that an exit lay to the west solidified by November 1: hence, as we have seen, the name, taken from that of the day—the Strait of All the Saints. Pigafetta said that verification came all the way back to the neighborhood of Cabo Froward, where *Victoria* and *Trinidad* remained at anchor.

> While in suspense, we saw the two ships with sails full and banners flying to the wind, coming toward us. When they neared in this manner, they suddenly discharged a number of mortars, and burst into cheers. Then all together thanking God and the Virgin Mary, we went to explore farther on.[16]

It seems unlikely that such a reconnaissance could have been completed so quickly. Perhaps, rather, the lie of the strait, which trends firmly to the west from that point, induced the optimistic mood. In 1523, the crown authorized the payment of 4,500 maravedíes to two survivors of the expedition "in reward for when they went ashore and the strait was revealed."[17] The lucky pair, Ocasio de Alonso, a crew member of the *Santiago* until it foundered, and Hernando de Bustamante, who served on the *Concepción*—as surgeon-barber, according to some records—obviously did something spectacular and decisive on shore. But we do not know what, where, or when. The temptation to link their escapade with the reconnaissance Pigafetta reported is hard to resist, but there is no warrant for it. A promontory toward the western end of the strait long bore the name of Campana de Roldán, Roland's Bell, exciting presumptions that another survivor, Roldán de Argot, a gunner on *Victoria*, must have been of the party or otherwise involved in a reconnaissance by land.[18] He returned to the strait with Bustamante on a follow-up voyage in January 1526 and went ashore to reconnoiter, "as they had been there before."[19] Roldán was not, however, among the recipients of the reward.

From the surviving descriptions, the point at which *San Antonio* departed was either around the widest part of the strait, where Paso Ancho meets Bahía Inútil, as that is where Admiralty Sound debouches to the south, or, more probably, a little further on, at about the southwest corner of Isla Dawson, where one can discern "three channels," such as the expedition's anonymous Genoese pilot recommended as a landmark.[20] According to an early tradition, Magellan asked San Martín to use astrological divination to determine the fugitive ship's whereabouts; the astrologer divined that she had fled with her captain in mutineers' hands, but Magellan "did not give much credit to that."[21] It is probable that San Martín dabbled in the occult: as we have seen, the reputation of a "judicial astrologer"—one, that is, who read the future from the stars—dogged him; and in the sixteenth century only a very blurred line divided sorcerers from scientists. But the story sounds like a fictional embellishment: Why would Magellan ask for a reading he was ill disposed to credit? *San Antonio* "did not reconnect with the other ships," nor did the prisoners know "whether she returned to Castile or was lost." Their ignorance is a guarantee of the good faith with which the captives testified. They did not expect to have to be judged against a rival, partisan narrative. "Magellan," the prisoners' account concludes, "proceeded with the ships that remained until he reached the exit."[22]

Maddeningly unclear is the problem of when the desertion happened. Pigafetta's sequence of events places it before the existence of a western outlet to the ocean was confirmed. That seems most unlikely. When Estevão Gomes, the *San Antonio's* pilot and, by common consent, the chief mutineer, tendered his dissenting advice about whether to continue with the expedition, the existence of a western exit seems to have been clearly established. Gomes's reasons for aborting the expedition made no reference to skepticism on that point. He said rather, as we have seen, that the Pacific, though accessible, was untraversible with the means the expedition commanded. The *San Antonio* was one of the ships that had reconnoitered halfway along the strait. Albeit not conclusive, the reconnaissance had improved the likelihood that the strait led somewhere promising.

Evidently, before *San Antonio* deserted, the presumption, if not the proof, that the expedition had found the fabled passage to the Pacific was well diffused and widely accepted throughout the fleet. It may seem odd to abandon an expedition at a speciously propitious moment. But the discovery of the strait may have constituted, for some mutineers, an extra incentive to leave: they could claim to have accomplished the essential part of their mission. They could hope for rewards for being the first to get home with the news. To expect consensus among them would be too sanguine. Mutinies multiply schisms. Divided opinions aboard *San Antonio* may not have been as various as they were across the whole fleet, but they were sufficient to make the ship dither and dawdle in the Atlantic. Should they head straight for home, or boost their credentials by rescuing the castaways from San Julián, or race for the Moluccas by "the Portuguese route" in continued defiance of orders? Every option had advocates.

. . .

Pigafetta relates what he and the crew of *Trinidad* did while they waited for the reconnoiterers to return. At first, they tried to shelter in a bay that was perhaps Bahía Inútil or maybe one of the coves near Cabo Froward, while a fierce storm blew them perilously "hither and thither about" and the advanced party was blown almost all the way back to the strait's mouth. During four or six days of enforced idleness, waiting in vain for the *San Antonio,* the men fished for "sardines," before setting out on a search for the missing vessel "in all parts of the strait, as far as that opening through which it had fled."[23] As ever, Pigafetta seems to have been in a world of his own, passionless except for his fierce devotion to Magellan, and strangely detached from horror and drama, even when it unfolded before his eyes. He could always find time to enjoy and allude to pleasure and do a little fishing. Though he could never describe anything vividly, his eye was always alert for beauty in seas and straits, even when other beholders responded with fear and loathing. Even the mutiny on *San Antonio*—of which, of course, he knew nothing for years after the event and of which his account is worthless—left him unmoved; the narration of it that he added to his text is curiously bloodless and matter-of-fact.

He provides, however, one plausible detail on the doings of the rest of the flotilla. *Victoria*, he mentions, doubled back—as far, he improbably declares, as the mouth of the strait[24]—to make the search for the missing ship sure but, finding nothing, erected "an ensign on the summit of a small hill, with a letter inside a pot placed in the ground" to inform the sister-ship of the remaining vessels' proposed course.[25] Naturally, Pigafetta wished to leave readers convinced that Magellan had done everything he could for his fellow seamen, but in principle the procedure he describes seems reasonable. The fact that the whereabouts of *San Antonio* were undetectable was bound to arouse disquiet in the rest of the fleet. The usual ways of locating missing elements of a convoy were to sound the guns on board and ignite smoke signals on shore. Failure raised a presumption that the errant ship was lost. Andrés de

San Martín told Magellan that he supposed *San Antonio* had fled for Spain.[26]

* * *

For what happened aboard the absent ship we have to turn, with cautious confidence in the reliability of the evidence, to the accounts of events that members of the crew provided to another inquiry, back in Spain, in Seville in May 1521. The transcript at our disposal is only partial, since seventy deponents were interrogated and it took half a day to get through each testimony. Only twenty-one interviews were complete at the time of the existing report. Special pleading ripples the testimony. The men insisted that they complied with orders until justified despair drove them to new expedients. The claim was obviously designed to mask what really happened: mutiny aboard *San Antonio,* with the captain stabbed and clapped in irons.

According to the summary prepared for Bishop Fonseca, who continued to head the Casa de Contratación, *San Antonio's* reconnaissance to the south led the ship to 54.5 degrees of latitude by an unspecified date toward the end of October. The next few days are a blank. The narrative resumes only when the witnesses reported for what turned out to be an unfulfilled rendezvous with *Concepción,* which, they assumed, must have rejoined Magellan. They searched for their companion ships—so they said—for four or five days,

> and as they did not find them, they agreed among themselves to turn eastwards, and to the said turn the said Álvaro de la [*sic*] Mezquita objected, [and] they came to blows, during which the said Álvaro de la Mezquita stabbed the pilot, Esteban Gómez [or Estevão Gomes in the native Portuguese version of his name], and the said Mezquita received another knife wound in his left hand, and in the end they seized the said Mezquita.

The exchange of wounds between captain and pilot became the subject of dispute, like a playground fight, over who really started it. The date given for the affray, October 8, is an obviously wild error, inconsistent with all other data, but November 8—well after the naming of the strait—would not be an impossible date. "And then," the narration by the Sevillan interrogator concludes, "they came straight here."[27]

After all that had befallen the expedition it is hardly surprising that the mutineers wanted to take their chance of escape. It was a risky but rational move. By being first back to Spain they could get credit for having discovered the strait, without suffering the miseries of prolonging the voyage. Gomes could put forward his own plan for a northwest approach to the Orient (though we have no evidence of when he formulated it), supported by proof of the failure of Magellan's venture. The mutineers could indemnify themselves against punishment by discrediting Magellan in advance of his return—in the unlikely event of his ever making his way home: Gomes knew the condition of the ships and the insufficiency of the stores. He suspected that the "Great Gulf" was unnavigably broad or at least untraversible with the means at Magellan's disposal. The interrogator's summation of the mutineers' expectations reflects Gomes's views:

> In the judgment and opinion of these men who have arrived here, the said Magellan will never return to Spain, because the route he was following they judge to be useless and unprofitable, and because he is unwilling to take the route via the Cape of Good Hope and the Isle of San Lorenzo [that is, via Madagascar and across the Indian Ocean by the Portuguese route] as he told them several times he would before retracting and telling them that he would stick to his planned course; because rather than go by the Cape of Good Hope and San Lorenzo he declared that he would scuttle the ships twice over.

The colorful language attributed here to Magellan is not confirmed elsewhere. Pigafetta's version of the threat is milder: Magellan swore only to prosecute his voyage as far as seventy-five degrees south, "where there is no summer night" (al tempo de la estate non ge è nocte).[28] As usual, however, the survivors' narrative is plausible. We have already seen examples of Magellan's equivocations about whether to hazard the Indian Ocean route if his hopes of finding a strait should fail (above, p. 87, 169). To do so would have been flatly against royal commands: the fleet was specifically barred from making incursions into the zone of navigation assigned by treaty to Portugal.

Whether such orders were intended to be taken seriously, or were included in instructions to Spanish expeditions for the sake merely of protecting the king from accusations of treaty breaking, is another matter. Not all seamen—not, indeed, most servants of the Spanish monarchy once they were out of reach of punishment—took orders from the king *au pied de la lettre*. "I obey orders but do not execute them" was the proverbial response, usually uttered only in mental reservation but universally, tacitly acknowledged. A largely unquestioned legal principle in Spanish tradition was that no obligation of obedience extended to unjust or unlawful orders.[29] Mental elision easily slid the limits of the exemption to orders that were impractical or exorbitant. On board ships, routine injunctions against, for instance, gambling and swearing were treated with the contempt they probably deserved. Issues on which war and peace depended might have commanded greater solemnity, but for most seamen the Cape of Good Hope represented an easy route to quick riches, as long as one could elude Portuguese patrols. It was more important to evade capture than obey orders.

If Magellan refused to take the easy way, therefore, it must—in the mutineers' minds—be out of fear of falling into the hands of the compatriots he had betrayed or, worse, out of secret collusion with Portugal. The interrogator's report on the mutineers continued with a bitter joke that rings true—a play on words on the hopefulness the Cape embodied:

> In their judgment, he was guilty of double dealing. In no way do they attribute good hope to him, and here, they say, we have better hope, on account of the fourteen months' delay while he tarried on southern shores, to say nothing of the evil and perverse accounting and billing rendered by His Highness's officials and captains, whom Magellan took with him, which has left the men shocked.[30]

The mutineers' charges therefore were that Magellan was financially dishonest, professionally incompetent, and politically treacherous. Further details of the version they gave of Magellan's alleged misdemeanors are discernible in other parts of the document. The cause of the discord between Magellan and some of his officers was the result, they claimed, of the captain-general's failure to comply with royal orders. With Cartagena, specifically, he fell out because Magellan failed to consult him with the deference due to a co-commander, appointed as such by the emperor. The accusers recalled or invented a vivid dialogue in which Cartagena demanded to know why Magellan departed from the course his orders laid down. As we have seen (above, p. 138), Magellan may have been prompted by the advisability of eluding Portuguese pursuers, or, more likely, he may have had a prejudice in favor of a southerly crossing point, or he may have been havering toward a decision to follow the Cape route after all; but the mutineers' version of his riposte to Cartagena captures his characteristic arrogance. "I know what I am doing," were words they put in his mouth. "Follow my lead. I do not have to account to you for my actions."[31] Cartagena's attempt to take over the expedition and assassinate his chief was occluded, in the mutineers' narrative by a version in which Magellan started the violence by placing his hand on his rival's breast and saying, "Consider yourself under arrest."

The report of yet another inquiry followed, conducted in Valladolid on October 18, 1522, when testimony was taken from the captain, master, and company of the only other ship to reach home, the *Victoria*. The questions put to the expedition's survivors reveal the mutineers' further complaints against their former commander. Why did he require the arrest of Luis de Mendoza and, when the arrest failed, order him to be killed? Did he promise some reward to the assassin? Why did he maroon Juan de Cartagena and the priest who died with him, and sentence Quesada, Mendoza, and other victims to death? Was it true that he favored Portuguese and members of his own family in promotions? Why did he spend so long in harbors on the South American coast, wasting time and provisions without replenishing supplies? All the mutineers' allegations were self-interested. Some were obviously false, such as that Magellan had not taken steps to restock larders on the South American coast (above, pp. 170–1). Others were tendentious. There was enough truth in them, however, to sway the inquiry in the mutineers' favor, to raise doubts of Magellan's fidelity and reliability, and to warrant further investigation.

The mutineers' strategy was broadly successful. Their sufferings in the strait had been hardly less grueling than the rest of their time under Magellan's command. It had been a serious error to sail straight home without trying to rescue the castaways from San Julián: it prompted suspicion that the mutiny was a product of panic by perpetrators devoid of proper compassion and sense of duty. Nonetheless, by making Gerónimo Guerra their official leader—a relative of the royal factor who was the biggest investor in the enterprise, and who held significant debts from the crown—they secured a favorable hearing for their case. When they reached Seville, Álvaro de Mesquita, the deposed captain of the ship, was instantly jailed, much to the chagrin of Magellan's father-in-law, who, the Sevillan official reported, "demonstrated great force of feeling, saying that Mesquita should be set at liberty and his captors jailed." Although Mesquita was eventually released and rehabilitated, the mutineers incurred little reproof and less punishment. The ringleaders all returned to royal service.

The conclusive point in the mutineers' favor was that events had proved them right: Magellan had chosen the wrong route.

. . .

When they emerged from the strait, most pilots seem to have been too exhausted to write a description. Beyond it, marking the extreme reach of the land, Cabo Pilar, black and desolate, passed unremarked in early accounts, though it rears nearly two thousand feet from the sea. The sight of dory and tunny chasing the shadows of flying fish distracted Pigafetta at the crucial moment.[32] John Narbrough, who drew up sailing directions through the strait for the Royal Navy in 1670, was exceptionally observant. "This Cape," he wrote,

> is a very remarkable land. It is high and steep with peaks on the hills and shows ragged rocks on the tops of the hills with a high couple [of] sugar loaf hill[s] at the north part. The Cape makes [a] bluff [visible] a league and a half north and south to a man's sight when he is four or five leagues [out] at sea. At the north corner of this Cape near the straits there stands [sic] two high peaked rocks, like the Needles at the Isle of Wight, but a great deal higher and bigger. One—the water runs round it; the other is on the edge of the hill. Also there is [sic] two rocks which lies [sic] in the sea, two or three cables length from that needle.

Narbrough's attention returned to sea level and became narrowly focused as danger revived: "There lies [sic] in the sea many craggy peaked rocks and broken ground near two leagues off. The shore [is] very dangerous from this point. . . . I call these rocks The Judges."[33] Magellan called the headland on the north shore "Cape Desired," to contrast the frustrations of the crossing, and the ocean before him he called the "Peaceful Sea" in gratitude after the storm-racked strait. Had he noticed the perils—ominous for the next stage of his journey—that caught Narbrough's eye?

The ill fortune that dogged Magellan's voyage was exceeded in every subsequent attempt to force the strait. Spain's follow-up expedition struggled through in 1526, but it cost four months of harrowing, enfeebling effort and the loss of three ships. The next attempt, in 1535, could not make way in the face of freezing winds, provoking mutiny and the murder of the commander. The strait's already gloomy reputation darkened. The stains of sin and misery deepened, which even the fiercest storms seemed unable to dilute. One vessel in a fleet of three completed the traverse in 1540, but the experience was so costly and laborious as to deter further attempts against the wind. In 1553 an expedition sent from Chile sought to make the journey in the opposite direction, only to run out of provender and patience in wasted, battered vessels. A follow-up flotilla in 1557 was dispersed by a storm: two vessels got lost in the maze of channels at the western end of the strait and never found the entrance; the captain of the third, Juan Fernández Ladrillero, claimed to have made his way to the Atlantic and back, but after all the other disasters, few were willing to believe him. No one tried again until Drake zoomed through in sixteen days in 1578. But he was the beneficiary of freakishly favorable weather, and his achievement proved unrepeatable for the rest of the age of sail. When navigation from the Atlantic to the Pacific resumed in the seventeenth century, seamen favored the storm-fraught route around Cape Horn in preference to Magellan's strait. Not until steamships arrived, with power sufficient to challenge the wind, did the strait become a modest highway of commerce. Even were it not so far away from all exploitable markets, it could not have served commercial interests in the age of sail.

After he had forced a way through, it was for Magellan—perhaps in the sense Elvis Presley's death popularized—"a great career move" to carry on without acknowledging failure. His strait led him to fame but did not lead the world to a new age, or toward globalization, or to substantial new knowledge, or to new commercial initiatives or possibilities. It did not lead Spain to the coveted dominion of the Spice Islands. It was not even a stage in a planned circumnavigation of the world: that plan probably arose only after Magellan's death. Heroic folly often guarantees fame. Had the captain-general succumbed to his pilots' advice and turned back, he could not have become what he became: the central character in his own romantic legend. His recklessness—Magellan's "normal"—grew more risky than ever, for he now had nothing to lose except the chance to conjure fame from failure. From henceforth, he lived without prudence and would die in a spirit of devil-may-care that befitted his life.

It is hard to imagine anything in fact or fiction more striking than the story of the passage of the strait: the fantastic setting, the twists of good and evil fortune that seemed to follow the kinks in the course of the waterway; the navigators' anxieties, the captain-general's ferocity and obduracy; the squalid escape of the mutineers and the inescapability of the force of circumstances that drove the rest of the expedition on. The most surprising feature of the escapade is perhaps that the life of the fleet went on regardless: the business of keeping the vessels afloat, the food stocks replenished, the men fed, the soundings taken—every few yards in channels where the depth of the bottom varied frighteningly. The eerie calm of Pigafetta's reminiscences might be the perfect counterpoint to the harsh realities of the fleet's experience of the strait, were it not for another, largely neglected text that exceeds its cold-bloodedness.

Every day, the phlegmatic pilot Francisco Albo (or whoever was responsible for the text at this point) kept up his relentless log. After thirty leagues

> we went south by southeast. . . . After that we went to southwest a matter of twenty leagues, and there we took the sun and we were in fifty-three and two-thirds degrees of latitude, and from there we returned to the northwest a matter of fifteen leagues, and there we anchored in fifty-three degrees of latitude. . . . After that we went to the northwest.[34]

And so it goes on. Because his job was to provide sailing instructions for the future, Albo noted the whereabouts of shallows and narrows and beaches and bays and the windings and zigzags of the strait; he sketched the landscape with a tiresomely workaday eye—"The chains of mountains are very high and covered with snow, with much forest"—but he never deigned to notice the human dramas unfolding around him. Of vanishing ships, rumored mutinies, or menace-laden exchanges between fearful navigators and their febrile commander he remained wordless, as aloof or unaware as the seals in their shallows and the penguins in their burrows.

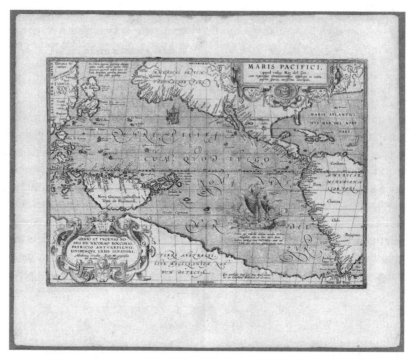

FIGURE 8. "I first encircled Earth with rapid flight/Magellan,my leader, through the new strait you sought/Duly VICTORIA is the name they ought/To call me. Wings for sails, and for the fight/Against the sea, glory's the prize I brought." Abraham Ortelius featured the image of *Victoria* in his map (Antwerp: Vrients, 1589) of "the Pacific Sea"—in Latin, the language of most of Ortelius's annotations—"commonly called," in Spanish, the "Mar de Zur" or South Sea. Tierra del Fuego appears as part of the great "Unknown Southern Continent" that geographers of the time postulated to balance the apparent disparity, which exploration revealed, between land and water on the surface of the Earth. More than three-quarters of a century after his death, Magellan's claims still influenced cartography: notice the unrealistically narrow dimensions of the Pacific, and the mysterious, still unidentified islands Magellan reported: San Pablo (here called Sancti Petri, just north of the Tropic of Capricorn), and Tiburones (or Island of Sharks, a little to the northwest). Courtesy David Rumsey Map Collection.

CHAPTER 8

The Unremitting Wind

The Pacific,
November 1520 to March 1521

Surely the isles shall wait for me, and the ships of Tarshish
first, to bring thy sons from far, their silver and their gold
with them, unto the name of the LORD thy God, and to the
Holy One of Israel, because he hath glorified thee.

—Isaiah 60:19

Although dauntingly vast on modern maps and, for a long time, imper-
meable to shipping from outside, the Pacific is the world's most reliable
ocean. The winds and currents flow more predictably than in the Atlantic, more consistently than in the Indian Ocean. El Niño disrupts the
regularity at odd intervals—typically twice in every decade. But for most
of the time navigable routes score the surface. Fierce but unsurprising
westerlies dominate the northern and southern extremities of the ocean,
south of about forty degrees south and between about thirty and fifty-
five degrees north. On either side of the wide belt of calms that lies north
of the equator, there are fairly constant trade winds, while equatorial
currents provide central avenues of access from east to west. The trade
winds echo those of the Atlantic, swirling in a counterclockwise system.
The northeast trades, blowing between about fifteen and twenty-five
degrees north, are among the world's most trustworthy winds. The
southeast trades, which bore Magellan, stalking a wider corridor, from
the equator to about twenty degrees south, are only slightly less so.

Along the western margins of the ocean, conditions are monsoonal,
with northerly winds in winter, reversed in summer. At the eastern rim,
the Humboldt Current carries ships northward from the frayed, fragmented southwest coast of Chile almost to the equator. It helped push

Magellan, when he emerged from the Atlantic, out of the zone of storms that threatened to drive his ships ashore. The California current, meanwhile, links the coast from southern Oregon to southern Mexico, flowing south along the shores off which the northeast trade winds spring.

It took until the late eighteenth century to decode completely this extensive but essentially simple system and explore the vast spaces between the corridors of wind and current. Magellan unlocked one crucial element: the path of the southeast trades, opening an avenue of access from South America to Southeast Asia. As far as we know, his was the first crossing of the entire span of the ocean. Indigenous mariners made long voyages across parts of the Pacific, but with neither the means nor the incentive to find the routes that could bind the ocean from coast to coast and link it to the rest of the world.[1] Their achievements remained limited, even if they sometimes spanned significant distances. In the Far North, for instance, scholars have postulated unrecorded crossings—most implausible, all unprovable—from China to America, dating back to the second millennium BC.[2] In the central and southern Pacific, by about 1000 BC, colonizers from Asia had reached as far as Fiji, Samoa, and Tonga. In the first millennium AD the Polynesian successors of these precocious argonauts sailed on, along roughly the same course, to Easter Island and perhaps to touch the Pacific coast of the South American cone.[3] Polynesian voyagers' peopling of remote islands demanded astonishing feats of seamanship. Their experience helps make Magellan's journey intelligible.

If you plot the affected islands on a map, the underlying rationale becomes obvious. All the journeys that led to the peopling of the South Pacific were made into the path of the southeast trade winds. Navigators set their prows against the wind—which confirms them as typical explorers of the age of sail (above, p. 41). It tells us something about their nautical technology, too, for the vessels must have been rigged with maneuverable, triangular sails to cope with the conditions. Occasional reversals of the prevailing wind—which are inevitable in any system, however regular—would help them along, without carrying them out of range of return. They sailed until they found something or ran short of provisions, in which case they could get home quickly by reversing the route. The wind that slowed the outward journey would hurry them home.

From the Polynesian heartland the Cook and Society Islands, Tahiti and the Tuamotus, as far as Mangareva, are all on the familiar trajectory, right in the path of vessels heading directly into the wind. Easter Island is a long way further out in the ocean but lies on a continuation of the straight waterway from the heartland through the intervening archipelagos. In further phases of expansion, Polynesians colonized islands as far off the wind as Hawai'i to the north and New Zealand and the Chatham Islands to the south. They did it not by aimless drifting but by purposeful exploration, probing beyond the wind, little by little, and sometimes ending up in islands so remote that it was impossible to maintain contact with the settlers' places of origin.[4]

. . .

The wind system that enabled native navigators to explore the Pacific also enabled Magellan to cross it. Reversing, in effect, the windward routes that the Polynesians pioneered, he picked up the trade winds off the South American coast. They hurtled him northwestwards.

Every consequent experience was surprising. Magellan had no inkling of the vastness of the ocean he had to cross: "more vast," Transylvanus averred, "than the mind of man can conceive" (vastius quam ullum humanum ingenium caperet)[5]—an exaggeration perhaps pardonable to the first minds obliged to modify it. Magellan could never have expected to sail so fast: there were days when pilots supposedly reckoned the fleet made up to seventy leagues[6]—probably over two hundred nautical miles, "nor were they ever able to depart from it" owing to "the force of the winds and gales."[7] Nor can Magellan have expected the paradoxical results. The very speed of the voyage emphasized the scale of the odyssey.

The size of the ocean condemned Magellan and his men to unprecedented physical and mental suffering: an almost intolerably long nonstop voyage of nearly four months without making land; the gnawings of starvation; the tortures of scurvy. The urgings of the wind magnified the misery. That is hard to understand now, when seagoers love the exhilaration of a speedy cruise, the feel of the wind at the helmsman's back and the sense of propulsion toward an ever-closer destination: these are among the pleasures of modern yachting. For Magellan and his men, there were no such consolations. Home receded, horrors multiplied, and anxieties quickened, while the force of the unremitting wind, sustained for months on end, menaced them with every seaman's greatest fear: that even if they survived hunger and disease or the menace of some unknown, looming reef or shoal or monster, they would never find a wind to take them back.

The story of the voyage is a kind of counterpositive to the usual sea saga, in which the enmity of the elements takes uncomplicated forms:

destructive storms, as in the Sinbad tradition, or deadly calms, such as befell the Ancient Mariner, or monstrous foes, like those of the leviathans that swallowed Jonah and disabled Ahab. Magellan's fleet more resembled the *Flying Dutchman,* condemned to sail for what must have seemed like forever without making port. The benignity of the weather was like a villain's smile. The following wind self-transformed from blessing to curse.

. . .

The start seemed good. West of Isla Carlos III, the Strait of Magellan is like a long, oppressively overshadowed corridor between dark, jagged cliffs. The sense of relief when you emerge into the broad channel, with the ocean before you, is almost irresistible. "Everyone was jubilant," as Ginés de Mafra exclaimed,[8] but the moment of release struck different men at different times: the Genoese pilot on the twenty-sixth of November and Pigafetta not until the twenty-eighth, while the narrator whose account Ramusio gathered felt they left the strait on the twenty-seventh.[9]

They were fifty-two degrees south of the equator, in the southern high summer. But the ocean that faced them was not yet "the Pacific." Even when the mariners gave it that new name, they did not necessarily think of the sea as previously unknown: Balboa had called it the "Mar del Sur"—the South Sea—and thought it was merely a "great gulf" of the Indian Ocean.[10] Although the length of South America exceeded Magellan's expectations, there is no reason to think he had changed his conception of it (above, p. 195) as a vast promontory of Asia. He and such of his men as believed him, therefore, thought they were nearing their objective.

At first, in any case, the Pacific did not seem pacific. The sea still heaved with hostility. The strong westerlies still prevailed, as they do as far north as forty degrees south. The fleet was on a lee shore that was uncharted but obviously dangerous, threaded with shoals and studded with rocks. The question of whether to turn west at once seems to have arisen, at least in some minds. Pigafetta guessed that had they done so, "we would have circumnavigated the world without finding other land" (haveresemo dato una volta al mondo senza trovare terra niuna):[11] he was right, though how the aperçu came to him is baffling. The current, however, drove northwards and it would make no sense to try to make way to seaward, except for the minimal requirements of safety, in the face of the wind.

So Magellan went with the flow. Or was he making, as some students suppose, for a quick run to the equator, from where he would turn to seek the Moluccas, which, as he knew, lay on the line?[12] Even if he was genuinely aiming for the Moluccas, it would not seem wise to lengthen the voyage by sailing along two sides of a triangle. Once Columbus pioneered the wind-riding methods that all subsequent oceangoing explorers from Europe followed, it made sense always to take advantage of winds and currents that took one roughly in the right direction, if one were lucky enough to find them; for "the wind bloweth where it listeth, and thou hearest the sound thereof, but canst not tell whence it cometh, and whither it goeth." In the event, the northward course suited Magellan. He had no difficulty in keeping the ships safely offshore—but with the added sense of security that comes from being in sight of land. The sea grew quieter the further north they went. It began to lull the expedition into complacency. "In truth," remarked Pigafetta, "it is very pacific," for in all the time they spent there "we did not suffer any storm."[13]

From December 1, as Albo's log records, the route remained northerly on average until the seventeenth of the month,[14] in the low or midthirties latitude south, when Magellan called for a course that wavered between north and northwest, trying to edge tentatively from the shore. From the eighteenth, land was out of sight. The moment was dramatic. Most of the crewmen whom the Portuguese later captured and interrogated recalled the way the fleet seemed suddenly to veer to the west, maintaining a fairly steady course to the northwest from that time on.[15] The wind dictated the direction of sail. It blew unceasingly. Herrera surely called on real seamen's experiences when he reported that "for days on end the mariners had neither to touch the tiller nor trim the sails, as if that vast ocean on which they sailed were a canal or a gentle river."[16]

It sounds idyllic, but the loneliness of the fleet, "in the midst of that open expanse," as Pigafetta kept saying, was unsettling. The crews were like the sailors in the Irving Berlin song who only "saw the sea." The infrequency with which they saw anything else seems uncanny. Even fish, for which Pigafetta was always on the lookout, were scarce, human life nonexistent. The rarity of sightings even of desert islands is remarkable and is the best guide to the course, which eluded all the archipelagoes that dot the ocean, including the Marquesas, the Gilbert and Ellis and Marshall Islands. It seems possible, as we shall see, that the fleet passed to the east of all of them. More than a month went by before the explorers caught their first glimpse of an island—at between fifteen and nineteen degrees south, according to rival pilots' reckonings or recall.[17]

Historians have wasted prodigious efforts (to which we shall return in a moment) in trying to identify the island and using it to fix the expedition's location and so elucidate its course. Magellan called the landfall San Pablo, presumably in honor of the feast of the conversion of St. Paul on January 25, as the wind bore them beyond it with no chance of anchoring. Another island—or what looked like one—appeared in the distance a few days later in shark-filled waters. They were at nearly eleven degrees south by Albo's reckoning—but in another entry in his log he places it at nine degrees of latitude beyond San Pablo;[18] so this is a case where his data cannot be trusted. Other estimates wavered so wildly—from nine to fifteen degrees[19]—that it is inadvisable to rely on any of them. The crew called the pair the Unfortunate Islands, perhaps in regretful allusion to the Fortunate Isles, the mythical paradise in the Atlantic that, in some ancient and medieval traditions, gave its name to what are more commonly called the Canaries. The name reveals the return or persistence of unquiet minds aboard Magellan's ships. Conditions would not allow the explorers to approach or land.

They were prisoners of the wind. And sickness had begun to kill: the tally of the dead recorded the first victim on December 23. Since then, five others had died, their names laconically annotated as dead of an unspecified malady. Additionally, Pigafetta had lost the companionship of Pablo, the Patagonian giant, who, when he fell ill, "asked for the cross and embraced it and kissed it many times" (domandò la croce abrassandola et basandola molto).[20]

. . .

It is fair to say they were lost, on an unexpected sea.

The discrepancies in the pilots' calculations of latitude are clues to the uncertainties of navigation in an unfamiliar environment, under a pole "not so starry," as Pigafetta put it, as in northern seas. The compasses' magnetism weakened. The absence of a star as reliable as the Pole Star drove pilots to choose, at variance with one another, to take their readings from celestial bodies that proved coy or skittish—hiding behind clouds or oscillating in the heavens. For readings of latitude, Francisco Albo's log was based, as we have seen, exclusively on sightings of the sun; so it is tempting to accept most of his figures as the best possible in the circumstances. Since, however, better-instructed pilots disagreed with him at almost every moment of the Pacific crossing, his contributions were not definitive.

Longitude remained the province of guesswork. On December 23, Andrés de San Martín attempted another of his readings of longitude by the lunar distance method. The result is unrecorded, but, like all such endeavors aboard a sailing ship, other than in a flat calm, it is unlikely to have been illuminating. The only other available method of testing for longitude was little better than guesswork: navigators could attempt to estimate the speed of the ship by timing their passage past a bit of jetsam or a feature of a coast or a rock. No accurate chronometry was available. So they had to count the seconds in their heads or use links in a chain, like rosary beads, to tell the tally. Maybe long experience taught them to do the job well. But they still had to extrapolate to get an average speed over a representative length of time in order to work out how far a ship might have sailed. Even in the Pacific, with constant winds and a steady course, it was impossible to be accurate because undetected currents and sudden hazards pushed ships off track, while magnetic variation, which could not be accurately measured, meant that the true course might waver from where the compass indicated. Pancaldo's log complains that his needles' fluctuations varied by as much as two quarters.[21]

An anecdote of Pigafetta's captures the prevailing uncertainty. "While we were in that open expanse," he wrote, Magellan accused the pilots of departing from the course he had laid down. They all denied it. "He answered them that they were pointing wrongly, which was a fact," and suggested that they were not making enough allowance for leeway and magnetic variation.[22] The experts aboard could not agree on their own whereabouts. There is little point in trying to plot their course today with a precision they could not attain or even imagine.

The effort to do so began in the Venetian workshop of the cartographer Battista Agnese in the 1540s. He produced dozens of magnificently drawn and decorated atlases, all slightly different, but all privileging Magellan's voyage and dedicating to it a world map in which the explorer's supposed route is picked out with a thread of silver leaf. Agnese's data on the voyage were good and his depiction honest. He omitted, for instance, the still-unexplored chunk of South American coast north of Magellan's reach. His depiction of the Strait of Magellan, though unfeatured in detail, follows descriptions that survive in sources from participants in the voyage. He shows what we now call the Southern Ocean as Pigafetta imagined it—circling the world like the rim of a bowl, with no intervening landmass of the sort with which most other mapmakers capped the Far South. He shows the Moluccas very prominently, just on the Spanish side of the Tordesillas antimeridian. He uses, in most of his maps, the name, rendered as "Latrones," that Magellan's men gave to the Marianas. The only other islands he depicts in the Pacific are labeled with the names of St. Paul (Isola San Pablo) and

Sharks (Tiburones). But the line that wobbles and waves across the ocean is evidently schematic and not meant to represent the real course, any more than the London Underground or New York Metro map is meant to show the real route of the railway lines between the stations, or a medieval portulan chart was meant to show distances to scale.

Historians more ambitious and less prudent than Agnese have tried to do better. Their consensus draws Magellan's course west of the Marquesas. The thesis is obviously problematic: it invites speculators to identify San Pablo and Tiburones with too many possible islands, all of which seem in the wrong place. San Pablo must have been a fairly substantial lump of land. It was first sighted on January 23 or 24. Unless Magellan named it for the eve of the next day's feast, it remained in sight until the twenty-fifth. Perhaps a day or so was wasted in a search for an anchorage: Albo said they took soundings but could find nowhere to land. The island must in any case have been fairly conspicuous—no barren islet, as Albo says it was well wooded.[23] Pitcairn has its advocates but seems much too far south.

Perhaps San Pablo and the shark-thronged island correspond with one or two of the Tuamotu group or some combination of the atolls that form the Malden and Line Islands, such as Vostok or Kiritimati or Caroline or Flint. All of these have their advocates, but the problem of how easy low-lying atolls are to sight in varying conditions make certainty impossible. In any case, they all seem too far north for the fleet to have reached them by the last week of January, and perhaps too far west for what had so far been a mainly northwesterly course since the fleet left the Chilean coast. They lie, moreover, in regions where islands are numerous, and where a navigator would be extraordinarily unlucky to have seen only two.

On the presumption that Albo's log can be discarded as fraudulent or as a late confection, made to deceive, claims have been advanced for Clipperton Island—barely a bump in the ocean—and Clarión, an outlier of the Revillagigedo Isles, well north of the equator;[24] these are far off any course reasonably inferred either from the surviving sources or from the effects of the winds. If Magellan sighted them, he must have been pursuing a track incompatible with the location of the Moluccas as specified in his memorandum of September 1519 (above, p. 79). He may have assured royal officials that the Moluccas were, relatively speaking, "not very far from" Spain's American colonies,[25] but he did not think they were so near as to make it worthwhile to hug the American coast before turning west in search of them.

As candidates for identification with San Pablo, islands in the Gambier group, which are south of the Tuamotus, but not so far as to be disqualified, fulfill most of the criteria. So—except that they are inconspicuously low coral island and atolls—do those along the eastern edge of the Disappointment Islands, such as Reao and Puka Puka, which are inside or almost inside the range of latitudes the pilots proposed for San Pablo. But unless Magellan made an unrecorded change of course after passing the Gambiers or the Disappointment Islands it is hard to see how he could have avoided all but one of the many isles and atolls to their north.

It is possible, therefore, that Magellan was in sight of two of the Marquesas Islands at the material time and even perhaps to their east. If his subsequent course took him north and east of the Line Islands, the absence of further landfalls is intelligible. A further reflection supports the suspicion: since he hoped and expected the ocean to be narrow, Magellan would have been ill advised to steer too far west, running the risk of overshooting his objective and trespassing on the exclusive domain of the Portuguese. It is also worth considering how the pilots allowed for leeway and variation. Especially after Magellan upbraided them for making insufficient provision, they may have overcompensated. A recorded course to the northwest may really have been closer to north by northwest. The unknowables are too many to justify unqualified assertions. We must abjure certainty and rely on imagination disciplined by the sources. Mine pictures Magellan just east of the Marquesas.

. . .

By the time the fleet reached the vicinity of the equator in the second week of February, at least two more dead had been cast overboard. Despite leaving the normal limits of the trade winds, the ships were still making good headway. There was an opportunity, however, to turn west with the South Equatorial Current for a current-assisted passage straight to the Moluccas. Why did Magellan not take it? The simplest explanation is that he missed it, and sailed on, without changing his northwesterly course, until he found the North Equatorial Current, in the last days of February,[26] having skirted the Marshall Islands without seeing them. In other words, he took the opportunity to swing westward when he could, and the delay was undesigned. But that seems unlikely, because he knew the Moluccas straddled the equator: he had said as much himself, in the memo he left for the king of Spain before setting sail. The fastest route, therefore, would have involved turning west as he approached the equator, rather than waiting until he had reached or passed it.

The Philippines, however, beckoned (above, p. 93). In any case, as a Moluccas-bound enterprise, Magellan's expedition was already a failure, and there was not much point in persevering with it, even had the commander been genuinely committed to it. Failure seemed encoded in the faulty preparations in Seville, prefigured in the quarrels and conspiracies in the Atlantic, blazoned from the gibbet at San Julián, threatened in the diminishing stores and gathering storms, preordained in the mutinies, defections, and deaths, obvious in the oncoming sickness. Above all, failure had become ever more clearly inescapable as the voyage lengthened. Even before Magellan entered the strait that led out of the Atlantic, he had taken an unconscionably long time to establish a laborious route that was commercially unviable and glaringly inferior to the Portuguese course to the Moluccas. When he got to the equator, therefore, it was reasonable—albeit contrary to the king's commands—to try a different expedient.

Leone Pancaldo thought his leader was deterred by the fear that the Moluccas would afford no fresh provisions:[27] maybe Magellan spun such a yarn, but he knew enough about the islands to realize that it was incompatible with the facts. Other explanations, which occurred to contemporaries or historians, are equally improbable. Magellan can hardly have thought he was already beyond the Tordesillas antimeridian, in Portuguese waters, as Portuguese propagandists later claimed, or that he had overshot the Moluccas:[28] even if either speculation were right, it would not explain his northward trajectory. That he might have been seeking Cipangu, the legendary land so named by Marco Polo and subsequently identified as Japan, seems a wild suggestion: Pigafetta, who was privy to Magellan's thoughts, believed they had already passed it, twenty degrees south of the equator.[29] Even by the standards of the imprecise geography of the time, the whereabouts of Cipangu were too vague to make it a practical target: organizers of an expedition in Magellan's wake in 1525 seemed to share Pigafetta's expectations of finding it well to the south of Japan's real position. They thought the fabled island might make a suitable station on the way to the Moluccas and so ease the path across the Pacific.[30]

The only clue to why Pigafetta placed the island so strangely far south is his conviction that he had also passed another legendary island, which he called Sumbdit Pradit, a name unexemplified in any other known source.[31] The ingenious—sometimes excessively ingenious but unquestionably brilliant—G. E. Nunn suggested nearly ninety years ago that the name is in part a corruption of "Serendib": a fabulous island "of gold" long renowned in Arab and Indian literature and often identified with Sri Lanka. The term *serendipity* in English, which Horace Walpole devised to mean a fortuitous and surprising discovery, derives from it. Nunn suggested that Pigafetta's term combined *Serendib* with *prada*, meaning "silver" in Malay languages.[32] The "Isles of Gold and Silver" persisted in myth and crop up speculatively in European maps of the Sea of Japan and surrounding waters in the sixteenth and seventeenth centuries.[33]

If Nunn's guess is right, Pigafetta and *a fortiori* Magellan were using information that came in part from Malay works or maps: that is entirely possible. As we have seen (above, p. 111), Francisco Rodrigues was among mapmakers who drew on Javanese originals while Magellan was in the East. There were plenty of opportunities to acquire Malay lore in Malacca, where Magellan lived. He had Enrique's knowledge at his disposal. His correspondent, Francisco Serrão, was in touch with Malay merchants and mariners by virtue of his residence in the Moluccas. The name of Cipangu was of course unknown to Malays, but the association with silver—a product that Marco Polo had linked with Cipangu and that was indeed a major export of Japan—was probably enough to suggest the connection in educated European minds.[34]

Finally, among his reasons for continuing northward, Magellan may have had mainland Asia in mind as a default destination if he made no prior landfall: Pigafetta says that the fleet's turn to the west, when at last the commander authorized it, was "so that we might approach nearer to the land of Cape Catigara" (per apropinquarse piú a la tera de Gaticara),[35] an ancient name for a Southeast Asian promontory of uncertain location; but there is no warrant in logic or the sources for thinking that it was Magellan's objective. In any case, he stuck to a northward course until at last deciding to turn west, in the last days of February, when he was, according to his pilots' estimates, between twelve and fourteen degrees of latitude north: again, the likelihood that he was thinking of the Philippines is hard to resist.

. . .

The commander's willingness to protract the voyage must have been for what, in his own mind, at least, was a good reason, because the plight to which scurvy and starvation had reduced his crews was obvious and urgent. Pigafetta's description may have been exaggerated to enhance the magnitude of his own martyrdom, the extent of his sacrifices, the miracle of his survival, the resolve of his captain, or the fortitude of the company. It is famous—one of the most quoted passages, perhaps, in the history of travel literature. But it is so vivid and convincing as to deserve another airing:

> We ate biscuit, which was no longer biscuit, but powder of biscuits swarming with worms, for they had eaten the good (it stank strongly of the urine of rats). We drank yellow water that had been putrid for many days. We also ate some ox hides that covered the top of the mainyard to prevent the yard from chaffing the shrouds, and which had become excessively hard because of the sun, rain, and wind. We left them in the sea for four or five days, and then placed them for a few moments on top of the embers, and so ate them; and often we ate sawdust from the boards. Rats were sold for one half-ducat apiece, if only one could get them.[36]

The line about the oxhide sheathing is especially suggestive, since the prospect of having nothing else to eat had been one of Magellan's threats to men whose morale had seemed to falter in the strait (above, p. 195). Gums swollen with scurvy would have made the horror of the menu intolerable. Pigafetta invoked horror again by describing the cannibalistic fantasies that overtook the starving and the scurvy-stricken. When at last the expedition reached land, "before we landed, some of our sick men begged us if we should kill any man or woman to bring the entrails to them, as they would recover immediately."[37]

Was the story just a flesh-creeping flourish? It seems hard to reconcile with what we have seen (above, p. 148) of how Tupi cannibalism, like that of other victim-peoples of European imperialism, served as a criterion for stripping communities of the protection of natural law or denying them the consideration due to fellow humans. Yet the notion that cannibalized innards might save scurvy sufferers' lives sounds scientifically verifiable and had some justification in European jurisprudence. Tripe of any species tends to require a lot of cooking to be masticable, but men too sick to chew could have sucked at the half-digested contents of human entrails to extract the nourishment. Plant food, relatively rich in vitamin C, and otherwise inaccessible to the patients, collects in the digestive tract to be slowly broken down and absorbed. The image of crewmen gorging on bloodily wrested human bowels may seem unappetizing and calculated to *épater les bourgeois*. Human flesh and blood, however, featured regularly in the European pharmacopoeia.[38] Seamen, moreover, knew "the custom of the sea." Necessity hath no law, and it was a well-established principle that, as long as the victims died willingly or accidentally, castaways and shipwreck victims could eat their own dead.[39] The moral context of Pigafetta's fellow travelers' cannibal appetite may seem less pardonable, but they were not asking their comrades to kill for their sustenance—only to bring them the spoils of victims of war, who would be dead, and therefore available for innocent cannibalization in any case. Gustatory cannibalism was taboo. Life-saving cannibalism was sacrifice.

Still, it is hard to find Pigafetta innocent of sensationalism. His description of the effects of scurvy seems restrained by comparison with his treatment of anthropophagy: "The gums of both the lower and upper teeth of some of our men swelled, so that they could not eat under any circumstances and therefore died. . . . Few remained well." The author was among them. "I, by the grace of God, suffered no sickness."[40] He mentioned his exemption, I suppose, as further evidence of God's favor to him (above, p. 131) and of his own miraculous survival and sanctification. Material circumstances played a part. His ship, as we have seen, was stocked with the life-saving foodstuffs—including the dried and sugared fruits and that precious little bottle of quince preserve (above, p. 111)—which elevated the diet of officers and privileged supernumeraries and may have had some antiscorbutic virtues.

. . .

The men were weak in body and subverted in morale by the time they made land. It was March 6, 1521, when they discerned two islands to the west: Guam and Rota, at the southernmost reach of what we now call the Marianas but that Magellan at first dubbed the Lateen Islands ("Islas de Velas latinas") because of the shapes of the sails of the shallow craft that came to meet him. Albo interrupted his usual featureless list of readings of latitude by describing the scene: two islands, "not very big" (no muy grandes), between which the fleet swung south and west, leaving each on either hand. Pigafetta, perhaps misled by the twin peaks of Rota, spoke of three islands, but no other can really have been in sight. Except between Guam and Rota, no approach was practical for the fleet or consistent with the evidence.[41] Ginés de Mafra recalled the lookout's cry, "¡Tierra!" Land ahoy! The wild joy that ensued overturned normal standards. "The men all cheered so madly that anyone showing restraint was deemed insane."[42]

"And thus," Albo continued,

We saw many small sails that were coming toward us, and they made so much headway that we thought they seemed to fly, and they had triangular woven sails, and they could sail equally well ahead or astern, making of the bow what had been the stern and vice versa at will.[43]

The islanders seem to have found the encounter less surprising than the mariners. The Marianas screen the Philippines to the east, at a far remove: Guam is over 1,500 miles from Manila. Unsurprisingly, therefore, little or no information about the archipelago or its inhabitants had reached Europe or the Portuguese outposts in Asia. Magellan, in consequence, was unprepared. On the other hand, the Marianas were not as isolated as Pigafetta supposed: the natives thought, he inferred, "from the signs which they made that there were no other people in the world but themselves."[44] Their trading relationships seem hardly to have extended beyond the Caroline Islands; yet, by transmission across intervening emporia from the Moluccas, rumors of Europeans and of what they and their ships were like had already alerted the inhabitants to what to expect.

The benefits of the stranger effect (above, pp. 142, 195) were unavailable. Among the Tupi of Cabo Santo Agostinho, Magellan had confronted people who were already acquainted with French and Portuguese visitors but whose cultural prejudices in favor of strangers had not yet been dispelled by familiarity or contempt. The giants of Patagonia had not yet received enough outsiders or seen enough European misbehavior to erode their inclination to be welcoming. The inhabitants of Guam knew Europeans only by report. Either that was enough to make them hostile, or they had no existing disposition to regard strangerhood favorably.

According to Pigafetta, Magellan hoped to land and acquire fresh food and water. Natives forestalled him by coming fearlessly alongside his ships in their little masted dugouts with outriggers, using their long paddles as tillers, as agile as dolphins.[45] Whether or not their initial intent was friendly, they were quick to take advantage of the starved and sickly torpor of the crew. Almost as soon as the fleet dropped anchor and before the mariners had a chance to furl the sails, natives severed the towrope of the flagship's little brig and paddled home with their prize. Grapplers came aboard and seized any goods they found to hand: "They sought us," said Albo, "in order to steal whatever they could from us" (Nos buscaban para hurtarnos cuanto podían).[46]

For a while, Magellan's men seem to have been able to do nothing but gape in astonishment at the boarding party's effrontery. "We could not protect ourselves," Pigafetta admitted, and, as Pancaldo confessed, "took no precautions."[47] The impression Pigafetta derived was that "those people are poor, but ingenious and very thievish," earning the archipelago the sailors' nickname "the Isles of Thieves" (Islas de los

Ladrones), which caught on with mapmakers and served as the official name until the 1660s. The shock of the thieves' success dispelled some of the crew's lassitude. Ginés de Mafra recalled a bout of fisticuffs on deck, to which the natives responded by throwing spears.[48] A few discharged firearms finally drove the interlopers away with their loot.

Magellan responded as usual: decisive and ruthless. The pillagers' presumption, in the words of the account Ramusio published, "cost them very dear."[49] The commander mustered able-bodied men—about forty of them, said Pigafetta—for a punitive raid. It gave Pigafetta, who joined it, an opportunity for hurried but surprisingly detailed ethnographic observations.

Some of his remarks served to justify his commander's violent retaliation. The natives lived, he said, "each one . . . according to his will." That is, they did not observe natural law and so did not qualify for protection under it. "They have no lords." They did not, therefore possess true sovereignty. The men's hair—tangled and long—and the women's, loose and dangling, betokened savagery. "They worship nothing"—which suggested to European minds a deficiency of reason and an offense against nature, but which also raised the prospect of easy conversion to Christianity. They all went naked, except that the women "cover their privies with a narrow strip" of the pith of the palm tree, and the men "wear small palm-leaf hats, as do the Albanians" (portano capeleti de palma como li Albanezi).[50] The comparison seems astonishing and makes one wonder what Pigafetta could have had in mind. Unless an error of transcription is responsible, he must have been alluding to the people of the Adriatic who still bear the same name or something like it in most European languages. They were familiar to Pigafetta, who, as we have seen (above, p. 130) knew the eastern Mediterranean well, where similar hats were widespread headgear since classical times, familiar to any scrutineer of pictures of ethnic types and regional attire. Their hats at the time were typically of wool, but their conical shape and the fact that they were worn close to the skull may have been the points of resemblance that prompted the remark.

If the Chamorros—as the people of Guam are called—failed to qualify for civilization by Pigafetta's standards, he found them praiseworthy in other ways. They looked the part of fully rational humans, "as tall as we, and well built, . . . olive-skinned but . . . born white." The women were "beautiful, delicately formed, and whiter than the men." Some of the natives' technology impressed Pigafetta, especially their boat building, which produced craft capable of darting in and out among Magellan's fleets ships and boats. He admired their houses, "all built of wood covered with planks," with wood-framed windows, beams for floors, and palm thatch for roofing. "The rooms and beds are all furnished with the most beautiful palm leaf mats." Women wove the same material into baskets and other housewares. "They sleep on palm straw, which is very soft and fine." They had an aesthetic sense of sorts, "staining their teeth red and black, for they think that is most beautiful," and anointing their bodies and heads with coconut and sesame oil.[51] To Renaissance men, these unctuous good manners were evidence of civilizability and perhaps of a predisposition to Christianization, reminiscent of Jewish customs described in the Bible and practiced on Christ in the house of Simon the Levite.

. . .

Despite forewarnings, Spanish military technology took the natives by surprise when battle was joined. "When," explained Pigafetta, "we wounded any of those people with our crossbows-shafts, which passed completely through their loins from one side to the other, they looking at it pulled on the shaft now on this side and now on that side, and then drew it out, with great astonishment, and so died,"[52] presumably in unawareness that by extracting the arrows they would unstem the flow of blood. "Others who were wounded in the breast did the same" (Et altri che erano feriti nel peto facevano el simile).[53] The natives' only weapon, according to Pigafetta, was "a kind of spear, pointed with a fishbone at the tip."[54] Clearly, Spaniards' crossbow bolts made deadlier missiles, and their steel blades were probably more efficient in close combat: the fleet carried one thousand lances and two hundred pikes.[55]

Firepower played no part in battle. Guns were only for show: they were not practical weapons in tropical climes and distant frontiers.[56] The volley that frightened marauders off the decks when the fleet first anchored at Guam was the sort of display they were good for; but early sixteenth-century handguns were cumbersome and inefficient. By the time an arquebusier had primed the pan, rammed shot down the muzzle, kindled a flame, and allowed the wick to burn down, a shock-armed assailant was likely to be upon him. Even if the shot discharged successfully, rather than exploding in the marksman's face, it was unlikely to hit anything as it splayed out of a smooth-bored barrel. Powder was volatile and would not function if damp. Once shot was fired it could not be replaced except by manufacture on the spot: Magellan's, like all Spanish expeditions, carried molds and lead for the purpose.[57] Powder was a wasting asset. The myth that technology is always progressive has occluded the uselessness of firepower on land in the establishment of early modern European empires.[58] It took generations of tactical change and technical improvement before firearms became efficient.[59]

In any case, Magellan's expedition was poorly equipped for battle on land, with only fifty-nine or sixty crossbows and fifty handguns to go round.[60] The numbers reflect, no doubt, those of men competent to handle these specialized weapons. There were no professional soldiers on board and only a handful of trained gunners. Some fairly large pieces of artillery were aboard the ships for engagement at sea with pirates or rivals. Most probably functioned as antipersonnel weapons or to disable ships' rigging. When the fleet set out, before the defection of *San Antonio,* which carried the largest share of equipment of all kinds, and the loss of whatever was irretrievable from *Santiago,* there were fifty-eight demiculverins capable of firing a shot perhaps of up to about nine pounds in weight; seven small-bore, long-barreled *falcones,* which fired shot weighing around three or four pounds; and, for lobbing larger balls, three short, fat-barreled lombards, bound with hoops of iron around cast barrels that tended to split. There were also three of the large sixteen-pounders known as *pasamuros,* or wall-busters. How many large guns belonged to the ships is unknown, as is the number of specialized deck-mounted pieces for repelling boarders.[61] The Portuguese factor in Seville told his king that there were eighty big guns, but that was probably a spy's exaggeration, designed to give his data more impact. *San Antonio* might have been big enough to carry a complement of up to about fifteen cannon, but in any case there was no means of deploying them on land, and the guns had vanished with the deserters over an Atlantic horizon. Most—perhaps all—of Magellan's guns were made of iron rather than bronze, which was more durable and efficient; though the proportions are unknown, some of the shot was of stone, which tended to shatter on impact and do little damage. In short, his firepower was negligible.

When the task force went ashore, Magellan "in great wrath" deployed his usual strategy: setting fire to a village and burning "some forty or fifty houses, together with many boats." Pigafetta expressed compassion for the native dead, condemned by their own bafflement when they tore crossbow bolts uncomprehendingly from their wounds, and for the weeping widows whom he saw gather in boats around the fleet and "who were crying out and tearing their hair, for love, I believe, of those whom we had killed" (gridare et scapigliarse, credo per amorede li suoi morti).[62] Magellan was not compassionately inclined. He had responded with similar terror tactics to supposed affronts from natives in Patagonia and was to try the same method in the Philippines.

At first glance, his behavior seems to instantiate the "Black Legend" of Spanish cruelty or, in this case, that of a naturalized Spaniard. Common sense ought to abjure the legend. Spaniards are no more morally inferior by virtue of being Spaniards than any other people by virtue of provenance. To assert otherwise is blatant racism. Spanish rule throughout Spain's global monarchy was—despite commonly adverse Anglo-Saxon attitudes—relatively less malign than most other European empires: "Si monumentum requiris, cirumspice." Native peoples survived and often prospered under the Spanish crown and maintain to this day their identities, traditions, languages, and numbers in most areas of former Spanish rule on the American mainland, whereas those in former British colonies were expelled or exterminated. The difference, however, has nothing to do with the respective moral qualities of Spaniards or Anglos. Spanish imperialism operated mainly in environments where native labor was precious—too precious for genocide—and the continuation of the existing, precontact economies was vital for the colonists' survival: too vital to disrupt. English settlements operated by and large in regions where natives were of little economic importance or utility and could be dispensed with in favor of European colonists or imported slaves.

Though Magellan has escaped the obloquy ignorantly lavished on other conquistadores in Spanish service, he was among the most ruthless of his kind. His apparently unreflective recourse to massacre, mayhem, and arson should not, however, be the subject of adverse moral judgments by people who live in other times and face different dangers. Terror is the tactic of the terrified. Almost everywhere Spaniards and Portuguese went in the world, they were outnumbered when and if hostilities started, and frighteningly ill equipped. They were far from home—over 13,500 kilometers as the crow flies in the case of Magellan in Guam—without hope of reinforcement, in dauntingly unfamiliar environments, where strange sicknesses threatened and unfamiliar food and drink tortured their bodies. Cultures they deemed savage surrounded and menaced them. They were often, in effect (though not in Guam), the prisoners of native "allies" who outnumbered them and who could oblige them to collaborate in traditional violence or in the creation of enemy-free utopias. In Patagonia and Guam, Magellan came not as a conqueror but rather as a supplicant or extortionist, in urgent need of supplies. He could not dally for protracted negotiations. Nor could he allow host communities to dominate him or, as in Guam, impose one-sided terms of exchange with impunity. His tactics were vile but rational. Terror was his best recourse.

It worked. Magellan retrieved the brig the natives had commandeered. He also obtained food—forty or fifty boatloads, according to Pancaldo,[63] perhaps in part by seizing it but evidently by intimidation or negotiating from strength. On March 9, swift departure seemed in order. According to Pigafetta, more than a hundred native craft saw the fleet off, following and teasing the ships "for more than one league," circling and outpacing ships under full sail. The party atmosphere, evidently, was not to honor the explorers but to fête their flight. Natives gestured with offers of fish as the fleet departed, "feigning that they would give them to us, but then threw stones at us and fled" (con simulatione de darnello, ma trahevano saxi et poi fugivano).[64] Was a biblical allusion in Pigafetta's mind? "Which of you," Christ asked his followers, "when a child asks for bread will give him a stone, or when he asks for a fish will give him a serpent?"

When the fleet skedaddled, prows turned, according to Albo, west and by south.[65] The nod in a southward direction was not an attempt or even a gesture toward recovering the latitude of the Moluccas. If Magellan really believed the Spice Islands were in the Spanish zone, he should have taken a sharper southerly course, as he had already traveled farther west, and closer to the Portuguese concession, than he had expected. Nor, obviously, does the slight shift south suggest an attempt to reach the Asian mainland. So what was the purpose behind the new course? Magellan was evidently following directions Chamorros gave him for reaching Yap or Palau, places with which they were in frequent touch. It is not usually prudent to follow directions from informants whose main objective is to be rid of you or send you to perdition, but Magellan needed any help he could get. To judge from recollections Maximilianus Transylvanus gleaned from the expedition's survivors, when they got home a year and a half later, Magellan met useful informants somewhere out at sea beyond Guam, in a canoe off a couple of desert islands—presumably, outliers of Yap. Communicating "by gestures and signs, as if the dumb were talking with the dumb," the explorers "asked . . . where they could get provisions, of which they were in great want." Their interlocutors directed them to "an island not far off, which was called Selani, which they almost showed with their finger [*quam et fere digito ostendebant*], and that it was inhabited, and that an abundance of everything necessary for life was to be found there."[66] A storm prevented them from finding "Selani." But they were now, at last, on the right road for the Philippines.

. . .

When Balboa crossed the Isthmus of Panama in 1513 and beheld his "South Sea," he assumed that it was no more than a narrow arm of the Indian Ocean, separating him from Asia. European navigators continued to underestimate its extent. Magellan would surely not have attempted the crossing had he known how wide the Pacific is. Pigafetta was uncharacteristically realistic in affirming, after traversing it, that the sea was too vast to cross safely; but he weakened the plausibility of his account by invoking miracles to explain the feat. "Had not God and His blessed mother given us such good weather we would all have died of hunger," he wrote, "in that exceedingly vast sea." He added another of his usual touches: a bit of boasting. "In truth I believe no such voyage will ever be made again."[67]

The effect was further to undermine his credibility and to dare successors to try it. They all underestimated the extent of the Pacific. Alvaro de Mendaña in 1595, making his second crossing, by which time he ought to have known better, mistook the Marquesas Islands for the Solomons, four thousand miles to the west. Even Pedro Fernandes de Quirós in the early seventeenth century, who saw more of the Pacific than any other man of his day, tended to log shorter distances than he actually sailed. In most atlases of the period, the Pacific looks little or no wider than the Atlantic. The vastness of the ocean defied credulity.

This blinkered vision represented the triumph of hope over experience. Early explorers, needing to recruit crews and lands, portrayed the ocean not as it was but as they wished to see it. The regularity of the winds and currents, which might reasonably have been expected to encourage navigation, may have inhibited it. It constricted Magellan and generations of subsequent European to the narrow corridors that winds and currents defined, and sped them past most of the ocean. Only by accident or mutiny did an expedition such as Grijalva's in 1536 stray outside the limits the elements prescribed to wander between the winds. No complete voyage back and forth across the ocean was effected until the navigationally inspired Augustinian Fray Andrés de Urdaneta guided a successful round trip from Mexico to Manila in 1564. Not until Captain Cook did any navigator from outside the Pacific systematically set out to chart the ocean beyond the reach of the winds. Even as late as the mutiny on the *Bounty*, in 1789, it was possible for Christian and his cronies to hide on an uncharted island. The environment of the ocean did not determine the history of its exploration, but, like every environmental influence in history, it limited the possibilities and conditioned the course of events.

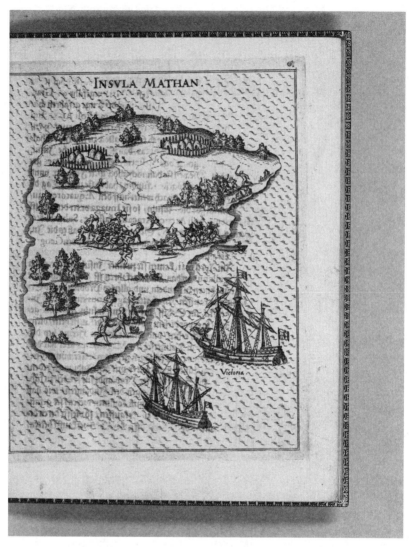

FIGURE 9. Loosely based on an earlier, Spanish engraving of 1601 from Herrera's account of Magellan's voyage, Hulsius's (v.s., p. 128) illustration of the death of Magellan shows the island of Mactan, with the *Victoria* offshore, while a landing party does battle with native defenders armed with clubs and bows. Magellan's native allies from neighboring Cebu take refuge among trees, unwilling to commit to a doomed fight. Hulsius's depiction of unclad inhabitants and of dwellings, crudely constructed and scattered irrationally behind irregular palisades, suggests a rudimentary state of civilization, at odds with the richness and sophistication of the natives' lives. The scene toward the bottom of the page, where natives attempt to entice Spaniards ashore with offers of food, may allude to the fleet's escape from Cebu, when those who accepted the rajah's delusive hospitality met their deaths, or perhaps to the scene that occurred on departure from the Marianas Islands, when hostile natives tried to lure the Spaniards back with gifts. Courtesy of the John Carter Brown Library.

CHAPTER 9

Death as Advertised

The Philippines,
March to July, 1521

But if thou wilt go, do it, be strong for the battle: God shall
make thee fall before the enemy: for God hath power to help,
and to cast down.

—2 Chronicles 25:8

Love of lists is one of those almost universal but rationally inexplicable
human tastes, like travel, quizzes, dancing, and fashion. Roy Plomley,
the deviser of *Desert Island Discs*, a BBC radio show, used to ask guests
to list the eight gramophone records they would want with them if cast
away alone on a desert island. Spin-off games I have played include
Desert Island Books, Desert Island Theatre, Desert Island Paintings,
and *Desert Island Dishes.* Historians should play *Desert Island Articles*,
in which they specify the scholarly articles they would most like to have
to hand in similar circumstances. Leonard Blussé's piece on Chinese
miners in eighteenth-century Borneo would be among my choices,[1]
because of the brilliance and vividity with which it conveys a universal
truth: identity arises outside the group. The miners whose correspondence Blussé discovered came from various parts of East Asia, and though
they were all subjects or former subjects of the Celestial Empire, they
never thought of themselves as Chinese until their Dutch employers'
habit of lazy generalization put the idea into their heads. No Native
Americans ever thought of themselves as such before Europeans introduced them to the notion of an American hemisphere and the possibility
that they might have something in common with each other. *Deutsch*
and *Dutch* are terms adopted from other peoples' words for foreigners.
Whites, not Blacks, made *Black* a designation.

Today our sense is strong that the Philippines—as we now call the islands where Magellan toured and died in 1521—are a single entity, not least because Filipinos have shared common political frameworks for nearly half a millennium, and because, until East Timorese independence, they were distinctive as Asia's only preponderantly Christian national community. Objectively considered, however, they inhabit a group of islands with nothing in common that collectively distinguishes them from others in their vicinity. The peoples Magellan found when he arrived did not think of themselves as other than mutually distinct, typically hostile communities.

Nevertheless, states do form as a result of initiatives from within—usually in consciousness of a shared sense of a common enemy, or in the course of widening family alliances, or as a result of the political imaginations of chiefs or rulers who enlarge their domains by conquering neighbors'. States nurture common identity. They may sometimes come into being as a result of it. When Magellan reached the Philippines, some islands, or pairs of islands at least, were, I think, poised on the brink of becoming states—rather like the Hawaiian islands or the Iroquoian confederacy on the eve of European intrusion in the eighteenth century, or the Mongol world at the elevation of Genghis Khan. That, at least, is what Magellan seems to have thought. His efforts to speed the process led to his death.

. . .

More than one island was in view when he made landfall, or at least on the following day, St. Lazarus's, the last Sunday before Palm Sunday, 1521, when he coined the name, "Islands of St Lazarus." Shallows kept the ships clear of the first island they approached. Pigafetta called it Samar. It is not clear, however, that it was the island that bears that name today. Samar lies on the approach from the direction of the Carolines, but a normal course there would take a vessel within sight of Mindanao, a large, highly visible island of which Magellan seems to have been unaware. According to Albo's log, the first land the expedition sighted appeared to the north, where the likelihood of bypassing Mindanao without making landfall would be puzzling.

The rest of the fleet's course among neighboring islands, however, is consistent. Balked at Samar, the ships turned south, Albo (or whoever was keeping his log) wrote, in order "to reach another small island, and there we landed . . . and this island is called Suluan."[2] Albo's account may have telescoped the events or confounded the places, as Pigafetta located the landing at an uninhabited island he called Homonhon; or perhaps the error was Pigafetta's—muddling the two islands, as is all too easy to do. Suluan, currently so called, is much smaller than present-day Homonhon. Both islets are in the right place, just south of Samar, and match Pigafetta's description: plenty of white coral, abundant fresh water, and coconut palms.[3] The absence of natives was perhaps a seeming advantage after the mishaps that had befallen the expedition in the Marianas.

In any case, everyone seems to have recognized the importance of arriving. For those for whom the Philippines were unacceptable as a final destination, the islands offered a halt for recuperation and reprovisioning. The presumption that the flotilla was on Spain's side of the Tordesillas antimeridian was unquestioned. There was no existing Portuguese presence to contend with. Pigafetta honored the occasion by specifying the full date—something he did only at moments he regarded as highly significant.[4] Albo paused for his own tribute: providing sailing directions to the islands—go west-northwest from the Strait of All Saints "and you will come upon them without deviating [*vais a dar en ellas justamente*]." The directions were misleading: they would have been more likely to lead to New Guinea. Albo also included a reading of longitude, though without claiming to have made it himself: the fleet, Albo opined, was 189 degrees west of Seville—an underestimate of nearly 40 degrees. He put the difference in longitude from the westernmost point of the strait at 106.5 degrees.[5] That is at least 10 degrees too little. Pigafetta placed the islands at 161 degrees west of the Tordesillas meridian: another error of at least some 40 degrees. So the figures—however obtained—tended to reinforce two of the prejudices on which the expedition was based: that the Spanish zone of navigation extended deep into Asia and that the ocean was traversibly narrow.[6]

The island at which the fleet was anchored afforded only modest supplies: nuts smaller than almonds and coconuts, "some good and some bad" (algune bonne et algune altre cative).[7] The men had to consume more of their own supplies. Their situation improved, however, when they attracted the attention of a boat bearing nine natives from a nearby island. Magellan was apprehensive at their approach and "ordered that no one should move or say a word without his permission."[8] But the expedition was back in the realm of the stranger effect. The islanders greeted Magellan "with great signs of joy."

They were "ornately adorned"—therefore encouragingly prosperous and aesthetically inclined—and gave an impression of being "reasonable men": that is, rational and therefore fully human and potentially redeemable. They exchanged courtesies—behavior that conveyed like-mindedness to anyone who, like Pigafetta, knew Castiglione's counsel for courtiers (above, p. 130). No common language was available, but gestures established a degree of communication, and Pigafetta found the natives "very pleasant and conversable." Magellan presented food and truck—some of the caps, bells, and mirrors the fleet carried for just such an eventuality, to which, according to Pigafetta, items of ivory and linen were added. The visitors reciprocated with fish, arrack, two coconuts, and what sound like plantains of a sort: "figs more than one span long and others that were smaller and more delicate."[9]

Despite the fresh supplies, two more scurvy victims were beyond help and died on the island; but the encounter involved four gratifying revelations for the explorers. First, when Magellan showed the visitors around the flagship and displayed samples of wares he hoped to acquire, they responded positively to his samples of gold, indicating that it could be acquired in a nearby island. They had gold ornaments of their own—armbands and, in the leader's case, gold earrings. To judge from remarks Pigafetta and Pancaldo made, there may have been gold or "signs of gold" on the very island on which the explorers were resting.[10] Second, a demonstration of the firepower of the expedition's *falcones* (above, p. 229) provoked "great fear" in the natives, who "tried to jump out of the ship." Fear was just the sort of emotion that might entail cooperation. Third, some natives, according to Pancaldo's account, indicated that they had seen men resembling the newcomers:[11] the names they gave for these strangers were unintelligible, but if the explorers understood correctly, they were surely referring to Chinese. From Europeans' point of view, proximity to China and access to Chinese trade were among the attractions of the Philippines. Finally, and for immediate purposes most auspiciously of all, the natives promised to return with more supplies.[12] When they did so, with renewed signs of pleasure, they brought "coconuts, sweet oranges, palm-wine," and at least one fowl. Magellan bought the entire stock.[13]

Who were the hospitable providers? As Magellan's Malay-speaking slaves could not understand them, they were presumably speakers of some ancient Austronesian language, perhaps the Waray tongue still used in the region or an archaic type of Samaran, fragments of the vocabulary of which are still recoverable, especially in old tales.[14] Pigafetta said that they came from Suluan, about twenty-five kilometers to the east of Homonhon. In a gesticulating crowd, where people are pointing in different directions in an attempt to communicate the names of various islands, it is easy to make a mistake; and it is not clear that Suluan was populated at the time. A larger island, such as Samar, seems more likely to have been the providers' home.

Pigafetta described people "living near" without making clear their relationship to the fleet's benefactors. He call them "Kaffirs, that is to say, heathen," as if the informants on whom he relied were Muslims, though it is hard to say how Muslims could have been present so far east and north in the archipelago at the time. Perhaps locals had picked the term up from Muslims on Mindanao or from Muslim traders and residents in Cebu, where the ruler had a counselor whom the explorers, when they reached that island, recognized as a "Moor."

The "Kaffirs" were fisher folk with boats "like ours," Pigafetta reports—presumably plank-built, not dug out—and weapons and shields inlaid with gold. They had olive-tinged skins, decorated with tattoos, and long black hair. They went naked "with a cloth woven from the bark of a tree about their privies, except some of the chiefs who wear cotton cloth embroidered with silk."[15] They smothered themselves with coconut and sesame oils "as a protection against sun and wind" (per lo solle et per il vento) and pierced their ears so generously "that they can pass their arms through them" (che portano le braci ficati in loro).[16] They were fat, recalling the potbellies often ascribed to the Negritos of the interior. Later, however, Pigafetta saw or heard of the latter and called them "black as Ethiopians." Albo concurred.[17] The "Kaffirs," in short, shared the culture Pigafetta described with many coastal-dwelling groups in the eastern Philippines.

After eight or nine days of recuperation Magellan began to cruise round the archipelago, as if looking for gold and for the best place to contract a robust alliance and establish a permanent presence. As the explorers entered the heart of the archipelago, they found people who could understand the Malay that Magellan's slaves spoke. Anchored off Limasawa on March 28, Holy Thursday, by Pigafetta's reckoning, they had another encounter with natives: eight boatmen who stood off to scrutinize the fleet. On this occasion, it was the latter who hesitated, but Magellan induced the visitors to accept gifts of "a red cap and other things tied to a bit of wood" as a float, which he threw out to them. Within two hours, two more boatloads of men arrived, bearing an official the expeditionaries took to be the local ruler, "being seated under an awning of mats." The "king," or "rajah" in the term natives used to designate paramount chiefs, approached the flagship. A slave addressed him. Presumably, the speaker was Enrique de Malacca or one Jorge, named in claims subsequently submitted to the crown by Magellan's heirs,[18] and called a "morisco" in a list of unpaid wages, where his salary, due for his sea service despite his servile status, is scheduled as payable to Magellan's estate.[19] The Sumatran woman, who may possibly also have been aboard (above, p. 99), would also, presumably, speak a Malay language and be available to interpret ("la esclava los entendia").[20] Whoever spoke for the explorers, the rajah understood, "For in those districts the kings know more languages than the other people."[21] Magellan bestowed gifts but tried to make a favorable impression by refusing "a large rod of gold and a basketful of ginger."[22]

The gesture might have been misunderstood, and it is hard to know why Magellan took the risk of rejecting his host's generosity. He may have wished to evince an air of quasi-divine needlessness. More probably, he was unwilling to show the depth of his interest in gold for fear of driving up the local price. As we have seen, one of Elcano's reasons for resentment was that the commander forbade his men to buy or barter for gold so as to avoid inflation (above, p. 94). Pigafetta confirmed that when a native offered "a crown with six points of solid gold" for six strings of glass beads, "the captain refused to let him barter, so that the natives should learn from the very beginning that we prized our merchandise more than their gold." He added that "the captain-general did not want to take too much gold, for there would have been some sailors who would have given away all that they owned for a small amount of gold, and would have spoiled the trade for ever."[23]

. . .

However canny Magellan's attitude to gold, he seems to have been undergoing changed sensibilities. The Monday of Holy Week had coincided with the feast of the Conception of Christ. On that auspicious day, Pigafetta experienced the miracle that shaped his account of the voyage and the rest of his life (above, p. 131), when he fell into the sea while fishing and was saved, as he thought, by God's providence. It was a kind of quasi-baptism by total immersion, a symbolic rebirth of the sort Christ specified as essential to salvation. Pigafetta's miracle coincided with enhanced religious priorities on the commander's part too, almost as if Magellan had also undergone a conversion experience. Long sea journeys are often transmutative. You don't have to sink full fathom five to undergo a sea change. Magellan and his men can hardly have been unchanged in the course of the longest voyage on record, under the constant strain of deadly menace, as stomachs shrank, gums swelled, pain sharpened, and no prospect of relief appeared.

One day early in April 1521, by which time he had transferred his center of operations to the island of Cebu, Magellan performed a miracle of his own—or one was attributed to him. In the dwelling of the ruler, he saw his host's grandson (or nephew in an alternate version)[24] in a fever, which, he was told, had been raging for two years. "But," according to one source that records this episode, Magellan

> told him to be of good cheer, and that he would immediately recover his health and former strength, if only he would become a Christian. The Indian accepted the condition, and, having adored the cross, he received baptism, and the next day declared that he was well, rose from his bed, walked, and took food like the rest.[25]

In consequence 2,200 of the chief's followers were "baptized and professed the name and religion of Christ."[26] Political repercussions ensued, if Pigafetta can be believed, as "the people themselves cried out, "Castile! Castile!"[27]

Biblical precedents echo in the language of the anecdote; acts of healing—medical, miraculous, or both in combination—stud the history of the establishment of the Spanish Empire and of explorers' escapes from peril. In Magellan's case, however, the story is surprising, even shocking, partly because the protagonist's newly found spirituality and newly aroused interest in spreading the Catholic faith in the islands seem at variance with his previous character and conduct. He had been to confession and communion when custom required and had made devout bequests in his will (above, pp. 109, 126), but these conventional pieties were no sort of preparation for thaumaturgic prowess. Shipmates quartered or marooned or driven to desertion or silenced in fear of their commander's vengeance are unlikely to have commended his spirituality or provided testimonials of his faith prior to this moment on their journey.

Even more oddly, the source I have just quoted for the story of Magellan's miracle—which a critical reader might dismiss, but for its provenance—is Maximilianus Transylvanus, who held no brief on Magellan's behalf, was close to some of the commander's enemies, and relied for his information about the voyage chiefly on Elcano: the former mutineer whose reasons for hating Magellan were magnified by his role as the last commander of the expedition and therefore in a sense his predecessor's rival for such glory or renown as the escapade might generate. What motive can Transylvanus or his informants have had for representing Magellan as a channel of divine mercy?

Of course, Transylvanus's patron and dedicatee was a bishop (above, p. 136), but a sophisticated one, whom miracle tales were unlikely to sucker. Transylvanus himself was something of a skeptic, unimpressed by "I know not what visions" the healed native might have related. The humanist's own analysis was that Magellan used evangelization as a means of lubricating his otherwise sticky strategy of control and conquest. To begin such a strategy, however, by promising a miracle was a stupendous bit of recklessness. I think we have to conclude that the story—shorn of miraculous elements—is true; otherwise it would scarcely have reached its reporter's ears or passed his scrutiny. Pigafetta confirms it, with minor discrepancies: in his version, a procession "with as much pomp as possible" preceded the healing, which was effected instantaneously; and Magellan offered his own life in expiation if his promised miracle should fail: "If that did not so happen they could behead him" (si ciò non se foce, li tagliassero lo capo alhora).[28]

Magellan took the risk, I think, because he was in a genuinely exalted mental condition. Every servant of the crown knew the importance Spanish monarchs always attached to Christianization as a means of legitimizing empire. Religious transports of a sort had overtaken Columbus and Cortés in times of adversity. They both came to prioritize the spread of Christianity when other, more materially rewarding objectives seemed to elude them. Success, as well as adversity, can induce religious awareness, especially if it follows harrowing trials such as Magellan had endured throughout the voyage up to this point. Now he had reached his objective. His force, though depleted by desertions and disease, was still viable. He was surrounded by gold—including, reputedly, nuggets "the size of walnuts and eggs"[29]—in a paradise full of food. The prospects for profitable trade were good: on arrival at Cebu he heard of the recent arrival of a Siamese junk, "laden with gold and slaves"—evidently the exports of the island—and met a Muslim merchant from that kingdom who had stayed to trade.[30] China, from where, it was said, merchants often came, was not far away.[31] Chinese bronze bells were visible and audible around him.[32] Everything seemed, as if in reward for suffering and sacrifice, to be going Magellan's way.

He had already shown the beginnings of reckless overconfidence in refusing the gifts of the local chieftain while he was on the island of

Limasawa. A reversal of fortune from bad to good, in short, was beginning to turn his head. Pigafetta, his closest confidant aboard his fleet, had become convinced of his own miraculous preservation to do God's will. Magellan shared the shift to transcendent values.

. . .

First signs of the shift appeared even earlier in a long series of more or less ritual transactions with the ruler of Limasawa, culminating in the celebration of mass and polite overtures about matters of religion. An exchange of gifts preceded a demonstration by Magellan of the virtues of Spanish armor and guns, which left the host "almost senseless" with amazement (Il ne restò casi fora di sè).[33] There were feasts of rice dressed with pork and fish in an elevated chamber under lanterns of inflammable resin wrapped in palm or banana leaves, military parades, and "a fencing tournament."[34] Deputed by Magellan, Pigafetta visited the ruler's house, where his entertainment began with everyone, including the Christians present, raising their hands skyward. "The king," as Pigafetta described the scene,

> had a plate of pork brought in and a large jar filled with wine. At every mouthful, we drank a cup of wine. . . . Before the king took the cup to drink, he raised his clasped hands toward the sky, and then toward me (at first I thought he was about to punch me) and then drank. I did the same toward the king. They all make those signs one toward another when they drink. We ate with such ceremonies and with other signs of friendship. I ate meat on Good Friday, for I could not do otherwise.[35]

It is an engaging and convincing picture of communication by performance, mimesis, and gesture in the absence of a common language, though one wonders whether Pigafetta can really have known what he was eating. No interpreter was present, and the interlocutors were "constantly conversing with signs" (sempre parlando con segni).[36]

The celebrations of blood brotherhood and of the mass seem to have been intended by the parties as complementary gestures of reciprocal amity. The first, if Pigafetta understood correctly, happened at Magellan's request: how he knew about the custom is unclear. Ginés de Mafra described the ceremony performed at Limasawa on Good Friday, March 29, as well as he could remember it.

> The lord of that island came to the ship and spoke very fairly to Magellan and to all of us, and our commander made peace with him according to the custom of the country, which is to let blood from the breast and to mix the

blood of both participants in a drinking vessel with wine and each drinks half of the mixture. This it seems is their rite of good friendship.[37]

In Pigafetta's account of a similar rite at Cebu, the ruler and Magellan exchanged drops of blood from their arms, while on third and fourth occasions, after Magellan's death, Pigafetta himself witnessed a royal blood-taking rite from palm and tongue and took part in a breast-gashing brotherhood ritual at Palawan.[38]

The mass that served as the newcomers' rite of solemnification of the alliance occurred on Easter Day, March 31. It is often called "the first mass in the Philippines," but the priests would have celebrated mass as soon as they were safely ashore, and the whole expedition would have turned out on Palm Sunday and Holy Thursday. The Easter mass was the first to be reported, because the occasion was significant and exceptional, on an inhabited island, with native leaders in attendance—including the ruler of Limasawa and a visiting ruler from a nearby island. A political alliance, moreover, was at stake. Ceremony, salutes, and solemnity were prominent. Veneration of the cross was included, although that was not a universal practice of the church on Easter Day, perhaps so that pagans, who were barred from communion, could take an active role. When the hour arrived, Pigafetta explained,

> we landed with about fifty men, without our body armor, but carrying our other arms and dressed in our best clothes. Before we reached the shore, six pieces were discharged as a sign of peace. . . . We went in marching order to the place consecrated, which was not far from the shore. Before the commencement of mass, the captain sprinkled the entire bodies of the two kings with musk water.[39]

Later, Magellan was to repeat this act, presumably intended to convey a sense of purification, when mass was celebrated for the wife of the rajah of Cebu: "The captain sprayed her and some of her women with musk rosewater; they delighted exceedingly in that scent."[40]

> At the time of the offertory, the kings went forward to kiss the cross as we did, but they did not make any offering. When the body of our Lord was elevated, they remained on their knees and worshipped him with clasped hands. The ships all fired their artillery at once when the body of Christ was elevated.[41]

Nothing betrays Magellan's state of exaltation more than his extraordinary self-transformation into an aspirant holy man. Despite the fact that priests were available, he took on himself the duty of explicating the gospel. His efforts began after the celebration of mass on Limasawa Island, when Magellan successfully sought the ruler's permission to erect a cross

> on the highest mountain, so that on seeing it every morning, they might adore it; and if they did that, neither thunder, lightning, nor storms would harm them in the least. . . . The captain-general also had them asked whether they were Moors or heathen, or what was their belief. They replied that the worshipped nothing, but that they raised their clasped hands and their face to the sky and that they called their god Abba. At this, the captain was very glad.

To worship nothing might have been a sign of indifference to natural law, as among the Chamorros (above, p. 227). In Limasawa, however, it was clearly not to be taken literally. The promising signs of a predisposition to become Christians—the clasped hands for prayer and the use, by coincidence that betokened grace, of one of Christ's names for his father—outweighed it. In any case the explorers knew that the natives did not worship "nothing." They were well aware of the prevalence of "idols." On Cebu, Magellan's healing miracle followed the natives' unsuccessful appeals to idols; he instructed the ruler's consort to keep an image of the baby Jesus, which he presented as a gift, "in place of her idols." Magellan's first sermon, if one may so call it, evidently showed confidence in his ability to command miracles, silencing thunder, outshining lightning, stilling storms, and anticipating his healing of the princeling in Cebu.

Pigafetta provides a further sample of his friend's technique as a evangelizer on board ship on Cebu, where Magellan addressed a conference of the ruler's relations and advisers. Having found them well disposed to his proposal for a peaceful alliance, Magellan felt encouraged to broach "arguments to induce them to accept the faith." He told them "how God made the heaven, the earth, the sea, and everything else, and how he had commanded us to honor our fathers and mothers, and that whoever did otherwise was condemned to eternal fire." This was a further example of Magellan's mood of daring, since he had just heard from the native delegates that "when the fathers and mothers grew old, they received no further honor, but their children commanded them." Magellan then reverted to his summary of the book of Genesis, "how we are all descended from Adam and Eve . . . and how we possess an immortal soul; and many other things pertaining to the faith." He made further odd and reckless utterances: "that we could not have intercourse with their women, . . . since they were pagans," and "that if they became Christians, the devil would no longer appear to them except in the last moment at their death." It is unclear which, if either, of these assurances acted as an incentive to convert.

Magellan showed, however, that he knew something, at least, of theology when he told his audience that "they should not become Christians through fear or to please us, but of their own free will."[42] Unfortunately, as we shall see, he was to show little appreciation of his own point when, later, he tried to threaten, bludgeon, and burn the people of the island of Mactan into avowal of Christianity. Meanwhile, his efforts to instruct the inhabitants of Limasawa and Cebu recurred at every opportunity: first at the ceremony of the baptism of the ruler of Cebu, when Magellan showed the congregation how to make the sign of the cross, enjoined them to immolate their idols, and ordered them to salute the cross morning and evening. and second, prior to his healing miracle, when he interrupted sacrifices on the sick man's behalf to tell the sacrificers "to burn their idols and believe in Christ."[43]

Despite his new mood and his belated discovery of Christian zeal, Magellan's efforts at evangelization were evidently connected with his political strategy. His policy started at a modest level: trying to forge alliances. It started with blood brotherhood and amicable rites and gift exchanges in Limasawa. At a further but still cautious phase, Magellan assured the ruler that the cross he erected would be for the natives' "benefit, for whenever any of our ships came, they would know we had been there, and would do nothing to displease them" and that "if any of their men were captured, they would be set free immediately on that sign being shown."[44] A Christian symbol, in other words, became a politically powerful sign. On what seems to have been the same occasion, encouraged by assurances of the rajah's "love," Magellan, through his interpreter, "asked whether he had any enemies, so that he might go with his ships to destroy them and to render them obedient to him." The offer, on this occasion, was politely declined, but it marked a new phase: from now on, Magellan was looking for opportunities to make a display of violence that would intimidate potential enemies, secure emergent alliances, and extend the power of his friends in the archipelago.

The offer to reduce other rajahs to the obedience of Limasawa represented Magellan's first approach toward an overall strategy of unifying at least some islands into a single state under a compliant ruler. That strategy would be extended with fatal results in the next island he made

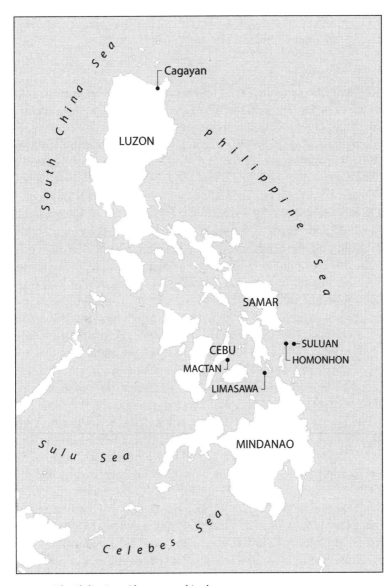

MAP 5. The Philippines: Places named in the text

for: Cebu, where, the people of Limasawa assured him, he would find "the largest" island "and the one with most trade."[45]

. . .

Throughout the time the expedition spent in the Philippines—a matter of nineteen or twenty weeks in all—Pigafetta had ample opportunity to make ethnographic observations, beyond those we have already reviewed concerning rites and religion. As usual, he compiled vocabularies, transcribing words with his customary good ear, catching a Visayan word for husked rice and recording terms for the language of Cebu, of which more than 80 percent correspond recognizably to terms still in use.[46]

He was drawn, of course, to customs that readers in Latin Christendom would find bizarre or amusing. He noticed betel chewing, "for it is very cooling to the heart, and if they ceased to use it they would die" (perchè rinfrescali molto el core. Se restasseno de usarle, moriebenno).[47] Carousing amused him, in, for instance, his description of how he was obliged to match the rajah of Limasawa cup for cup, and how a young prince, when invited to join the feast, "became intoxicated as a consequence of so much eating and drinking" (per tanto bere et mangiare diventò briaco). When the time came to leave the island, departure was delayed after a feast at which the rajah entertained a neighbor, because "the kings ate and drank so much that they slept the entire day. Some said to excuse them that they had taken somewhat ill."[48]

Sexual customs demanded attention. Pigafetta seems aware of being offered sexual hospitality: he records Magellan's refusal of an offer to supply women (above, pp. 245–6) and says he was made to dance with three "completely naked" houris at an entertainment, where the musicians and dancers were "very beautiful and almost as white as our girls and almost as large."[49] The experience should have embarrassed a knight of St. John, vowed to chastity. His observation that "the women loved us very much more than their own men" (le donne amavano asay più noi que questi) is credible: a side effect of the stranger effect.[50] Strangers' exoticism makes them attractive; their independence of existing local ties makes them relatively unproblematic as partners. When the people of Cebu finally turned against their guests, some members of the expedition thought it was in revenge for their excessive indulgence in the sexual opportunities they encountered.[51]

Pigafetta "very often" asked men on Cebu "both young and old to see their penis, because I could not believe" the way they pierced their members with "a bolt as large as a goose quill," adorned with hooks for bells.

Other sources confirm the existence of this quaint custom.[52] "In the middle of the bolt," Pigafetta continued, "is a hole through which they urinate. . . . They say that their women wish it so, and that if they did otherwise they would not have intercourse with them."[53] Best left in Gibbonian obscurity are his descriptions of the effect on women's vaginas and of the technicalities of procedures in which women take the initiative, while "the penis always stays inside until it gets soft, for otherwise they could not pull it out."[54] Only slightly less colorful are Pigafetta's accounts of the simulation of intercourse with a dead rajah prior to burial, and of pig sacrifices, in which priestesses worshipped the sun, danced, blew on trumpets, ate fire, and marked participants' foreheads with hogs' blood.[55]

Pigafetta did not, however, let his eye for the curious and picturesque distract him from the main point: most of the people of the islands were obviously civilized and rational, with true polities and promising dispositions. "Those people live in accordance with justice, and have weights and measures. They love peace, ease, and quiet."[56] By joining the rites, the feasts, the dances, and even, for two days before the fleet left Limasawa, the rice harvest, Magellan's men showed that they recognized the validity of indigenous culture.[57] The quiet, however, was about to be disturbed, the ease disrupted, and the peace shattered.

. . .

The first signs that the explorers were a cause of dissension accompanied preparations to leave Limasawa for Cebu. The destination was the choice of the rajah, who wanted to guide Magellan there for purposes of his own. His relations with his counterpart in Cebu were amicable and evidently deferential. After Magellan's parades, drills, demonstrations of weaponry, and offers to place his power at his friends' disposal, the temptation to secure the newcomers as allies in internecine wars was obvious. At first, the rajah's neighbor and guest (who, to judge from the way Pigafetta transcribed the names, was probably from Cagayan in Luzon or from Mindanao) was enlisted as a participant, but he temporized, made excuses, and withdrew on the grounds that "he should have his rice harvested and other trifles attended to" (facesse coglire el rizo et altri sui menuti.[58] Only the ruler of Limasawa accompanied the fleet on the next leg of its journey, traveling in his own craft, struggling to keep up with the visitors' ships.

They arrived in Cebu, after delays to avoid overnight sailing and to allow the rajah to catch up, on Sunday, April 7.[59] The fleet put out all flags and fired a salute to impress the crowd that lined the quays.[60] Ginés de Mafra blamed Magellan's "pigheadedness" for the diplomatic fracas that ensued when the leader refused to pay the customary toll.[61] Magellan's threats to "send so many men that would destroy" Cebu and all other enemies, "as our handkerchiefs wipe off the sweat," were an inauspicious way of opening negotiations. He benefited, however, from the good luck that had brought a Muslim merchant from Siam to court, who was able to warn Humabon, rajah of Cebu, that Europeans were dangerous. "If they are treated well," he said in Pigafetta's report, "they will give good treatment, but if they are treated badly, they will deliver bad treatment and worse, as they have done in Calicut and Malacca."[62]

Thanks to the mediation of the rajah of Limasawa, the incident passed off without violence, but rather with exemptions from tribute mutually conceded, and ended with the ceremony of blood brotherhood. To negotiate the terms of an alliance, a meeting on the flagship took place the same night. Magellan presided in a red velvet chair, while his guests—including the rajah of Limasawa, the Muslim merchant, and various chiefs and officials—sat on leather. The only point that emerges clearly in the surviving accounts is that "if they became Christians he would leave them a suit of armor" (sie deventavano Christiani, gli lassarebe una armatura)—presumably openly coveted by the native negotiators. Magellan swore on his faith as a knight of Santiago "that he would give them peace with the king of Spain. They answered that they promised the same."[63]

The rajah of Cebu was willing to placate Magellan by accepting baptism: he wanted the newcomers' help in his own wars and was disposed to accommodate them in a way that involved no cost to himself. Pigafetta understood his acquiescence to imply subordination to the Spanish monarchy. "I and my vassals," he claims Humabon said, "all belong to your sovereign" (Io et li mey vasalli semo tucti del tuo signiore).[64] Other chiefs, however, were less willing to oblige. Magellan persuaded them to comply by threatening that "unless they obeyed the king as their king, he would have them killed and give their possessions to the king."[65] Consciously or unconsciously, he was warping the nature of Filipino rulership, turning a presiding paramount into a despot.

Mass became a daily ritual on shore, which Magellan attended without fail, and at which the rajah and members of his family were expected to endure more of Magellan's indoctrination—he "told the king many things regarding the faith" (diceva molte cose della fede).[66] The captain exploited the occasions for political purposes, to consolidate the changes he was contriving in the polity: concentration of local power in the rajah's hands, and recognition of Spanish sovereignty. "Before mass one day," for instance, Magellan renewed his promise of fidelity to the rajah, again invoking his honor as a knight and drawing his sword "before the image of our Lady," with the assurance that "if anyone so swore, it would be better he should die rather than break such an oath."[67]

On this occasion, he demanded oaths exceeding all former obligations: from the "chief men" he required that they "swear to be obedient" to the rajah and that they kiss his own hand: obviously they would be unaware of the solemnity, in Christian practice, of such a gesture of fealty. To Magellan, it was a sign that made clear his own aspiration to lordship: the first evidence of the self-image he had pursued in his negotiations with the king of Spain (above, p. 104). As for the rajah, Magellan insisted that he "declare that he would always be obedient and faithful to the king of Spain" (d'essere sempre hobediente et fidelle al re de Spagnia).[68] The promises Magellan made on his own part were at least as extravagant: that henceforth the rajah would "more easily conquer his enemies than before" (que vincerebe più facilmente li sui nemisi che prima) and that in the future Spain would supply him "with so many forces that he would make him the greatest king of those regions."[69]

All seemed ready for the next stage of Magellan's program: the extension of Cebu's mastery to other islands. To understand what went wrong, we have to change focus and look ahead to the inquiry that Spanish investigators undertook more than a year later, when the expedition was over and they were untangling the evidence of the events that led to Magellan's death.

. . .

Did they have a murder mystery on their hands? For a while, the authorities in Seville thought it was possible. As evidence accumulated of mutinies aboard Magellan's ships, and the silence lengthened beyond the time when the captain-general, by rights, should have accomplished his mission and sent reports home, rumors seemed believable that he might have been the victim of some dastardly deed. When the survivors of the *Victoria* arrived in port, exhausted, wasted, and sick, in October 1522, they found officials of the Casa waiting for them.

"How," read one of the questions in the interrogation that ensued, "did the indios kill Captain Magellan?" Up to that point the question must have seemed unthreatening to those who faced it. It reflected the narrative they were eager to share. The sting came in the clause that followed: "Some," the questionnaire continued, of those who remained in the East or arrived aboard the *Victoria*, "say that he died in a different way."[70]

Magellan died in battle against native defenders when he attacked the island of Mactan: that much was confirmed in the investigation.[71] But in every other respect mystery remains. The proceedings of survivors in Seville, the interrogation of others who fell into Portuguese hands in Ternate, and the reminiscences gathered over the next years and decades are full of contradictions. Not only did different deponents give mutually incompatible accounts, but some of them recorded self-contradictory recollections or inventions at different times.

Historical forensics can help exclude falsehoods or assertions made unreliable by the self-interest of the attestators or their absence from the scenes they reported. When the chaff has been sifted, an inquirer usually looks to logic, common sense, and knowledge of human nature to identify the truth. Sherlock Holmes's maxim comes into play: when the impossible has been eliminated, whatever remains, however improbable, must be true. In the case of Magellan's death, there are only improbabilities to choose from.

The conflict of evidence begins with Magellan's reasons for launching his attack in the first place. No one claimed that the initiative was other than entirely his own. Pigafetta even suggested that he chose the day of battle, Saturday, April 27, "because it was the day especially holy to him" (perchè era o giorno suo devoto), though he does not say why.[72] No account suggests that Magellan had, by any reasonable standard, adequate provocation. Part of his quarrel with the ruler of Mactan arose over his demands for tribute in the form of the supplies his expedition desperately needed. One witness said that the captain-general felt insulted and cheated by the offer of one bushel of rice and a goat.[73] According to the narrative by Pancaldo, the Genoese pilot, who alleged that the dispute over provisions was Magellan's only reason for taking action, the people of Mactan countered a request for three goats, three pigs, and three measures (fardos) each of rice and millet: they offered only two items in each of the requested categories, on the grounds that "they had no more to give."[74]

Victuals were crucial, but politics as well as provender were at stake. Magellan had resorted to violence before—in Patagonia and the Marianas—but for revenge, or supplies, or kidnap victims. The livestock of Mactan was not food: it was symbolic of power. Magellan could not afford to be weak in negotiations. He had to conceal the parlous condition of his ships and men under a display of bravado. He was always a chancer in the calculation of risk. We have seen how he was wounded in Calicut, ventured his life in the Maldives, defied justice in Morocco, provoked the king of Portugal, abandoned his country and family for a new life in Castile, and embarked on his great voyage against all reasonable expectations of success. An entirely circumspect gambler might not have bet the entire expedition against a couple of pigs and goats, but it was perhaps not inconsistent for Magellan to do so.

The problem of how to extend promising alliances to the rest of the archipelago might have been solved or shelved in various ways. Magellan's chosen strategy resembled that practiced at about the same time by Cortés in Mexico: to treat one ruler as paramount and to try to govern the others through him. In Mexico, Cortés chose Cuauhtemoc, the chief of Tenochtitlan: it proved to be an infelicitous election, and the chosen one was disposed of by judicial murder when the experiment seemed to fail. Magellan was similarly rash in backing the chief of Cebu.

He was committed to his alliance with Humabon. They had sworn blood brotherhood. The rajah had reciprocated by assenting to the ritual quid pro quo Magellan had demanded: baptism, when he took King Carlos's Christian name as his own, and formal submission to the king of Spain.[75] Cebu was Magellan's most pliable ally, but in many ways it was a poor choice of base: an island small and unripe for elevation to power above its neighbors. Magellan chose it in ignorance, on the self-interested recommendation of the rajah of Limasawa (baptized as "John" at the same ceremony). He was stuck with it. He could not afford to leave a garrison, with his force depleted and debilitated. But if he left Cebu to its own devices in the rajah's war against Mactan, defeat might undo all he had achieved so far.

It is worth recalling that in order to make the most of his contract with the crown (above, p. 104), Magellan needed to gather as many islands as possible under his wing: only those in excess of six would bring him the rewards he wanted.[76] He believed there were five islands in the Moluccas (above, p. 21). Even if he successfully subjected them all, and Limasawa and Cebu were counted to his credit, he would have a total of only seven. Ginés de Mafra believed that Magellan's anxiety to conquer Mactan arose from the terms of his commission.[77] The suspicion is credible: Magellan needed islands he could keep for himself, where rulers would kiss his hand and the king confirm him as governor. His temptation to spread his conquests farther is understandable.

Magellan therefore tried to impose Humabon's rule on other rajahs, as if he were a mediator of the superior rights of the king of Castile, some of which, under the contract, along with some of the profits, would revert to Magellan. In retrospect it seems obvious that the policy was nonviable, but it may not have seemed so at the time. Fragmented polities, as we have seen, sometimes show a tendency to consolidate. The fusing of chiefdoms into states is a common theme of history. Magellan's program, however, had two fatal flaws, which he was hardly in a position to appreciate: other communities were unwilling to submit to Cebu; and Mactan, where Magellan proposed to start his campaign, was a bigger, more powerful entity, with a larger number of men under arms. Presumably Magellan did not know that when he undertook to conquer it.

He boosted his delusive confidence by burning villages to coerce one or two—the sources are at variance—subordinate chiefs. Witnesses hostile to Magellan claimed that he gained little by his high-handedness, as his allies in Cebu consumed most of the provisions he seized.[78] Pigafetta, the knight of St. John, who might have been expected to know about Christian standards of chivalry, disagreed. He made village burning seem like a successful sort of piety, the effectiveness of which could be measured in baptisms: everyone in Cebu was baptized within eight days. "On a neighboring island we set fire to another village" (bruzassemo una vila, . . . la quale er in una ysola vicina), as if *pour encourager les autres*.[79]

Witnesses differed over what exactly Magellan required of Mactan and why his demands met rejection. Those with mutiny on their consciences tried to undermine his reputation by claiming that Lapulapu, rajah of Mactan, professed himself willing to defer to the king of Spain, or to the Christian God, but not to his own counterpart on Cebu. The implication is obvious: Magellan was more interested in implementing his own policy in his own interests than in serving the king of Spain. The claim was obviously self-interested. Yet to me it seems credible: it is always easier to profess allegiance to some source of authority too distant to matter than to submit to a neighbor whose pretensions might be vexatious. One of the advantages perceived by many subject communities of the Spanish monarchy in the New World was the comfortable distance from which the king of Spain was obliged to issue his commands: in the unlikely event that these would be relevant by the time they crossed the ocean and reached their destinations, they could usually be safely ignored. Officials commonly greeted them with rituals of submission: kissing the documents hallowed by royal provenance and placing them on their heads in token of subjection before filing them away with the traditional formula: "I obey but do not comply."[80]

In any case, whatever formulas Magellan sought, subordinate rulers refused to grant. "The people of Mactan," as one witness at the Seville inquiry put it, "did not want to kiss the hand of the said king of Cebu."[81] A show of force, another torched village—Magellan may genuinely have believed that Mactan would be easy to intimidate. Yet he had access to plenty of accurate data, in Cebu and in the course of negotiations with other polities.

One of the strongest themes of the surviving evidence is the unanimity with which all witnesses who refer to the matter insist that the rajah of Cebu and some of Magellan's own officers tried to deter him from immediate recourse to violence. Ginés de Mafra, whose recollections, as we have seen, were inconsistently reliable, was the most circumspect: "The king of Cebu," he reported, "told Magellan not to concern himself, because with time the recalcitrance would subside"; the rajah, moreover, "would be able to sort it out himself, because his counterpart in Mactan was his brother-in-law."[82] It is unclear where the chronicler Herrera got the details he added in his account, but he quoted, in indirect speech, advice he attributed to Juan Serrano:

> that Magellan should not attempt the campaign he proposed, not only because no good could come of it but also because the ships would be left inadequately defended and liable to easy capture; and that if he insisted on going ahead he should not take part in person but appoint a substitute.[83]

Herrera also understood that Magellan's native allies advised against the adventure.

The whole story sounds fictional: the hero flawed by hubris, the fates gathering to precipitate his downfall, the drama of a doomed encounter. If, to reconstruct the next phases of the story, one weaves the witness testimonies into a single narrative, the sense of heightening tension and impending disaster accumulates, as if the events belonged to romance rather than to history. Like García Márquez's famous novel about a murder in Colombia, the chronicle is of a predictable death—a death advertised but unstoppable.

• • •

Magellan rushed into battle against Mactan as unwarily as a knight at the tilt. He was heedless, according to the united weight of surviving testimony, of the strength of his opponents. The numbers of men Mactan deployed cannot be accurately computed. In the heat of battle or the flow of narrative, observers and chroniclers may have exaggerated them. But every witness put them in the thousands, whereas Magellan did battle with a handful of companions—reckoned by different witnesses at between thirty-eight and "about sixty."[84] He could hardly have mustered more without enlisting native auxiliaries, as most of his crew were too weak from sickness, hunger, or exhaustion to take part in fighting.

He also underestimated the enemy's technical and tactical capacity. His crossbows were effective only as long as the bolts lasted: and the Mactanese had the good sense to stay at maximum range until the Spaniards' supply was exhausted. Arquebuses were deployed but were ineffective: rendered useless by the volatility and shortage of powder, its exposure to damp as the task force waded ashore, the slow rate of fire that limited matchlock weapons, and the general inaccuracy of smooth-bored firearms (above, p. 228). The defenders had, by most accounts, longer spears than the Spaniards, abundant stone missiles, poisoned darts, and iron-tipped lances acquired from Chinese traders.[85] The route the invaders had to traverse was dug with pits (concealing, by some accounts, deadly staves), which would impede advance and imperil retreat.

Magellan made the assault in vessels that could hardly have been less fit for purpose: launches too deep in the keel for the shallow shoreline, which foam-washed rocks protected, but too light in construction to carry heavy artillery. In consequence, his boats had to stand out to sea, too far to bring fire to bear on the battle. The beachhead he sought was unsuitable, obliging his men to advance waist-deep in water to get to dry land. His approach, in short, gave the enemy every advantage of place.

As if all this were not foolhardy enough, Magellan's preparations climaxed in a chivalric gesture of almost unbelievable obstinacy. Ever since Roland supposedly refused to blow his horn at Roncesvalles, the notion that it is noble to refuse help in combat had become a topos of knightly literature. Primaleon, the hero of the romance Magellan had alluded to at San Julián, repeatedly refused help in facing the foe. The story of David and Goliath and the Homeric tradition of battle as a setting for single combat fortified the tradition. According to Ginés de Mafra, Magellan continued, exceeding Roland's valor, to refuse to allow his allies from Cebu to enter the fray, even when he was defeated and dying. They did so only to rescue the wounded when Magellan himself was dead.[86] The author of the Leiden manuscript confirms that "while he lived he did not want his ally the king to help him with the men who were there for the purpose, saying that with God's help the Christians were enough to conquer the enemy's entire mob."[87]

The terms in which Magellan exhorted his men may be unrecoverable. But early sources provide versions that have consistent features. Elcano and the survivors whom Transylvanus and Oviedo interviewed agreed that Magellan recalled feats of arms in which two hundred Spaniards had put hundreds of thousands of *indios* to flight. What precedent might have been in his mind? A battle in Juvagana, say both chroniclers; but that was an obvious error—a name Spanish writers attributed to an uninhabited island in the Philippines or the Carolines.[88] A persistent tradition has made Magellan cite Cortés. But how could news of battles in Mesoamerica have reached the Pacific? It is not entirely impossible. Cortés's first reports arrived in Spain after Magellan left Seville. The Casa de Contratación did not send his message on to the court until November 7.[89] By that time, Magellan was beyond reach. In the interim, however, a sketchy version of Cortés's deeds might have reached Magellan's fleet aboard a pinnace, along with other mails that continued to arrive at least until the fleet had passed Tenerife. Transylvanus and Oviedo were working in the tradition of Herodotus, without hesitating to put a suitable speech, of their own invention, in a hero's mouth; but they could hardly have invented two such similar versions independently. They must have got it from Elcano, or another survivor whom they both interviewed. Perhaps Magellan offered a general appeal to heroic tradition and the locations the chroniclers mentioned were adumbrations of their own.

In short, the story of Magellan's death, as the witness statements and early historical tradition represent it, is a tissue of literary topoi. Some of the earliest chroniclers fell in unhesitatingly with the romantic agenda. Peter Martyr spoke of Magellan's evil star and blamed his death on a moral flaw: "his greed for spices" (Suae cupiditati aromatatrice finem imposuit).[90] Herrera attributed his death to excessive daring and a needless compulsion to tempt Fate.[91] For Pigafetta, Magellan's sacrifice was Christlike—of "a good shepherd who refused to abandon his flock," defying his wounds to protect his comrades' retreat to their boats. In describing a protracted agony reminiscent of the passions of hagiographical tradition, most accounts depicted the victim's resistance to a multiplicity of wounds—in the legs, the neck, the crown of the head—before he finally succumbed.

What, then, really happened? If the story of Magellan's death sounds like a romance, it is not only because of the inventiveness of the witnesses and the literary proclivities of the chroniclers. Survivors of battle do romanticize to make their memories bearable: Paul Fussell demonstrated that in one of the great works of modern literary criticism, *The Great War and Modern Memory,* showing that even the most supposedly realistic poets occluded horror by wrapping it in artifice. But in Magellan's case it is impossible to resist the conclusion that he was the original author of the saga of his own demise. He crafted his death to suit a narrative he composed in his own mind before the event, imagining a knightly consummation in a battle sanctified by crusading ideals, ennobled by echoes of the tradition of Roland, and elevated by acts of heroism that were, by mundane standards, merely reckless. He died as advertised.

. . .

The other fatalities were few. Peter Martyr thought seven fellow combatants fell at Magellan's side, with twenty-two wounded (cum septem sociis trucidatus, vulnerati duo et viginti).[92] There were six, according to the Genoese pilot and seven in the sources Herrera ransacked, eight (plus four natives of Cebu, caught by friendly fire) according to Pigafetta.[93] Cristóbal Rebelo (above, p. 119) died alongside him, as did six others, according to the official tally of the dead.[94] Antón Escobar, one of the mutineers of San Julián, died of his wounds two days later. A gunner from the Victoria, similarly wounded, did not succumb until September 1. Andrés de San Martín was another fatal casualty, according to Barros, who might be expected to know, though, as we shall see, other sources suggest his demise occurred a few days later.[95] Magellan's slave Enrique de Malacca was among the wounded, recuperating pathetically, Pigafetta claimed, under a blanket. Pigafetta tells a story, credible in view of what we know of the treatment of slaves in Magellan's household (above, p. 76): apparently suspecting malingering, Duarte Barbosa, the captain-general's cousin by marriage, threatened Enrique with flogging and return to his mistress's abusive keeping when they all got back to Spain.[96] That was unconscionable: Magellan had freed Enrique in his will (above, p. 51); if the charge was true, the slave's resentment was justified.

Although, as Pigafetta admitted, "many of us were wounded," the low death count prevented the explorers from realizing, at first, the magnitude of their defeat. The assumption that they would flee from the Philippines was instantaneous, to judge from the way the trade goods displayed on shore were immediately loaded aboard ship. But the urgency of their plight did not impress the leading men of the expedition. Magellan's promises to the rajah of Cebu had proved empty. The Spanish war machine had failed. Wherever the stranger effect broke down, as it had in the Marianas because of the natives' peculiarities of culture and foreknowledge of their hosts, or in Mactan, because of the demands of regional politics, Spaniards faced defeat. The newcomers had been humiliated. The fortunes of war had shifted significantly against Cebu and in favor of Mactan. The best policy, from a perspective in Cebu, was to turn against the explorers and try to recoup good relations with the rulers of other islands.

It should have been obvious, therefore, that the alliance with Cebu was at an end and that the friendship of the former allies could no longer be assumed. Yet on May 1 a party, perhaps twenty-eight or thirty strong, accepted Humabon's invitation to a feast, at which the rajah would present them with jewels he had promised as a gift for Magellan; the intended recipient was dead, but they might make a suitable offering to the king of Spain. Juan Serrano, whom most accounts represent as taking over Magellan's place, led the shore party. San Martín, according to Pigafetta, went with him, as did João Carvalho and Gonzalo Gómez de Espinosa. Pigafetta remained aboard ship, nursing a leg swollen by poison from Mactanese missiles.

His account of what transpired is too vivid to ignore. Carvalho and Espinosa became suspicious when they saw the beneficiary of Magellan's miracle separate a priest from the other Spaniards, as if to protect him from a plot. Scarcely had they returned to the fleet "when we heard loud cries and lamentations. We immediately weighed anchor and discharging many mortars into the houses, drew in nearer to the shore." Juan Serrano, bloodied and coatless, appeared on the shore, begging his comrades to cease fire, "for the natives would kill him." The rest of the shore party had already been massacred, except, perhaps, for Enrique de Malacca. Serrano pleaded to be ransomed, but Carvalho warned against the risk of sending another boat ashore. The latter's advice was not disinterested: he would be next in line for overall command.

Serrano,

> still weeping, asked us not to set sail so quickly, for they would kill him, and said that he swore to God that at the day of judgment he would demand his soul of João Carvalho, his comrade. We immediately departed; I do not know whether he died or survived.[97]

The official record shows him among the presumed dead of the "killing by treachery" (mataron a traición), along with Duarte Barbosa, Magellan's cousin by marriage, who had taken over as captain of the *Trinidad* and had allegedly threatened Enrique de Malacca with reprisals for malingering, and twenty-five others, including Enrique himself and Andrés de San Martín.

The fate of Enrique is a mystery that eludes elucidation. Pigafetta, who rarely agreed with Juan Sebastián Elcano, concurred in blaming the slave for the massacre. Defecting to the natives' side, according to their assumptions, he induced Humabon to turn against the explorers and seize their goods. Yet there is no need for recourse to conspiracy theory. It was in Humabon's interest to rid himself of his troublesome guests and right his account, if possible, with the fellow rajahs whose autonomy Magellan's policy had menaced.

. . .

The fleet was now free to resume its original mission and seek the Moluccas, though it would take another six months to get there. Would Magellan have done so, had he lived? His interpreter told the Moorish royal adviser on Cebu that the Moluccas were the fleet's next destination, but Magellan told the rajah of Cebu that his next trip would be home to Spain, from where he would return with reinforcements.[98] He had always kept the Philippines in the fore of his mind. Yet it was surely inadvisable to ignore the orders of the king of Spain entirely. Whether Magellan lived or died, the voyage had long lost all chances of success. The route he pioneered toward the Moluccas was obviously impracticable. The disasters in Mactan and Cebu merely confirmed the fact. Having come so far, however, the fleet had no reason to leave its original mission unfulfilled.

In some ways, by his act of showy heroism, Magellan had made a "great career move." We can picture him on the shore, wading, with bare legs bleeding and swelling from his wounds—inching backwards, sinking into sand, as the harassed task force retreated, with the darts and spears and stones flailing the sea and stinging their limbs. Battle, especially if you are wounded and losing with the enemy closing in, is not a perfect environment for calm contemplation of the future. The past is more likely to flash before your mind's eye. Magellan's prospects—if he had a chance to envisage them—were dire. He had disobeyed orders. His mission was ending in commercial failure and military defeat. If he got home, the mutineers and deserters would gather to denounce him, like buzzards over bones. His high-handedness affronted shipmates and alienated former friends. By condemning or killing the nominees of Bishop Fonseca, he had earned the enmity of an impregnably powerful interest at court. He was confronting the collapse of his hopes of a storybook fadeout to a life based on fictional models. It was a good moment for revised priorities. He had little to lose but his life. When he fell, he evaded the costs of failure and laid the foundation of a legend.

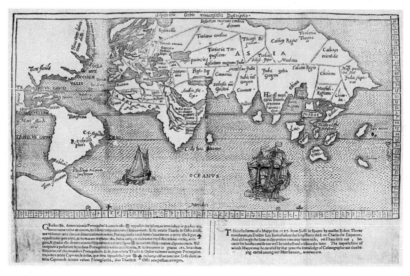

FIGURE 10. A map drawn by Robert Thorne, an English merchant in Seville, who was active in trade with the Levant as well as with the Spanish and Portuguese worlds, in 1527. According to Richard Hakluyt, the English propagandist for empire who published this version in 1582, Thorne sent the map to the English ambassador in Spain. Thorne showed the Tordesillas meridian on the longitude of the Cape Verde islands, showed the Tordesillas line itself a little over twenty degrees to the west, and included indications of the rival locations Spanish and Portuguese cosmographers assigned to the antimeridian, remarking that what he called "the islands of Tarshish and Ophir"—apparently identifying the Moluccas and Philippines with King Solomon's sources of wealth—lay in the disputed area. The influence of Magellan's voyage is obvious: Thorne shows San Julián somewhat south of its real position and includes the Strait of Magellan under the name the explorer gave it: "Strictum Omnium Sanctorum" or Strait of All the Saints. From *Divers Voyages touching the Discovery of America and the Islands adjacent: Collected and published by Richard Hakluyt, Prebendary of Bristol, in the Year 1582* (London: Hakluyt Society, 1850). Courtesy of the Hakluyt Society.

CHAPTER 10

Aftermath and Apotheosis

The World, from 1521

By terrible things in righteousness wilt thou answer us, O
God of our salvation; who art the confidence of all the ends
of the earth, and of them that are afar off upon the sea.

—Psalm 64/65:5–6

When I was a small boy, I was at a terrible school, where we had to run
around a heath every afternoon. I resented it, not only because I was
slow and feeble, but also because even my fastest classmate proved
nothing except that he could, at the cost of a lot of pointless effort, be
first back to where we started.

Circularity has continued to annoy me ever since. Global circum-
navigations impress me only as monstrous recurrences of the same folly
as haunted my childhood. They are interesting not because they are
admirable but because they are odd. Being "first" to do something valueless is generally an overappreciated achievement in a world irrationally obsessed with records. I recall another moment in my early youth,
when I heard A. J. P. Taylor, one of the most commercially successful
academic historians of all time, interviewed on—I think—the BBC
about the first manned flight to the moon.

"What's the historical significance of this moment?"

"None," he replied.

The interviewer seemed bamboozled. Taylor was almost right: the
US had invested disproportionately in a political gesture, which showed
that Soviet Russia was no longer ahead in the "space race" but which
added nothing to knowledge or happiness.[1] I feel the same about the
first circumnavigation of the world. It did not matter.

Yet no sooner was it accomplished than it generated huge pride and huge praise. Juan Sebastián Elcano, who led the stragglers home, received a coat of arms emblazoned with a globe and the legend "Primus Circumdedisti Me": *You Circled Me First*. Almost as soon as the survivors got home, Transylvanus hailed their ship as worthier of fame than the Argonauts' because it had completed the circuit of the Earth ("In Orientem penetrans, rursum in Occidentem remeavit").[2] Oviedo concurred.[3] So did Peter Martyr, who thought the event was notable because circumnavigation was unprecedented.[4] The voyage of the *Victoria,* opined Antonio de Herrera y Tordesillas, appointed chronicler of the Spanish Empire in the last years of the sixteenth century, was "the most admirable thing any man has ever seen since the creation of the world."[5] More admirable than the incarnation of God? The claim was obviously, tellingly extravagant.

Magellan got more than a collateral share of the misplaced glory. Pigafetta promoted his dead patron on the grounds that "no other had so much natural talent nor the boldness nor knowledge to sail around the world, as he had almost done."[6] Magellan had done no such thing. He had not even come close, or even envisaged it. Yet his apotheosis began. Christian Jostmann, one of the best of his modern biographers, associates the takeoff in his legend with nineteenth-century romanticism.[7] As we shall see, the celebration of the explorer achieved a new intensity and universality in that period, but it had begun with his death and had spread with Ramusio's publication of Pigafetta's narrative. It accelerated toward the middle of the century with Agnese's maps (above, p. 219). Gonzalo Fernández de Oviedo summed up the view of many contemporaries: although Magellan never reached "the Spicery, the praise of it is due to him alone, and to him this great voyage and discovery are ascribed."[8] In 1589 Ortelius accompanied his image of the *Victoria* (above, p. 212) with immortalizing doggerel:

> I circled first the world with flying, billowing sheets:
> Magellan! Through the strait thou led'st me to new feats.[9]

Johannes Stradanus and Theodore de Bry, whose engravings fixed images of exploration and imperialism in late sixteenth-century minds, depicted Magellan in armor—like the knight he was—with a celestial sphere at his side, as if he comprehended the cosmos. Neptune and a rather portly mermaid are in attendance, surrounded by romanticized Patagonians, with Apollo bearing a lyre, as if to demonstrate deeds worthy of memorialization in verse.[10] Magellan's portrait crowns those of select circumnavigators in a Dutch narrative of the first voyage around Cape Horn in 1615.[11] Engravings of him in various heroic poses multiplied in the following century, often with a globe as a prop: in Holland, for example, in *De Nieuwe en Onbekende Weereld* of Arnoldus Montanus in 1671 and in Fredrick de Wit's *Atlas* of 1675, where (if Magellan is indeed the intended subject) he appears at his strait, with natives improbably smelting gold; or in France by Nicolas de Larmessin, with dividers and a scroll, in 1695.[12]

Even those who think highly of round-the-world voyages should not credit Magellan for undertaking one. He never intended it (above, p. 126). He would have been foolish if he had, as his entire undertaking was based on the assumption that the shortest route ("nuevo y más breve camino") to and from his destination lay across the Atlantic.[13] Equally valueless is the suggestion that he might have completed the circuit of the globe inadvertently, having made an unrecorded trip eastward to the Philippines, prior to sailing there in the opposite direction.[14] There is no evidence of such an earlier voyage. Similarly, claims advanced on behalf of one of Magellan's slaves, on the grounds that he might previously have visited the Philippines, or even that he might have been born there, or have returned to his birthplace on some undocumented later voyage, have no evidence to support them, although they have some amusement value in assigning the achievement to a subaltern figure, of non-Western provenance and dark pigmentation.[15] It is good to see a slave getting credit usurped by a master. In any case, disputes over who should be credited with this meaningless "first" are like power struggles in a bunker: a waste of time, effort, and emotion.

Nor does anyone who shipped with Magellan seem to have envisaged returning to Spain by completing the circuit of the Earth. Rather, the route was a desperate expedient, tried when other options seemed impossible. The only contrary evidence, as far as I know, was Pigafetta's speculation (above, p. 216) that the fleet could have continued westward around the world from Cabo Deseado. That, however, could have been an afterthought, occurring only when he wrote up his account after the circumnavigation had been achieved. The allure of putting "a girdle round the Earth" has remained seductive—and, for most of the time, diminishingly meaningful—ever since.[16]

Circumnavigation contributed, no doubt, to what Joyce Chaplin calls "planetary consciousness"[17]—awareness of the possibility of world-spanning connections, of the graspable world that Carlos Borja felt between his hands (below, p. 267). But it had no scientific consequences.

Everyone who thought about it already knew the Earth was round. There are surely more flat-earthers in our world than in Magellan's. Francisco López de Gómara, the humanist chronicler to whom contempt came easily, was suitably curt: "Any reasonable man, even if ill educated, can see the error of those who once thought that the world is flat, and no more need be said."[18] If anything, educated people overestimated the Earth's sphericity, assuming that it was a perfect sphere, on the grounds that God could hardly have been so blundering an artificer as to produce a world out of true. The demonstration of the real shape of our planet as an oblate spheroid awaited the speculations of Newton and the work of La Condamine and Maupertuis in the eighteenth century.[19] Stefan Zweig's paean to circumnavigation is so overwritten as to be self-undermining: "With the circling of our planet, pursued in vain for a thousand years, humankind acquires a new intimation of our capabilities, discovering ourselves, revealing our own greatness, increasing our joy and our worth in proportion to the greatness of the size of the world."[20] People can succumb to hyperbole, but its emptiness is obvious.

The nearest thing to a scientific revelation from the voyage Magellan initiated came when expeditionaries expressed surprise, toward the end of the voyage, to find that their own count of days was one day behind that of people who had stayed at home. Peter Martyr of Anghiera made a great fuss about the paradox, "which," in the version of his first English translator, "will fill the readers with great admiration, especially those who think they have the wandering courses of the Heavens beforehand";[21] but in a world where "the sun that bids us rest is waking our brethren 'neath the western skies" the solution occurred readily enough to whoever gave it thought, and might have been expected in advance of experiment. Magellan's own scientific expertise was slight. Pigafetta praised his skill in navigation but offered no evidence except the commanders' reproaches to his navigators (above, p. 219). Magellan's memorandum on the location of the Moluccas was all wrong and in any case was probably indebted to Faleiro's work rather than to his own.

. . .

Ironically Magellan has attracted adulation for what did not concern him—the circumnavigation of the world—but none for his real achievement: crossing the Pacific and demonstrating its vast extent. His voyage should have had a valuable effect on the prevailing world-picture, as it proved that the world really was much bigger than Columbus and the cosmographers he influenced had claimed.

Circumstances, however, combined, to cheat Magellan of credit. His pilots and navigators had to perpetuate the myth of a narrow sea or "gulf" (above, p. 195) between Asia and America in order to shore up Spain's dubious claims that the Moluccas lay on the Spanish side of the Tordesillas antimeridian. The range of exploration and the proliferation of globes that could be handled favored the spread of a sense that the world was graspably small; as Carlos Borja expressed it, evidently struggling for something to say in a thank-you letter for his uncle's gift of a globe in 1566: until he grasped its image in his hands, he had never realized how small the planet is.[22] Magellan's transpacific voyage demonstrated the real immensity of the Pacific, but the most widely diffused accounts of the journey encouraged cartographers to show it as navigably narrow. All sixteenth- and seventeenth-century world maps squeeze the Pacific into a seductively, traversibly narrow slimness.[23] In Spanish imperial geopolitics, the colonies on either side of the ocean were always "one"—close enough to be treated as a unit.[24] As we have seen, voyagers in Magellan's wake consistently underestimated the distance they had to traverse, even in some cases on repeat voyages (above, p. 232). The Jesuit polymath José de Acosta, who wrote what he called a "moral" history of the New World toward the end of the sixteenth century, thought the greatness of the *Victoria*'s voyage was that it demonstrated that the Earth was subject to man, "because he could measure it."[25] Perhaps in a sense Acosta was right: in principle, the length of a journey around the Earth was measurable, but no practical means were available to do the job.

. . .

In the immediate aftermath of Magellan's death, the flotilla was literally without direction. When they fled Cebu, the survivors' first and overwhelming concern was to get fresh provisions. They regrouped off the nearby island of Bohol, with its famous landscape of humpy hills. There were not enough men left to man three ships; so they scuttled *Concepción* and redistributed her cargo and crew.[26] Most of the officers had died at Cebu. Whether by election or emergence, Carvalho assumed overall command, while Gómez de Espinosa took over the *Victoria*. Giving evidence years later, Espinosa omitted mention of Carvalho and said that he "was chosen" leader in Magellan's place (fue elegido en su lugar):[27] perhaps a pardonable oversimplification.

Fending off "great hunger," they went island-hopping, heaping elephants' backs with presents for the rajah in Palawan and capturing junks off Brunei.[28] Sometimes they enjoyed lavish hospitality, but they never acquired supplies for the long voyage still ahead. The ships needed caulking. The tale of famine interrupted by feasts continued. Tension mounted, while complaints against Carvalho's leadership mounted; another mutiny can be inferred, with Gómez de Espinosa taking over the flagship and Juan Sebastián Elcano the *Victoria*. The mutual resentment of Spaniards and Portuguese played no recorded part in the reshuffle, as if hatred had subsided in the course of common sufferings and in the face of shared challenges, but rumor blamed Carvalho for taking gold for himself in ransom for the captured junks, instead of bartering for food, and for monopolizing the services of kidnapped slave women (above, p. 143). Only a captured cargo of coconuts brought temporary relief.[29]

It took six weeks, according to Pigafetta, to career the ships off the coast of Borneo, taking them to mid-September. Here at last the men could reprovision. Thoughts revived of continuing to the Moluccas, but no one aboard knew how to get there.[30] They wandered, following the unreliable directions of captured wayfarers or the counsel of kidnapped native pilots, until on November 6, 1521, "we discovered four lofty islands." A native pilot—the only one they had at the time, according to Pigafetta—identified the find as part of the Moluccas. "Therefore we thanked God as an expression of our joy and discharged all our artillery."[31] Reflection on the hopelessness of exploiting such a long route as they had traversed might have modified the joy.

Accurate readings of longitude, had they been able to make them, might also have had a depressing effect, as the ships had left Spain's permitted zone of navigation. Since the loss of Andrés de San Martín, it is not clear that anyone competent in the science of longitude was aboard. Albo's log resumed data on longitude when the fleet reached Borneo. The figure given for what he called Brunei—201 degrees and 5 minutes "from the line of demarcation"—is roughly correct if the Tordesillas line is what Albo meant and if we locate it about 46 degrees west of Greenwich. A reading of this sort would have shown the explorers that they had got well into the Portuguese zone; such would not be the case, however, if Albo, despite the wording of his log at this point, were calculating from his normal point of departure, the meridian of Seville. In any case, since the explorers were also beyond the Moluccas, any suspicion that Borneo, or part of it, was in the Portuguese sphere need not have been a cause of despair—even had the reading of longitude attracted confidence. It surely did not. No seaman trusted estimates of longitude. Even latitude, as we have seen, was a matter of rough measurement and rougher guesswork (above, p. 151). Albo's latitude put

Brunei (if he had in mind the place that now bears that name) much further north than in reality—about a degree and a quarter too far. It should have been possible for a pilot to do better than that on land.

In the direction of the Moluccas, Albo provided his next reading for "the island that is on the equator": it sounds like Moti (if that is the right identification for what he calls Motil), which he placed slightly too far south, exactly on the line. Albo, however, gives two contradictory values for "the island on the equator" at different points in his log: the first, perhaps intended for Ternate if not Motil, at 190.5 degrees west of the Seville meridian; the second at 191.75 degrees west. The math may seem tiresome, but it is vital to understand it if we want to know where the explorers thought they were and, in particular, whether they thought they had strayed into Portugal's privileged sphere. There is no justification for assuming that Albo's measurements were made from the Tordesillas line: here, his language unequivocally indicates Seville as the starting point.[32] Both his readings therefore place the islands emphatically on the Spanish side of the Tordesillas antimeridian by a good 30 degrees or so. Pigafetta's assignment of Tidore to "161 degrees from the line of demarcation" is equivalent to a longitude of about 200 degrees west of Seville[33]— significantly further than Albo (and at least 70 degrees short of the true figure) but within the same order of magnitude. As Albo also says that the archipelago is aligned from northeast to southwest, it is theoretically possible that part of it, in his mental map, might have been much closer to the Portuguese side. The widespread assumption, on the other hand, that Albo located the Moluccas in Portugal's domain is false.

. . .

Ternate was the best place to do business, as Magellan knew from Francisco Serrão's letters. But as Serrão was dead and the likelihood of a friendly reception in Ternate was slight, Elcano and Espinosa set up shop in Tidore to exchange their truck and cloth for cloves, nutmeg, and mace. No single island could supply the amount for which they had the capacity and, at local prices, the means to acquire. So the ships at anchor became a sort of emporium, frequented by merchants from all over the archipelago.

The Spaniards made treaties, most notably with the sultan of Tidore: Elcano was still clutching the precious documents when he got back to Spain. The ships ended up with more cargo than they could safely carry; so they got the ruler of Tidore to agree to warehouse part of it. They left their surplus truck and other rubbish behind. The list makes depressing

reading, including great heaps of unwanted textiles, four dozen putres-
cent hams, eight dozen combs, 150 dozen hawks' bells, seventy-five
dozen red caps, a small pair of scales, a big copper oven, ten broken
guns, two broken anchors, and five dozen squeakers or whistles, made
in Paris, that had been stowed by Mendoza as his personal quota of
trade goods. Five hundred sewing needles remained as a present for the
sultan of Tidore, who presumably did not want them. A saddle of velvet
and silk was left for the same beneficiary.[34]

The ships were in poor condition: *Trinidad* too wormed and leaky to
make the voyage home. Elcano was therefore deputed to captain the
Victoria back to Spain at once, while her sister vessel underwent repairs.
The Spaniards learned that Portuguese squadrons were hunting them
and had tried to intercept them before they left the Atlantic. The question of how best to elude the pursuers on the way home had to be
confronted.

The obvious course was to follow what was surely Magellan's original plan and to try to return to Spain by the way they had come. The
strength and regularity, however, of the winds that had brought the fleet
across the Pacific precluded that. Nonetheless, the only way to keep out
of the Portuguese zone, and, therefore, both to follow the king's orders
and to evade pursuit, was to set sail to the east. Central America, or
Darién, as most documents called it, was the nearest friendly destination, where Spanish colonies were already present. The main problem
was that of finding westerlies: in the latitudes of the Moluccas none
were available and the currents were adverse.

Presumably on the basis of consultations with local seafarers,
Gonzalo Gómez de Espinosa resolved to attempt the eastward run in
the *Trinidad,* heading northeast in search of a favorable wind. "I prayed
God," one of his shipmates, Juan Bautista de Punzarol (more properly,
perhaps, Giovanni Battista da Punzarolo in his native Italian), later
recalled, "that the ship could be ready in fifty days to go to Darién." In
fact, it took four months before the *Trinidad* was ready to put to sea. A
sketch of her in full sail, but low in the water, appears in Diego Ribero's
world map of 1529, accompanied by the legend, "This is the ship *Trinidad,* which, seeking a way to the Southern Sea, but encountering adverse
winds, ascended to latitude forty-two degrees north, and from there
returned to Maluco once again, because she had already spent six
months at sea and was leaking water and lacked provisions."[35] Not
only did the voyage shatter Magellan's record for nonstop seafaring: it
also demonstrated the worthlessness of the project of a Spanish route to

the Moluccas. Even had it been possible to reach the islands via the Atlantic, the return voyage would have to traverse Portuguese waters. No one found a practical way of crossing the Pacific from west to east until the 1560s (above, p. 232).

The misery of the first attempt to do so fills Gómez de Espinosa's reports to the king, made—partly in the form of his letter of January 12, 1525, from his prison in Portuguese-held Cochin, and partly as testimony to an inquiry in Valladolid in 1524.[36] The voyage took seven months. In spite of relentlessly struggling northward to forty-two (according to Espinosa's letter) or (according to his testimony) forty-three degrees of latitude, the searchers never found the wind they sought. They endured a storm that lasted twelve days, while most the crew "grew weak because they had nothing to eat."[37]."

Of about fifty-five crew members, only twenty or so were still alive and aboard when they got back to the Moluccas. "Because the crew of the said ship were sick, and many of them were dead, and there were insufficient hands to man the ship," Espinosa appealed to the Portuguese in Ternate to help him back to Tidore.[38] To Espinosa's professed astonishment, "because this witness took it for certain that the said Moluccas belonged to the royal crown of Castile and fell within the limits and lines and demarcations thereof and because this witness had taken possession of them in his Majesty's name," the crew were arrested. When they were taken to Ternate and put to work building the Portuguese fort, Espinosa refused, responding in the spirit of the River Kwai. "This witness told Antonio de Brito, Portuguese, who was their leader, that if he had to put a stone in place, it would be in the name of the king of Castile."[39] Ginés de Mafra, on the other hand, admitted to taking part in the work "for fear that they would kill him if he refused" (de miedo que lo matasen sy lo dease de hacer).[40] After being shunted around the Portuguese East and imprisoned for years, while more men died, the last four survivors made it back to Spain amid sufferings we have already described (above, p. 134).

. . .

According to Pancaldo's evidence to the inquiry of 1524, the decision to try to send *Victoria* and *Trinidad* home by different routes was a rational division of risk "so that one ship or the other might reach Spain and give news to his Majesty."[41] In consequence of the decision, the *Victoria* was bound for Spain via the Cape of Good Hope, through the Portuguese zone that the king had expressly forbidden the explorers to infringe. António Brito, the Portuguese commander on Ternate, had no reason to commend his Castilian enemies, but he acknowledged that "in view of the state of the ship, wasted and worm-eaten, and of the ill will of the crew toward the captain, and further of the many traps that await Spaniards in the East and the difficulties that I myself shall not fail to put in their way," the *Victoria's* return voyage would be as demanding as the outward one.[42] He presumably picked up evidence of dissension in the crew from his interrogation of the survivors of the *Trinidad*. His information was probably valid. For unknown reasons, for instance, Pigafetta had transferred to the *Victoria* for the homeward voyage, and it is hard to imagine that he, as Magellan's champion, and Elcano, as Magellan's detractor, were mutually cordial.

The horrors the crew of the *Victoria* experienced were almost as harrowing as those of her sister ship, but less protracted and rather less deadly: only four survivors from the *Trinidad* got home to Spain, whereas eighteen men were still alive aboard the *Victoria* when she reached Seville, and thirteen shipmates whom the Portuguese had captured were returned soon afterwards. Leaving Tidore on December 21, 1521, the ship stopped for repairs at Alor Island; after reconnoitering the fabled sandalwood markets of Timor, she struck into the open ocean. Meanwhile, Pigafetta continued to gather ever more sensational and salable stories, allegedly by native report but evidently from ancient Greek and Indian myths: of pygmies whose huge ears dangled to their ankles, of an island of women whom wind impregnated, and of birds so big that they could hoist elephants in their talons.[43] The most pressing concern was to elude Portuguese ships—which meant skirting far to the south, to the very edge of the roaring forties, which would have blown them back off their course and represented, for them, the southernmost limit of navigation.[44] On April 30, when all provisions save rice were exhausted, they had barely begun to climb back to hospitable latitudes.

They rounded the Cape of Good Hope about the middle of May and struggled north, casting their dead into the sea, and quarreling over whether to throw themselves on Portuguese mercy. "The ship was leaking badly," according to Pigafetta, but some "more desirous of their honor than of their own life determined to go to Spain living or dead."[45] A sort of democracy had taken over the direction of the ship. Eventually a majority of those still alive opted to pause at the Cape Verde Islands in the hope of finding fresh food and water.[46]

They were aware of the need to keep secret their infringements of Portuguese waters; but, having no cash, they tried to pay for provisions with cloves. The landing party, perhaps thirteen strong, was arrested. The ship badly needed repair as well as supply, but vowing that they would rather founder at sea than fall into Portuguese hands—that at least was what Elcano said, perhaps in an attempt to gild the survivors' shame or embarrassment with specious heroism—those aboard the *Victoria* abandoned their shipmates.

They were too few and feeble –maybe twenty-four souls at this point in the voyage—to make much speed. They did not clear the Cape Verde Islands until late July. They had, by their own reckoning, to go as far north in the Atlantic as the *Trinidad* in the Pacific to find a westerly wind,[47] uttering the same imprecations against the cold, before struggling into Sanlúcar on September 6, 1522, more than three years after they had left the same port in Magellan's company.

. . .

The voyage had been an unmitigated failure. It was amazing that anyone got home alive; and to return with any sort of cargo was enough, perhaps, to make commodity marketeers salivate. The common opinion, however, that the expedition made a cash profit is false. That opinion derives from a tally drawn up by the Casa de Contratación in 1537 in the course of a legal wrangle with Cristóbal de Haro over the reimbursements and other sums due to him for his investment in the expedition (above, p. 108).[48] The figures deployed at that time showed total expenditure of 8,334,335 maravedíes. But that was an underestimate. The costs itemized during the preparation of the fleet add up to 8,480,684 maravedíes.[49] Many other costs were incurred subsequently, for messages dispatched to the fleet after it set sail (above, pp. 136–7), arrears of pay, pensions, disbursements to widows and other heirs, and the administration of the many inquiries and lawsuits that followed the return of mutineers and survivors. Staggeringly, nearly 400,000 maraviedíes went on costs of accounting. Of the money the crown received from the liquidation of cargo, 2,793,157 maravedíes were divided among five religious houses and forty survivors, their heirs. The heirs of selected fatalities made other costly claims. The Casa de Contratación reckoned the yield at 8,680,551 maravedíes. Almost all of that came from the sale of the spices the *Victoria* brought to Spain; clearly, however, the yield did not cover all the costs. Even at the Casa's nominal figure for cost, the total excess of income over expenditure for the voyage was only 346,216 maravedíes—less than the amount spent on drawing up the accounts.[50]

It is the role of public expenditure accountants to put the best possible gloss on official profligacy. In the case at present under consideration, even if we admit their calculations as valid, the values that underlay them were warped and wicked. The Casa officials' estimate of 350,000 maravedíes or so of profit would represent a yield of about 25,000 per life lost of members of the expedition—less than a year's wages of one of the pilots (above, p. 112), to say nothing of the suffering of the crews, the villages they burned, the havoc they wrought, and the deaths they inflicted in Patagonia, the Marianas, and the Philippines.

At the time, some people thought it was all worthwhile—people sufficiently influential to send another expedition in Magellan's wake in the hope that experience and improved management might cut the costs and increase the profits. In July 1525, Juan Sebastián Elcano was packed off to pilot the voyage with seven ships under Francisco García Jofre de Loaisa, who, like Pigafetta, was a knight of St. John. The voyage was even more disastrous than Magellan's and only served to confirm that the route he had pioneered was worthless. The Strait of Magellan proved hard to find, even for Elcano. The relatively easy passage the first fleet experienced was unrepeatable: conditions reverted to normal; the wind virtually stoppered the strait. It took seven weeks to get through it. By then, only four ships were still operative. Storms dispersed them. Two more vessels vanished, while a third survived with eighty men aboard by fleeing to an American port.

On the surviving ship, deaths mounted. When the survivors reached the Moluccas, their ship foundered. If they did not die fighting against the Portuguese or in the service of indigenous chiefs, they eventually fell into Portuguese hands. Perhaps half a dozen got back to Spain.[51] Meanwhile, efforts to locate the Moluccas from American ports showed that it was fairly easy to reach the islands, impossible—in the prevailing state of knowledge—to get back. In 1529 the Castilian crown cut its losses and ceded its claims over the Moluccas to Portugal for the almost risibly modest sum of 350,000 gold ducats at 375 maravedíes to the ducat, or 1,312,500 maravedíes.[52]

. . .

Failure is often an ingredient of heroic status. The tragic hero of Greek tradition is a victim of change from good fortune to bad. The Dunkirk Spirit is unavailable except to the defeated. Japanese hero worship reveres "the nobility of failure." The values of romanticism are those of the *Blaue Blüme,* unplucked because unreachable, or of the lover on Keats's urn, who "wilt never have thy bliss," or of the impossible dream or "the high that proved too high, the heroic for Earth too hard." In espousing a death in imitation of Roland's, Magellan seems to have subscribed to tragic values. Yet it is for spurious success, not noble failure, that he is revered. It is hard to see him as exemplary in any conventional sense, surely not as "good," even by the patchy standards of other secular heroes. His apologists blame his accusers for branding him with charges of treachery, mendacity, recalcitrance, evasion, cruelty, egotism, bloodshed, and crimes—"a cargo of lies," decried by Samuel Eliot Morison, "to cover mutineers' misdeeds."[53] When the San Antonio brought the first escaped mutineers home, Peter Martyr was shocked at their "lamentable charges against Magellan. We believe that disobedience so grave shall not go unpunished."[54] Yet the charges—and others not officially brought against him—were all true. Magellan did flaunt royal commands by failing to share his planned route with his subordinate officers and navigators (above, pp. 137, 139) and by bypassing the Moluccas (above, p. 221). He did exceed the proper limits of his authority as commander. He did usurp sole power in his fleet. He did commit murder and judicial murder (above, pp. 164, 166). He did waste royal resources when he wintered in Patagonia. As a commander he failed in his duty: his preparation for the voyage was inadequate, his execution negligent. He took needless risks, which ultimately led to starvation, sickness, and defeat. He managed his men poorly, provoking—or failing to forestall—mutinies.

His conduct toward the indigenous people he encountered was such as in most other standard-bearers of European imperialism usually incurs odium today. Columbus, whose comportment was, at worst, equivocal and at best highly appreciative of Native American cultures, attracts extraordinary obloquy and irrational vengeance for offenses he never committed or conceived. Enragés topple his statues, blacken his memory, and, in my own university, occlude murals that glorify him in order to protect them from onlookers and onlookers from offense. Statues of St. Louis are in peril in the very city named after him because he led a crusade, as are those of St. Junípero Serra in California because his beneficence, love, and sufferings for his Native American congregants were insufficient to assuage modern wrath. Churchill's, Washington's, and Jefferson's are among the monuments denounced as commemorating the vices of their times rather than the virtues of their persons.

Magellan's behavior was at least as bad as that of any of these. Nonetheless, statues of or monuments to him in Lisbon and Sabrosa stand without, as far as I know, attracting objections from or causing offense to the local bien-pensants, or from more traditionally inclined Portuguese who recall his treason. Those in Mactan and Guam, where he launched invasions and burned natives' homes, or Cebu and other parts of the Philippines, where he tried to found an empire, or Punta Arenas in Patagonia, where his native victims had cause to regret his coming, are undefiled, except by the memory of his own excesses. The only Magellan monument I know to have been destroyed was in Manila, in 1945, in an accident of war, not a postcolonial protest. The survival, beyond approval of the dead, of statues, street names, tombstones, plaques, and other *lieux de mémoire* always makes me glad because I think that to understand our present we must embrace the whole of our past, including the bits we dislike; the inconsistencies, however, puzzle me.

If the truth were known or generally acknowledged, Magellan would surely attract at least as much obloquy and outrage as any other former hero now deemed politically incorrect. His escape owes something, I suppose, to his reputation as a scientific explorer. As we have seen, he does not deserve it. Yet he has become a standard recourse for scientific naming projects. Dwarf galaxies first recorded in Pigafetta's narrative were named for him in 1678.[55] Craters on the moon and Mars bear his name, respectively since 1935 and 1976. An asteroid was dubbed 4055 Magellan in 1985. The Magellan telescopes adorn the Las Campanas Observatory in Chile. NASA launched the Magellan Mission to Venus in 1989. Species of penguin and woodpecker in South America and of butterfly in the Philippines share the explorer's name.

The number of scientific awards named after Magellan attest to the assumption that he achieved something scientifically commendable. The habit of treating Magellan's name as authenticating scientific work started in 1786, when Jean-Hyacinthe Magellan persuaded Benjamin Franklin that the American Philosophical Society should administer a prize, the Magellanic Premium, for major contributions in navigation, aviation, astronomy, and science generally. "Nititur in ardua virtus" is the motto of the prize: it is true, no doubt, that virtue gleams amid hardship, but Magellan demonstrated that vice can be conspicuous too. The popularity of Magellan's soubriquet with organizers of university competitions is particularly remarkable as evidence of ignorance about the explorer in the loftiest circles, because universities are peculiarly vulnerable to students' woke sensibilities. The Magellan Prize, for instance, "offered through the generosity of the descendant family of the pioneering navigator, Ferdinand Magellan," is awarded in each year to "the best student beginning graduate studies in the University of Oxford in the language, literature, culture or history of the Portuguese speaking world."[56] Washington & Jefferson College publicizes its Magellan Project as

> a unique . . . project of possibilities. When you pursue a Magellan Project, you choose to take the lead in your educational experience. W&J College provides funding so that you can set sail in the summer months, exploring a passion, an interest, an internship, a study abroad opportunity, a research trip, or all of the above. You're out of the classroom, learning the ropes of whatever you've set your sights on. It could be right around the corner or right around the world.[57]

Here, the strangely unmeasured language of bright young university bureaucrats ticks as many Magellan-related boxes as possible: setting sail, exploring, passion, opportunities abroad, science, ropes and world-girdling. South Carolina University advertises an even more bouncily recommended "Magellan Ten Scavenger Hunt":

> OUR's Magellan Programs just turned 10. That's a big deal! Help us celebrate this major milestone—and win prizes—by taking part in the #Magellan10 Scavenger Hunt. . . . To participate, take a photo inspired by our daily #Magellan10 prompt and tweet it at us as fast as you can. . . . The first tagged photo . . . will win a Magellan Programs t-shirt and an Office of Research tote bag![58]

Less inappropriately, but still surprisingly, in Chile (where there is a "Universidad de Magallanes") the government bestows "the Strait of Magellan Award for Innovation and Exploration with Global Impact" to "explorers of the XXI century, who embody the same spirit of those who accomplished the epic achievement that connected the world: 500 years ago."[59] And "what," asks the website of a New York club,

> do General Douglas MacArthur, author James A. Michener, oceanographer Captain Jacques-Yves Cousteau, Senator John Glenn, newsman Walter Cronkite, religious leader Dr. Norman Vincent Peale, astronaut Dr. Sally Ride, ocean explorer Dr. Robert Ballard and balloonists Bertrand Piccard and Brian Jones have in common? They are among an elite group . . . whom the Circumnavigators Club has honored with the coveted Order of Magellan in recognition of their contributions to world understanding."[60]

Though it seems astonishing that an award for "world understanding" should be named for a failed conqueror who burned villages and coerced and killed people, the Circumnavigators Club of New York can be forgiven for promoting the notion that circumnavigation matters. Do many of them feel, like Mark Twain, at the end of his own round-the-world trip, proud "for a moment"?[61]

Commercial organizations, too, exploit Magellan's fame, with varying degrees of justification. The "global" resonance that attaches to his name, despite its unfitness, is evidently a marketable commodity and accounts for why a company that makes GPS devices calls itself Magellan Navigation, Inc., and why the business that manages my pension has a global investment fund called Magellan, to say nothing of a defunct Internet search engine of the same name.

A Canadian manufacturer is called Magellan Aerospace. Magellan is also the name of a firm that seeks "to provide . . . a fantastic motorcycle holiday." Magellan Vacations Inc. runs Magellan Luxury Hotels. You can find the Magellan guides online. The same name graces or disgraces a cruise ship that offers, incongruously, trips to see the Northern Lights. If you go to the Gold Coast Railroad Museum in Miami-Dade County, Florida, you can see the Ferdinand Magellan Railcar, which carried US presidents around their country in the 1940s and '50s. You can buy a 135-foot yacht named after him if you have appropriate amounts of money and sense.[62]

Magellan continues to accumulate evidence of adulation. "Following its postponement due to the COVID 19 pandemic," we hear that "The Tall Ships Races 2020 will return as The Tall Ships Races Magellan-Elcano 500 Series 2021." That is understandable. Less so is the fact that there is a company called Magellan Health ("Leading the community to healthy, vibrant lives"). The association with health seems unfortunate in view of the lives lost to disease on Magellan's ships and their role in spreading pathogens; yet other references echo it. "Our product," reports Dr. Abhiram Dukkipati, founder of Magellan Life Sciences, "is a protein-based sweetener that addresses the global sugar reduction problem. It tastes like sugar with none of the calories of sugar. Moreover, it is stable under extreme temperature and pH conditions, providing flexibility from a formulation perspective."[63]

Magellan is the name of a racehorse, a container ship, a Venezuelan baseball team, Magellan Petroleum Corporation, a former progressive metal rock band, a family of video games, an auction house in North Yorkshire, and a light fitting by Kuzco Lighting, "composed of multiple shades surrounding a single orb that emits a diffused light or in clusters."[64] Magellan House is the name of a care home in East Grinstead and 2,947 square feet of self-contained creative workspace in a dock in Leeds. "Magellan Outdoors Waterproof Leather Landman Snake Hunting Boots" are, presumably, named after the explorer; presumably, however, "Magellan's EveryWear Camisoles and Women's Full Cut

Briefs" are not. Still, it is an impressive tally of arrogated endorsements. Nowadays, when sponsors and advertisers regularly cancel the contracts of celebrities who commit offenses against political correctness, the unabrogated ubiquity of Magellan's name is strong evidence of how favorably people perceive him. Yet most use or abuse of his name defies the facts or strays into irrelevance. Magellan's immunity from the revulsion other paladins inspire seems insecurely founded on mistaken notions.

. . .

Of course, Magellan had admirable qualities. No one can indict his courage, or the resolution, dauntlessness, and perseverance that sprang from it, or the vividness of the way he imagined himself and pursued the future he pictured. Pigafetta praised his indifference to hunger and his constancy in the face of storms.[65] His skills in leadership were expensive, in terms of the lives his opponents lost, but he did manage to lead his men further from home than anyone, as far as we know, had ever gone before. And although he did not circumnavigate the world, he did get halfway round it. He was, as the cabin boy, Martín de Ayamonte said (above, pp. 135, 153), the seamen's preferred leader, a kind of demagogue or popular tyrant, representing the common man against Cartagena's court-appointed coterie. Magellan's own loyalty was negotiable, but he could inspire deep fidelity in others. His courage edged into recklessness, and he drove his expedition to disaster when the rational, prudent course was to turn back for home. But "the high that proved too high, the heroic for Earth too hard," are, according to a respectable code, "music sent up to God by the lover and the bard."

Good qualities cannot explain Magellan's resistance to dethronement. The rule of the people who want to tear down monuments, desecrate shrines, and besmirch renown is that no goodness can purge the taint of association with imperialism, slavery, incontinent bloodlust, unjust discrimination, or any of the commonplaces of the past that we now recast as crimes.

In a former age, however, Magellan was a good fit for the spaces myths make for heroes. He suited the Romantic era, when supermen were saviors and great men, including Carlyle, Hegel, Nietzsche, and William James, agreed on the glory of their kind. "Universal history is at bottom the history of the Great Men."[66] Amoretti's translation of Pigafetta, published in 1800, molded Magellan to the age.[67] The Hakluyt Society was founded in the 225th anniversary year of Magellan's death. In August of the following year, the society's ruling council voted that the vignette of the *Victoria* should be imprinted on the cover of all volumes. As we shall see, the image was a good choice of symbol for exploration in general, but the choice is further evidence of Magellan's uncontested status.

Marguerite Cattan has exposed the way nineteenth- and early twentieth-century writers crushed Magellan's life into a standard heroic narrative, patterned on epic, saga, sacred history, and hagiography.[68] The first step in what may not have been a conscious project was to emphasize mysterious elements in the story, such as the unknowns of Magellan's birth and early life. Mysterious provenance is a heroic topos. Episodes of adversity follow—starting with that of the orphan nurtured far from home in Magellan's case, and the narration of *mein Kampf* against rejection by men and desertion by Fortune. Magellan provided plenty of grist: the underappeciation that drove him from his homeland; the instances of sagacity and daring that enlivened his battle against alienation in distant climes; his purgation of suffering amid the travails of the great voyage; the peripeties of fortune; the demands of command, which deepened the hero's loneliness; heroic moments, frankly invented and expressed in purple prose.

Nineteenth-century biographers liked to differentiate Magellan from ordinary mortals by attributing to him presumed intimations of his own foreordained course, such as his sense of being "predestined for high deeds and immortal glories" (predestinado para altas empresas e glorias immortales) or his nursing ambitions of greatness from boyhood.[69] Rather than being the act of treachery contemporaries condemned, Magellan's defection to Spain was turned by later biographers into a Rubicon-like crossing. His most successful apologists in Portugal, Queiroz Veloso and the Vizcomde de Lagoa, wanted to rehabilitate Magellan as a Portuguese hero.[70] They succeeded: Magellan appears among the Portuguese navigators who cluster around the Infante Dom Henrique in a vast sculpture in Lisbon. The right to celebrate his voyage and commemorate his death has been fiercely contested between Spain and Portugal, where media organs have pestered me and, I suppose, other students of the subject to endorse the claims of one country or the other.

Outside the Iberian peninsula in the same period, historians failed to question the romance of the great man. "His faults, if faults they were, were those of strength," averred F. H. H. Guillemard, for whom Magellan exemplified the aristocratic virtues that—along with physical beauty, muscular Christianity, and sporting prowess—were among the qualities overvalued in England's old universities. Guillemard was a professor of geography at Cambridge and by no means devoid of critical intellect, which, however, seems to have deserted him. Magellan "bears," he wrote, "a name of untarnished honor. There is no singular story against him, nothing to hide or to slur over; no single act of cruelty in the age of cruelties."[71] The truth lies somewhere—or at different points and different times—between Guillemard's flawless paragon or Morison's matchless combination of "audacity, wit, and ability" and the "more or less . . . ruthless thug" Joyce Chaplin beheld in her history of circumnavigation.[72]

Twentieth-century sensibilities did not dispel the heroic tradition. Two curious examples illustrate its persistence. E. F. Benson, whose life of Magellan appeared, without scholarly pretensions but with deft use of existing works, in 1929, was no historian but a master of all literary trades, whose hallmark was accuracy of social observation, in devastating comedies of manners. Yet he could not behold Magellan with the same deadly glare that illumined his contemporaries' foibles. Magellan, for him, was one of the "great lamps of human enlightenment." His achievement was "supreme" and all his excesses were excusable.[73]

Even more surprising was the lack of judgment in the biography that Stefan Zweig wrote in exile in England in the late 1930s. A Jew who was a pacifist, a victim of persecution, and a refugee from Nazism might have been wary of national heroes and resistant to the superman myth of German idealists. Yet Magellan's apotheosis was never more fantastic than in Zweig's pages. How can the author really have believed that Magellan was "sworn to oppose needless bloodshed, truly the antithesis of the butchery of all the other conquistadors"?[74] Zweig's language echoes the propaganda of the flawless "men of destiny" who were his political enemies, embarked on "the holy war of humanity against the unknown . . . on wings of genius, . . . stronger than Nature herself, destined to be the unique one."[75]

The book, which has appeared in dozens of translations, has rarely been out of print and is, I suppose, the most widely read life of Magellan to this day. Instead of revulsion, it has inspired imitation. Most biographies since Zweig have repeated the heroic shibboleths. One biography—uncritical in every sense, yet widely cited—borrowed Pigafetta's tribute to "so noble a captain" for its title.[76] Of Samuel Morison, a redoubtable scholar, who devoted so much space to Magellan in his history of *The European Discovery of America* that his work amounts to a biography, we have seen enough to realize that he was spellbound by his subject's claims to greatness and protested too much in Magellan's defense.[77] His investment in heroics exceeded his discrimination in determining facts.

Most other biographers are best overlooked in the interests of kindness. In the twentieth century only two strove to deliver both objectivity and genuine scholarship: one near the beginning of the century, the other close to its end. Jean Denucé, whose study appeared in 1911, mastered to near perfection the material available at the time; his is still the most recommendable work on the subject (although his readers will detect, in the present book, points at which I correct errors he made and dissent from some of his judgments). Even he, however, could not shake off the heroic tradition, succumbing to the temptation to exaggerate ludicrously in calling Magellan "beyond contradiction, the greatest of navigators ancient and modern" (sans contredit, le plus grand des navigateurs anciens et modernes).[78] Denucé exonerated the hero from culpability in responding bloodily to his enemies. Even more strangely, he commended Magellan's generosity to fellow seamen, apparently because of a list of backlog wages, still unpaid in 1537 and due to 108 men who served on the expedition. (There are 107, according to the enumeration, but Francisco Albo and João Carvalho are squeezed into a single item.) No fewer than 31 of the men on the schedule are listed as in debt to Magellan, in sums ranging from a few hundred maravedíes in some cases—a mere 68 in the case of Andrés de la Cruz, a cabin boy—to substantial amounts: up to 25,034 maravedíes in the case of Francisco Albo.

It is not clear, however, how or why such debts were incurred or whether the figures include interest. Some of the larger sums were evidently the result of advances to supernumeraries recruited by the commander, such as Gonzalo Hernández, explicitly mentioned as lacking the king's commission, who was owed 15,190 maravedíes, the Portuguese Alonso de Ebora (10,600), and Alonso Coto (17,186). It is impossible to say how much charity, if any, was involved on Magellan's part.[79]

For objectivity, a rival to Denucé's work did not appear until 1992, when Tim Joyner, in retirement from his career with the National Marine Fisheries Service, published *Magellan* in a genuine attempt to make the subject seem human.[80] Underacknowledged because, I suppose, of the writer's lack of professional qualifications, it was a commendable effort, not least in recognizing the extent of Magellan's failure. But the author lacked the contextual knowledge, historical sensibility, humanistic discipline, and factual command the task demanded. He tried so hard to restrain value judgments that he ended with a picture of Magellan too ill defined to be convincing or even coherent. Tradition, moreover, continued to shape his narrative along classically tragic lines, portraying what is still, at least in glimpses. a recognizably heroic downfall as an outcome of hubris. Since then, although shabby, showy work has attracted praise and sales, only Christian Jostmann has managed a creditable biography of Magellan. Though he calls his book a popularization, scrupulous scholarship underpins it, albeit unfortunately without the support of a critical apparatus, and the author, immune to the magic of great men, proclaims a "demythologizing" project.[81] Yet the curse of the heroic tradition persists: the book is a narrative of adversity, triumph, and tragedy in the career of a crusader, cursed with a chancer's addiction to risk.[82]

. . .

The heroic mask hides the dynamism of Magellan's character and the real nature of the models that influenced him. Hubris and adversity are part of a truthful picture; but to understand their roles and how they dovetail with other elements of Magellan's life, we have to discard literary patterns and look back over what really happened. Early orphandom is a bad start for anyone in life. On the psychoanalyst's couch, Magellan would have to say how he felt about it. His feelings are inaccessible, but the experience of dislocation and removal from his provincial home to a distant and unfamiliar court is a fact. We should shut out the temptation to imagine Magellan in the royal palace, nursing resentment or accumulating vexation amid boys richer and nobler than himself, like a social misfit in a snooty boarding school. But we can say for certain that he learned two things: a sense of entitlement, such as all noblemen and royal pensioners had, and the chivalric ethos that informed his attitude to the world.

If I am right in the inferences I have drawn from the evidence of his readings in chivalric romance, fiction shaped his future: he could identify with Primaleon and other protagonists, down on their luck, for whom far-flung deeds of derring-do were the only means to social ascent or recovered birthright. Portuguese courtly education was specifically geared to preparing young men for service in the empire, as conquests and commerce stretched ever-longer, ever-feebler fingertips eastwards. Magellan embraced the opportunity. His first hopes of a fortune foundered on the shoals of the Maldives. His demand for an increased stipend failed. His equivocal record on campaign in Morocco seemed to condemn his career to stagnation. But knight-errantry beckoned: first, across the Castilian border, where he could join the country's most prestigious knightly order, then in a vocation as an explorer—which, after the church and war, was the commonest route to reward in early modern Iberia.

At no point in his life did his character stop developing. Some constant traits were in place before he set out from Seville: he had shown courage and leadership in the Maldives, prowess in wars around the rim of the Indian Ocean, resourcefulness and defiance of enemies in Morocco, arrogance in making demands of the king of Portugal and in abjuring allegiance when he felt rejected. His vulnerability to resentment, outrage, and scorn toward others were already obvious. His record was full of daring that edged into recklessness. In negotiating with Aranda and with the agents of the Castilian crown he showed canny calculation and an awareness of when to make concessions—not so much in a spirit of compromise as in calculation of how easily concessions could be reversed or ignored at need.

His ambitions are readable in the terms of his contract with the king: he wanted to rule a fief of his own on the fringe of the world—as far as possible from effective royal interference, in a fade-out reminiscent of the dénouements of those tales of seaborne chivalry on which he modeled his life. Like Columbus, when he bade for patronage he was willing to change his pitch to suit the potential sponsor. The king of Spain wanted a short route to the Moluccas; so that was what Magellan proposed and that was where he focused his explicit goals. But Francisco Serrão had already set up there in an enviable degree of autonomy, power, and wealth. It made better sense for Magellan to look further, to the Philippines, which gleamed with gold and offered untouched prospects for self-aggrandizement. His ambitions did not include circumnavigating the globe: there is no evidence that he ever thought of such a thing, and, if he had, it would not have cohered with his other objectives.

In terms of character development, the expedition he led can be divided into two phases: the first, dominated by the perils of the Atlantic, the disillusionment of elusive success, the hardships of the voyage, and the menace of mutiny. He responded with ruthlessness previously unexemplified in his life but consistent with his audacity and ambition. He eliminated opponents by three of the most elementary means imaginable: stabbing, strangling, and stranding. Ginés de Mafra recalled him as "pensive" amid manic mood swings along the Patagonian coast, alternating hilarity, whenever a possible strait opened, with depression when hopes faltered.[83] Other dissidents obliged by deserting. By the time he emerged from what would come to be called Magellan's Strait, his demands for obedience and even assent reached proportions consistent with paranoia.

In the second phase of the voyage, Magellan never fully recovered a sense of reality or a willingness to be constrained by practical considerations; he drove the fleet—unhesitatingly, in defiance of the king's orders and of his men's interests and well-being—beyond its declared objectives, probably in order to get to the Philippines. Now, however, the elements of sea and wind worked to change him unprecedentedly. Hardship sharpened, as the fleet faced starvation and sickness; yet the following wind and his awareness of nearing the lands he sought induced a new atmosphere of freedom, perhaps of irresponsibility, enhanced by the challengeless, vast, unpopulated, and "pacific" sea. God seemed at last to favor him, and perhaps to spare him for great deeds: mere survival was evidence in an expedition on which most men died. A mood of religious exaltation overtook some expeditionaries: in Pigafetta the mood was explicit, after his apparently miraculous escape from death. In Magellan it can be inferred from his actions in the Philippines, where he set up as a miracle-working healer with a hot-line to God, and as a sort of street preacher, haranguing the natives in a theology he barely understood himself, sidelining the expeditions' chaplains in his self-appointed role of evangelist. Religion was never his lodestar: his grasp of Christianity was superficial and did not stop him from deploying terror and arson against enemies, or provoking a war that was unjustifiable in the terms St. Augustine laid down. He reverted to his long-standing values in the end. He leaped into battle and resigned himself to death in imitation of long-standing chivalric heroes.

What about the problem I broached at the beginning of this book: that of why Magellan, his backers, and shipmates were willing to take outrageous risks against all reasonable chances of success? Here, 120,000 words or so later, we may be no nearer to a solution; but an approach to one lies through an examination of the exploring itch in general. We need to see Magellan in the context of the history of exploration as a whole.

. . .

"It is a strange serpent," said Lepidus. The crocodile—had they met—would doubtless have thought Lepidus equally strange.

Every species is unique, but humans think themselves unique in a unique way. Most such self-congratulation is unwarranted. Primatology and paleoanthropology show that we share with other creatures, in varying degrees, the capacities for culture, cognition, and communication that we once thought peculiar to ourselves. Maybe a fully objective observer—beholding us from the cosmic crow's nest, at an immense distance of space and time, or from another universe—would define us as creatures unique only in thinking ourselves unique.

The galactic onlooker would also surely notice that on our little planet we humans are distinctive in two modest but puzzling ways. First, we occupy a staggering range of environments: apart from the bacteria we carry with us, no other land-based species has spread to more of the Earth. Second, despite our unparalleled diffusion around the globe, human communities are in touch with each other, whereas populations of other creatures know only those near them. Explorers have led us into both these conspicuous forms of oddity—beating the paths that peopled the planet, finding the routes that reunited sundered cultures.

Why? Is exploration an innate or acquired impulse? A rogue feature—a weird genetic mutation, perhaps—or a common human trait unusually concentrated in some people? Is it a response to material circumstances or the property of a "spirit" or the result of some transcendent quality of mind or soul? Are explorers driven or drawn?

Maybe there is no such thing as a typical explorer. Maybe the search for a universal explanation of what makes exploration happen is doomed to failure, like expeditions to El Dorado or Shangri-La. At different times, in different cultures, and in different individuals, we can detect no uniform exploring profile but contrasting motives and pressures leading to similar results.

If we go back to the first explorers, whose existence we can postulate but not document, we confront the most puzzling of all cases: that of people who forsook their environment of origin in the Rift Valley of East Africa to lead tiny groups—small, biddable bands or just a family or two—across unfamiliar terrain to destinations that demanded unprecedented adaptations. Something of the sort happened perhaps a million years ago or so, among communities of *Homo erectus*. About one hundred thousand years ago, long after *H. erectus* had vanished from the Earth, individuals of our own species traveled in his wake and ended by vastly exceeding his range.[84]

Our fellow apes can help us understand how extraordinary the proceedings were. Chimpanzees, for instance, stay in the forests to which they are well adapted, but they often break with their communities of origin to venture into what for them is the unknown—beyond the control of their tribes, in territory often patrolled by hostile bands. The commonest reason for such secessions is sexual frustration. Young males, whom elite rivals exclude from opportunities for mating, have two options: mounting a sort of coup against the existing alpha male and his cronies or seeking fulfillment elsewhere. The ancestors of mountain gorillas were surely escapees from restricted opportunities, perhaps of various sorts, in the lowlands. The impulses that drove the first human venturers beyond their homelands were—if the analogy is valid—of similar kinds. Internal competition for females or food or other resources, or fear of rivals at home, or the rigors of war or of lower-level forms of violence, or the results of defeat could set them off and keep them going.[85]

Material gain or exigency justified most of the risks we know about: people to enslave or trade with advantageously; land to cultivate or cull; minerals to mine; conquests to exploit for tax or tribute; exotica cultivable only in new environments; allies to deploy in war or rebellion.

When did less crudely calculated motives, such as scientific or religious zeal, intervene? As so often, with early examples of almost everything now seen as launching Western traditions, Herodotus is the source of the first known examples: in the fifth century BC he recorded travelers who, starting along established trade routes, pursued them further than the merchants to report on the mist-enshrouded seas of the boreal Cimmerians, or the dreamlike interior of Asia, where ghosts and monsters dwelt,[86] or West African lands, reportedly of beasts "with shaggy bodies, whom our interpreters called gorillas."[87] A couple of hundred years later, Buddhist *jatakas* relate unreliable tales of journeys to spread the faith; well-documented cases follow, from about the fourth century AD, of Buddhist missions from China to India to gather texts or to Southeast Asia to make converts.[88] Irish monks whose leather-bottomed curraghs explored North Atlantic islands, perhaps as far as Newfoundland, in late antiquity and the early Middle Ages were practicing a form of penitential self-exile in which wandering, not arriving, mattered. "Is not God," one abbot remarked, "the pilot of our little craft? . . . He directs us whither He wills."[89] Some of the Norse navigators who established colonies in the same region a little later were expulsees from their homelands, compelled to wander as convicts for crimes or losers of quarrels.

All the motives discernible so far seem to have coalesced in exploration's reputed "great age" of European out-thrust to much of the rest of the world from the fifteenth century to the eighteenth. The material greed, the imperial impulse, the missionary justification, the scientific curiosity: all are familiar to every reader of the literature. But what about that elusive exploring "spirit"?

I think it was there. You see it documented, to begin with, in two kinds of fiction: first came hagiography, which elevates seaborne wanderers, such as Brendan and Eustace, to the rank of saints, hallowed by their sufferings and their self-exposure to God's will in form of wind and wave. Romances of chivalry followed—late-medieval light reading, which, as we have seen, was the common reading matter of the age and which Magellan shared. Fiction has a habit of anticipating reality, especially if readers take it seriously and model their lives on it. That is what happened with the romances of seaborne derring-do that inspired explorers. The navigators who brought Atlantic archipelagoes into the ambit of European knowledge, commerce, and settlement in the fourteenth and fifteenth centuries were often disreputable cutthroats, but they gave themselves storybook names, such as Lancelot and Tristram of the Isle.[90] Cartographers gave their discoveries (and the speculative islands that they scattered over their maps) names derived from Arthurian myths.[91] When members of the entourage of a Portuguese prince reached and rounded the West African bulge in the 1430s, he spoke of the natives they found there as "homines sylvestri"—the "men of the woods" of chivalric legends.[92]

Much has been written of the religious zeal of Columbus and Cortés—but they came to it late, as most of us do, and as Magellan did, when disappointment had embittered them and the failings of human patrons made them confide in God. Their initial quests were for social ambition—to follow the plot of a chivalric hero to fame and fortune. One of Cortés's men likened the sight of the Aztecs' city to the revelation of an ogre's castle in the best-selling knightly novel *Amadis of Gaul*.[93] Columbus was better read in romance and hagiography than in geography. Historians were long baffled by his insistence, against the evidence, that he had been the first aboard his flotilla to sight the New World. The explanation is that in the *Romance of Alexander,* in a Spanish version of the medieval rewrite of the story of Alexander the Great, the world-conqueror travels to India by sea and "proclaims that of all his crew he was first to see the land." Columbus, whom Alexander's image obsessed, was representing himself in his hero's likeness.[94]

Chivalric inspiration persisted throughout the "great age." In the very year of the publication of *Don Quixote,* Pedro Fernández de Quirós, the Pacific explorer who discovered the island he called La Austrialia del Espíritu Santo, celebrated his achievement by knighting every member of his expedition, including his cooks, and clothing them in blue robes as chevaliers of "the Order of the Holy Spirit."[95] The last great explorer in the service of the Spanish crown, the Welsh-nationalist renegade Juan Evans, whose mission in 1796 preceded that of Lewis and Clark in search of a route to the Pacific via the Upper Missouri, was consciously trying to replicate the fictional achievements of Prince Madoc, the hero of a chivalric legend of the twelfth century.[96]

"Exploring spirit" became part of the mental equipment explorers needed in the Western tradition. The romantic quest—pursued in defiance of practicality to appease the soul rather than to feed stomachs or fill pockets—captivated explorers from the Enlightenment onwards. When Captain Cook vowed to exceed scientific demands by going "not only farther than any man has been before me, but as far as I think it possible for man to go," or when Pierre-Louis Moreau de Maupertuis dreamed of "chasing God in the immensity of the heavens," or when Meriwether Lewis broke into uncharacteristically exalted fervor for the unknown, or when Burke and Wills set off to cross Australia, or Scott to conquer Antarctica, in what can fairly be described as a cavalier spirit, they were taking part in a tradition of conscious adventure-seeking that started with knight-errantry, continued with romantic self-consciousness, and led, all too often, to disaster.[97] Without the successes that sometimes happened against the odds, the reconnected, joined-up, globalizing world we inhabit might never have been attained.

Magellan's part in the story is slighter than his adulators have claimed. But his role was representative. What was admirable in him was what is admirable in all explorers, irrespective of their defects of character, or disappointed ambitions, or malign effects on the peoples they met or environments they touched. Magellan wanted wealth, power, and fame; but those are commonplace roots of evil. People whose motives are merely squalid pursue them in business or politics or showmanship or crime. Evil is unredeemable only if its motives are material. Magellan's motives were better than that. The spirit of adventure can and often does mislead its followers into rapacity and rapine. But they deserve some credit for answering its call. Where would we be without it? Where would we be without them?

A Note on the Author

Felipe Fernández-Armesto's awards for work in maritime and imperial history include the World History Association Book Prize, Spain's Premio Nacional de Investigación Geográfica, the Caird Medal and the John Carter Brown Gold Medal. He is a Vice-president of the Hakluyt Society. In 2016 the King of Spain recognised his services to education and the arts with the award of the Gran Cruz de la Orden de Alfonso el Sabio.

His previous publications include the critically acclaimed *Out of Our Minds, A Foot in the River, 1492, Millennium, Pathfinders* and *Food: A History*. He occupies the William P. Reynolds Chair at the University of Notre Dame, where he is a professor of history and, concurrently, of classics and of the history and philosophy of science.

Notes

PREFACE

1. I. López-Calvo, "Magallanes, Elcano, la marca España y los resabios de nostalgias imperiales," *Revista Communitas* 3, no. 6 (July–December 2019): 56–67, provides a formidable list of commemorative events.

CHAPTER 1. THE GLOBE AROUND MAGELLAN

1. M. de Navarrete, ed., *Colección de los viajes y descubrimientos que hicieron por mar los españoles* (Madrid: Imprenta Nacional, 1837), 4:3–8.

2. Quoted in M. Lucena Giraldo, "Renaissances, Reformations, and Mental Revolutions," in *The Oxford Illustrated History of the World*, ed. F. Fernández-Armesto (Oxford: Oxford University Press, 2019), 273.

3. P. Burke, L. Clossey, and F. Fernández-Armesto, "The Global Renaissance," *Journal of World History* 28 (2017): 1–30.

4. The global reach of the Little Ice Age is disputed (J. T. Houghton et al., *Climate Change 2001*, Working Group I, Intergovernmental Panel on Climate Change [Cambridge: Cambridge University Press, 2001], sec. 2.3.3: "Was There a 'Little Ice Age' and a 'Medieval Warm Period'?," https://web.archive.org /web/20060529044319/http://www.grida.no/climate/ipcc_tar/wg1/070.htm), but ice-core evidence from Antarctica and Southern Hemisphere glaciers and other proxies seems decisive (Inka Meyer and Sebastian Wagner, "The Little Ice Age in Southern South America: Proxy and Model Based Evidence Past Climate Variability in South America and Surrounding Regions," *Developments in Paleoenvironmental Research* 14 [2009]: 395–412).

5. James Belich is engaged in an investigation of this problem. See J. Belich, "The Black Death and the Spread of Europe," in *The Prospect of Global*

History, ed. James Belich (Oxford: Oxford University Press, 2016), https://doi .org/10.1093/acprof:oso/9780198732259.003.0006.

6. H.H. Lamb, *The Little Ice Age: Climate, History and the Modern World* (London: Routledge, 1995); B.M. Fagan, *The Little Ice Age: How Climate Made History, 1300–1850* (New York: Basic Books, 2001).

7. C. Pfister et al., "Daily Weather Observations in Sixteenth-Century Europe," *Climatic Change* 42 (1999): 111–50.

8. The scholarly consensus is shifting in favor of acknowledging the congruence of plague and cold. See M. Eisenberg and L. Mordechai, "The Justinianic Plague and Global Pandemics," *American Historical Review* 125 (2020): 1659 and the references given there; F. Fernández-Armesto, "The Perils of Environmental Truthfulness, 106 (2021), 41–65"; *Antonianum*, from proceedings of the conference "Il Patto Educativo," chaired by G. Buffon and I. Colagé.

9. On plague in general, see W.H. McNeill, *Plagues and Peoples*, 3rd ed. (New York: Doubleday, 1998). On the Black Death, see M.W. Dols, *The Black Death in the Middle East* (Princeton, NJ: Princeton University Press, 1977); O.J. Benedictow, *The Black Death, 1346–53* (Woodbridge: Boydell Press, 2004); Monica H. Green, ed., *Pandemic Disease in the Medieval World: Rethinking the Black Death* (Kalamazoo, MI: Arc Humanities Press, 2015) [special issue, *Medieval Globe* 1, no. 1 (2014)].

10. Bertrand and Hélène Utzinger, *Itinéraires des danses macabres* (Chartres: Garnier, 1996); James C. Clark, *The Dance of Death in the Middle Ages and the Renaissance* (Glasgow: Jackson, 1950).

11. E.D. Williamson and P.D.F. Oynston, "The Natural History and Incidence of *Yersinia pestis* and Prospects for Vaccination," *Journal of Medical Microbiology* 61 (2012): 911–18; Mark Achtman et al., "Microevolution and History of the Plague Bacillus, *Yersinia pestis,*" *Proceedings of the National Academy of Science* 101 (2004): 17837–42.

12. F.M. Snowden, *Epidemics and Society from the Black Death to the Present* (New Haven, CT: Yale University Press, 2019), 80, takes a different view, but see G. Parker, *Global Crisis: War, Climate Change and Catastrophe in the Seventeenth Century* (New Haven, CT: Yale University Press, 2017), xix–xxiii, 1–23, 69.

13. W.J. MacLennan, "The Eleven Plagues of Edinburgh," *Proceedings of the Royal College of Physicians of Edinburgh* 31 (2001): 256–61, www.rcpe .ac.uk/journal/issue/vol31_no3/T_Eleven_Plagues.pdf.

14. E.A. Heinrichs, "The Plague Cures of Caspar Kegler: Print, Alchemy, and Medical Marketing in Sixteenth-Century Germany," *Sixteenth-Century Journal* 43 (2012): 417–40.

15. L. Luiz de Oliveira, *Viver em Lisboa: Século XVI* (São Paolo: Alameda, 2015), 40; T. Rodrigues, *Crises de mortalidad em Lisboa, séculos XVI e XVII* (Lisbon: Horizonte, 1990), 16.

16. J. de Villalba, *Epidemiología española o Historia de las pestes, contagios, epidemias y epizootias que han acaecido en España desde la venida de los cartagineses hasta el año 1801* (Madrid, 1802); V. Pérez Moreda, *La crisis de mortalidad en la España interior (siglos XVI–XIX)* (Madrid: Siglo Veintiuno, 1980), 248–49.

17. António dos Santos Pereira, "The Urgent Empire: Portugal between 1475 and 1525," *Journal of Public Health* 4, no. 2 (Winter 2006), www.brown.edu /Departments/Portuguese_Brazilian_Studies/ejph/html/issue8/html/apereira_ main.html.

18. A. W. Crosby, *The Columbian Exchange* (New York: Greenwood, 1972).

19. J. F. Pacheco, F. de Cárdenas, and L. Torres de Mendoza, eds., *Colección de documentos inéditos relativos al descubrimiento, conquista y organización de las posesiones españolas en América y Oceania,* 42 vols. (Madrid: Quirós, 1864–84), 2:373.

20. J. N. L. Biraben, "Essai sur l'évolution du nombre des hommes," *Population* 34 (1979): 13–25.

21. S. von Herberstein, *Notes upon Russia,* ed. R. H. Major, 2 vols. (London: Hakluyt Society, 1852), 2:42.

22. P. Freeman, *Out of the East: Spices and the Medieval Imagination* (New Haven, CT: Yale University Press, 2008).

23. F. Fernández-Armesto, *Food: A History* (London: Macmillan, 2001), 177–78.

24. J. C. Brown, "Prosperity or Hard Times in Renaissance Italy?," *Renaissance Quarterly* 42 (1989): 761–80; C. F. Beckingham and G. W. Huntingford, eds., *Some Records of Ethiopia* (London: Hakluyt Society, 1954).

25. Robert S. Lopez, "Hard Times and Investment in Culture," in *The Renaissance: A Symposium* (New York: Metropolitan Museum of Art, 1953), 19–32.

26. T. Goldstein, "Geography in Fifteenth-Century Florence," in *Merchants and Scholars: Essays in the History of Exploration and Trade,* ed. J. Parker (Minneapolis: University of Minnesota Press, 1965), 9–32.

27. R. H. Major, *India in the Fifteenth Century* (London: Hakluyt Society, 1957), 10.

28. Major, *India,* 6.

29. Major, *India,* 9.

30. Major, *India,* 30.

31. Major, *India,* 23.

32. Major, *India,* 11.

33. N. Conti, *The Most Famous and Noble Travels,* ed. N. M. Penzer (London: Argonaut, 1929), 169.

34. W. E. Washburn, "A Proposed Explanation of the Closed Indian Ocean on Some Ptolemaic Maps of the Twelfth-Fifteenth Centuries," *Revista da Universidade de Coimbra* 33 (1985): 431–41.

35. J. H. Parry, *The European Reconnaissance* (London: Macmillan, 1968), 68.

36. C. F. Beckingham and G. W. B. Huntingford, eds., *The Prester John of the Indies,* 2 vols. (Cambridge: Hakluyt Society, 1961), 2:369.

37. A. Blázquez y Delgado-Aguilera, ed., *Descripción de los reinos, puertos e islas que hay desde el Cabo de Buena Esperanza hasta los Leyquios* (Madrid: Torrent, 1921).

38. M. N. Pearson, *Port Cities and Intruders: The Swahili Coast, India, and Portugal in the Early Modern Era* (Baltimore: Johns Hopkins University Press, 1998).

39. Philippe Beaujard, "Progressive Integration of Eastern Africa into an Afro-Eurasian World-system," in *The Swahili World*, eds. Stephanie Wynne-Jones and Adria LaViolette (London: Routledge, 2018), 375.

40. Jonathan Walz, "Early Inland Entanglement in the Swahili World, c. 750–1550 CE," in Wynne-Jones and LaViolette, *Swahili World*, 388–402; Chapurukha M. Kusimba and Sibel B. Kusimba, "Mosaics: Rethinking African Connections in Coastal and Highlnd Kenya," in Wynne-Jones and LaViolette, *Swahili World*, 403–18.

41. *O livro de Duarte Barbosa*, ed. Maria Augusta da Veiga e Sousa, i (Lisbon: Ministério da Ciencia e da Tecnologia, 1996), pp. 15–17. I owe this reference to the kindness of Professor Sanjay Subrahmanyam.

42. M.L. Dames, ed., *The Book of Duarte Barbosa*, 2 vols. (1918, 1921), 1:29.

43. Pearson, *Port Cities and Intruders*, 119.

44. Pearson, *Port Cities and Intruders*, 8.

45. R.E. Dunn, ed., *The Adventures of Ibn Battuta: A Muslim Traveler of the 14th Century* (Berkeley: University of California Press, 2005), 231–40.

46. J.M. dos Santos, "Kalu Muhammad Hilali, Sultan of the Maldives (1491–1528)," *Archipel* 70 (2005): 63–75.

47. C.E.B. Asher and C. Talbot, *India before Europe* (2006), 107 (translation modified).

48. Asher and Talbot, *India before Europe*, p. 77.

49. G. Bouchon, "Les Musulmans de Kerala à l'époque de la découverte portugaise," *Mare Luso-Indicum*, 2 (1973), 3–59.

50. G. Bouchon, *Mamale de Cananor: Un adversaire de l'Inde portugais* (1507–28) (Geneva: Droz, 1975), 16–23, 13–37.

51. G.P. Badger, ed., *The Travels of Ludovico di Varthema* (London: Hakluyt Society, 1863), 135–36.

52. M.N. Pearson, *Merchants and Rulers in Gujarat: The Response to the Portuguese in the Sixteenth Century* (Berkeley: University of California Press, 1976), 67–73.

53. N. Tarling, ed., *The Cambridge History of Southeast Asia*, vol. 1 (Cambridge: Cambridge University Press, 1992), 483.

54. A. Cortesão, "As mais antigas cartografia e descrição das Molucas," in *A viagem de Fernão de Magalhães e a questão das Molucas: Actas do II Congreso luso-espanhol de História Ultramarina*, ed. A. Teixeira da Mota (Lisbon: Junta de Investigações Cientificas, 1975), 53.

55. G. Bullough, ed., *The Lusiads in Sir Richard Fanshawe's Translation* (Carbondale: University of Illinois Press, 1963), 329.

56. Tarling, *Cambridge History*, 409.

57. R.A. Donkin, *Between East and West: The Moluccas and the Traffic in Spices up to the Arrival of Europeans* (Philadelphia: American Philosophical Society, 2003), 3–13.

58. Donkin, *Between East and West*, 149, 156.

59. W.W. Rockhill, "Notes on the Relations and Trade of China with the Eastern Archipelago and the Coast of the Indian Ocean during the Fourteenth Century," *T'oung Pao* 16 (1915): 61–159, 236–71.

60. L.F. Thomaz, "As cartas malaias de Abu Hayat, Sultão de Ternate, a El-Rei de Portugal," *Anais de história de Além-mar*, 4 (2003): 412.

61. Leonard Y. Andaya, *The World of Maluku: Eastern Indonesia in the Early Modern Period* (Honolulu: University of Hawaii Press, 1993), 47–9, 80–1.

62. A. Cortesão, ed., *The Suma Oriental of Tomé Pires*, 2 vols. (London: Hakluyt Society, 2010), 1:213–16.

63. B. Laufer, "The Relations of the Chinese to the Philippine Islands," in *European Entry into the Pacific: Spain and the Acapulco Galleons*, ed. D.O. Flyn and A. Giráldez (Aldershot: Ashgate, 2001), 55–91.

64. Laufer, "Relations of the Chinese," 59–63.

65. J.L. Phelan, *The Hispanization of the Philippines* (Madison: University of Wisconsin Press, 1965), 15–16.

66. Laufer, "Relations of the Chinese," 64.

67. Cortesão, *Suma Oriental*, 133–34.

68. Arnold J. Toynbee, *A Study of History*, vol. 1 (Oxford: Oxford University Press, 1987), 570; Lopez, "Hard Times."

69. J. Serrão, "Le blé des îles atlantiques: Madère et Açores aux XVe et XVIe siècles," *Annales* 9 (1954): 337–41.

70. F. Fernández-Armesto, "Atlantic Exploration before Columbus: The Evidence of Maps," *Renaissance and Modern Studies* 30 (1986): 1–23.

71. A. Rumeu de Armas, *La conquista de Tenerife* (Santa Cruz de Tenerife: Aula de Cultura, 1975).

72. F. Fernández-Armesto, *Columbus* (London: Duckworth, 1996), 19–21, 49.

CHAPTER 2. THE EDUCATION OF AN ADVENTURER

1. J.M. Garcia, "Fernando de Magallanes y Portugal," in *Congreso Internacional de Historia: "Primus Circumdedisti Me,"* ed. C. Martínez Shaw (Madrid: Ministerio de Defensa, 2018), 95–110.

2. X. Castro, J. Hamon, and L.F. Tomaz, eds., *Le voyage de Magellan*, 2 vols. (Paris: Chandeigne, 2007), 1:312–15. See also A. Morais Barros, *A naturalidade de Fernão de Magalhães* (Oporto: Afrontamento, 2009), and I. da Silva Dantas, *Entre memórias: A questão da naturalidade de Fernão de Magalhães* (Braga: Universidade do Minho, 2012).

3. J.T. Medina, ed., *Colección de documentos inéditos para la historia de Chile*, 14 vols. (Santiago: Ercilla, 1888–1902), 2:374; Vizconde de Lagoa, *Fernão de Magalhãis: A sua vida e a sua viagem* (Lisbon: Serra Nova, 1938), 1:90–91.

4. A. Baião, "Fernão de Magalhães: A problema de sua nacionalidade rectificado e esclarecido," *História e memorias da Academia das Sciencias de Lisboa*, n.s., 14 (1922): 25–81.

5. Lagoa, *Fernão de Magalhãis*, 1:91.

6. "Uma quinta de viñas y castañelas y tierras de pan sembrar, radicadas en tierras de Guyan, término de la ciudad de Puerto de Portugal, que el Comendador Fernando de Magallanes, capitán de Sus Altezas, hijo legítimo de Rodrigo de Magallanes y de Alda de la Mesquita, difuntos, vecinos que fueran de la

citada ciudad de Puerto de Portugal, hace a su hermana Isabel de Magallanes."
F. Arias del Canal, "Magallanes," *Revista Hispano-Americana* 4 (2004): 31.

7. Medina, *Colección de documentos inéditos*, 2:376.

8. J. Gil, *El exilio portugués en Sevilla: De los Braganza a Magallanes* (Seville: Sevilla Fundación Cajasol, 2009), 167–69; "Magallanes en Sevilla," in *Magallanes y Sevilla*, ed. E. Vila Vilar (Seville: Universidad de Sevilla, 2019), 45–46. On the claim that "Pedro" was the name of Magellan's father, see A. Baião, "Fernão de Magalhães: Dados inéditos para a sua biografía," *Archivo histórico portugues* 3 (1905): 308–9.

9. Baião, "Fernão de Magalhães: A problema."

10. J. Denucé, *Magellan, la question des Moluques et la première circum-navigation du monde* (Brussells: Hayez, 1911), 98; Medina, *Colección de documentos inéditos*, 2:293–95.

11. M. Villas-Boas, *Os Magalhães: Sete séculos de aventura* (Lisbon: Estampa, 1998).

12. On the queen's household, see I. Carneiro de Sousa, *A rainha D. Leonor (1458–1525): Poder, misericórdia, religiosidade e espiritualidade no Portugal do Renascimento* (Lisbon: F. Gulbenkian, 2002), 167–203, 841–88.

13. A. da Silva Regio et al., eds., *Documentos sobre os portugueses em Moçambique e na África central,* 9 vols. (Lisbon: Instituto de Investigação Científica Tropical, 1962–89), 1:110.

14. R. Costa Gomes, *The Making of a Court Society: Kings and Nobles in Late Medieval Portugal* (Cambridge: Cambridge University Press, 2003); H. Baquero Moreno and I. Vaz de Freitas, *A corte de Afonso V* (Gijón: Trea, 2006).

15. A. H. R. de Oliveira Marques, *Portugal na crise dos séculos xiv e xv* (Lisbon: Presença, 1987), 305.

16. S. Humble Ferreira, *The Crown, the Court and the Casa da Índia* (Leiden: Brill, 2015), 9.

17. "Itinerarium," ed. L. Pfandl, *Revue hispanique* 48 (1920): 80–81.

18. "A corte de Portugal / Vimos bem pequena ser / Depois tanto enobrecer / que nao há outra igual / na Cristandade a meu ver." *Miscelânia*, quoted in J. P. Oliveira e Costa, *D. Manuel I, 1469–1521: Um príncipe do Renascimento* (Rio de Moura: Círculo-Leitores, 2005), 259.

19. Oliveira e Costa, *D. Manuel I*, 49.

20. A. Disney, *History of Portugal and the Portuguese Empire*, 2 vols. (Cambridge: Cambridge University Press, 2012), 2:89.

21. Humble Ferreira, *Crown*, 62–63.

22. Humble Ferreira, *Crown*, 159.

23. P. E. Russell, "Arms versus Letters: Towards a Definition of Spanish Humanism," in *Aspects of the Renaissance: A Symposium*, ed. Archibald R. Lewis (Austin: University of Texas Press, 1967), 45–58.

24. P. K. Liss, *Isabel the Queen* (New York: Oxford University Press, 1992), 24.

25. M. H. Keen, *Chivalry* (New Haven, CT: Yale University Press, 2005), 16.

26. R. Kaeuper, *Chivalry and Violence in Medieval Europe* (Oxford: Oxford University Press, 1999).

27. F. Fernández-Armesto, *Before Columbus: Exploration and Colonization from the Mediterranean to the Atlantic* (Philadelphia: University of Pennsylvania Press, 1987), 221–22.

28. F. Fernández-Armesto, "Naval Warfare after the Viking Age," in *Medieval Warfare: A History*, ed. M. Keen (Oxford: Oxford University Press, 1999), 240–22.

29. Fernández-Armesto, *Before Columbus*, 182–98; P. E. Russell, *Prince Henry "the Navigator": A Life* (New Haven, CT: Yale University Press, 2000).

30. F. Fernández-Armesto, *Columbus* (London: Duckworth, 1996), 3, 17; F. Fernández-Armesto, ed., *Columbus on Himself* (London: Folio Society, 1992), 197–98.

31. "Digas tú, el marinero, / que en las naves vestías, / Si la nave o la vela o la estrella / Es tan bella. / Digas tú, el caballero / que las armas vestías, / si el caballo o las armas o la guerra / es tan bella." *Poesía de Gil Vicente*, ed. D. Alonso (Madrid: Cultura, 1941), 28.

32. F. Fernández-Armesto, "Sea and Chivalry in Late Medieval Spain," in *Maritime History*, vol. 1, *The Age of Discovery*, ed. J. B. Hattendorf (Malabar, FL: Krieger, 1996), 123–35.

33. W. Worcestre, *Itineraries*, ed. J. H. Harvey (Oxford: Oxford University Press, 1969), 390.

34. W. Lambard, *Archaionomia* (London, 1568), fol. 55; G. A. Williams, *Madoc: The Legend of the Welsh Discovery of America* (New York: Oxford University Press, 1987), 55.

35. *Historia de la linda Melosina*, ed. I. A. Corfis (Madison, WI: Hispanic Seminary of Medieval Studies, 1986), 53 (chap. 23); J. d'Arras, *Mélusine*, ed. C. Brunet (Paris, 1854), 121; J. Goodman, *Chivalry and Exploration* (Woodbridge: Boydell, 1988), 57.

36. "E sy Dios vos da ventura que vos podays conquistar tierra, governad vuestras personas e los vuestros segun la natura e condiçion de cada uno. E sy ves que vos son rreveles, guardad que los humilies e que seays señores. No dexes perder nada de los derechos de vuestra señoria. . . . Sobre vuestros subjetos tomad vuestras rrentas e derecho syn mas los pechar sy no por causa justa." *Historia de la linda Melosina*, ed. Corfis, 54.

37. F. Fernández-Armesto, "Colón y los libros de caballería," in *Cristóbal Colón*, ed. C. Martínez Shaw and C. A. Parcero Torre (Burgos: Junta de Castilla y León, 2004), 115–28.

38. J. Martorell y M. J. de Galba, *Tirant lo blanc*, ed. M. de Riquer (Barcelona: Seix Barral, 1970), 124; *Don Quijote* 2.52; J. Goodman, *Chivalry and Exploration*, 67.

39. *Don Quixote* 2.42.

40. *The first booke of Primaleon of Greece Describing the Knightly Deeds of Armes, as also the Memorable Aduentures of Prince Edward of England: and Continuing the Former Historie of Palmendos, Brother to the Fortunate Prince Primaleon* (London: Cuthbert Burby, 1595), 12, https://quod.lib.umich.edu/e/eebo/A10109.0001.001/1:3.1?rgn=div2;view=fulltext.

41. M. Reeves, *Joachim of Fiore and the Prophetic Future* (London: SPCK, 1976).

42. A. Milhou, *Colón y su mentalidad mesiánica el ambiente franciscanista español* (Valladolid: Casa-Museo de Colón, 1984); J. L. Phelan, *The Millennial Kingdom of the Franciscans in the New World* (Berkeley: University of California Press, 2021).

43. D. Pacheco Pereira, *Esmeraldo de situ orbis,* ed. G. T. Kimble (London: Hakluyt Society, 1892), 16–17; Geneviève Bouchon, "Trade in the Indian Ocean at the Dawn of the Sixteenth Century," in *Merchants, Companies and Trade: Europe and Asia in Early Modern Era,* eds. Sushil Chaudhury and Michel Morineau (Cambridge: Cambridge University Press, 1999), 42–51, p. 48; Jean Aubin, ed., "Lettre de Baba Abdullah à D. Manuel (début 1519)," *Mare Luso-Indicum,* 2 (1973), 201–11.

44. L. F. Thomaz, "L'Idée impériale manueline," in *La découverte, le Portugal et l'Europe,* ed. J. Aubin (Paris: Fondation Goulbenkian, 1990), 35–103.

45. E. G. Ravenstein, ed., *A Journal of the First Voyage of Vasco da Gama* (London: Hakluyt Society, 1898), 48.

46. B. de Las Casas, *Historia de las Indias,* ed. el Marqués de la Fuensanta del Valle and D. José Sancho Rayón, 5 vols. (Madrid: Aguilar, 1927), 4:376–78 (bk. 3, chap. 101).

47. G. Fernández de Oviedo, *Historia natural y general de las Indias occidentales,* ed. J. Amador de los Ríos, 14 vols. (1851–55; repr., Asunción: Guarania, 1944–45), 2:9.

48. G. Pereira, ed., "De Lisboa a Cochim," *Boletim da Sociedade Geografica de Lisboa* 17 (1899): 355–65.

49. Pereira, "De Lisboa a Cochim," 367.

50. G. Correia, *Lendas da índia,* ed. M. Lopes de Almeida, 4 vols. (Porto: Lello, 1975), 2:28. But cf. above, p. 66.

51. A. da Silva Rêgo, ed., *As gavetas da Torre do Tumbo,* 12 vols. (Lisbon: Centro de Estudos Ultramarinos, 1960–77), 10:359.

52. J. Aubin, "Albuquerque et les négotiations de Cambaye," in *Mare Luso-indicum: Études et documents sur l'histoire de l'Océan Indien et des pays riverans à l'époque de la domination portugaise,* 2 vols. (Geneva: Droz, 1971), 1:16; Sanjay Subrahmanyam, "The Birth-pangs of Portuguese Asia: Revisiting the Fateful ´Long Decade' 1498–1509," *Journal of Global History* 2 (2007), 278.

53. Baião, "Fernão de Magalhães: Dados inéditos," 306.

54. J. de Barros, *Decadas da Asia,* 24 vols. (Lisbon: Regia Officina, 1777–88), 4:396 (Dec. II, bk. 4, chap. 3).

55. Baião, "Fernão de Magalhães: Dados inéditos," 307.

56. Correia, *Lendas da índia,* 2:28, 625; F. Lopes de Castanheda, *História do descobrimento e conquista da India pelos portugueses,* 7 vols. (Lisbon: Rollandiana, 1833), 3:17 (bk. 3, chap. 5).

57. Barros, *Decadas da Asia,* 4:374–75 (Dec. II, bk. 4, chap. 1).

58. R. A. Bulhão Pato and H. Lopes de Mendonça, eds., *Cartas de Afonso de Albuquerque,* 7 vols (Lisbon: Academia de Ciências, 1884–1935), 1:287–89; J. M. Garcia, *A viagem de Fernão de Magalhães e os portugueses* (Lisbon: Editorial Presença, 2007), 22.

59. M. N. Pearson, "Goa in the First Century of Portuguese Rule," *Itinerario* 8 (1984): 36–57.

60. G. V. Scammell, "Indigenous Assistance in the Establishment of Portuguese Power in Asia in the Sixteenth Century," *Modern Asian Studies* 14 (1980): 1–11.

61. R. M. Loureiro, "Los años portugueses de Magallanes," in Vila Vilar, 28; Garcia, *Viagem*, 23.

62. A. Cortesão, ed., *The Suma Oriental of Tomé Pires and the Book of Francisco Rodrigues*, 2 vols. (London: Hakluyt Society, 1944),1:55–57.

63. G. P. Badger, ed., *The Travels of Ludovico di Varthema* (London: Hakluyt Society, 1863), 224.

64. Cortesão, *Suma Oriental*, lxxv.

65. T. F. Earle and J. Villiers, eds., *Albuquerque: Caesar of the East. Selected Texts* (Liverpool: Liverpool University Press, 1990), 81.

66. "Que fosse bem recebido, ca sendo o nosso nome era espantoso entre aquelles povos." Barros, *Decadas da Asia*, 5:584 (Dec. III, bk. 5, chap. 6).

67. Lopes de Castanheda, *História do descobrimento*, 3:257 (bk. 3, chap. 52); Garcia, "Fernando de Magallanes," 103–4; P. Valière, ed., *Le voyage de Magellan raconté par un homme qui fut en sa compagnie* (Paris: Gulbenkian, 1976), 26–27. Fernández de Oviedo, *Historia natural y general*, 2:15, places Magellan in the Moluccas to buy a slave-interpreter who served on the great voyage, perhaps confusing what he calls "Maluku" with Malacca. Elsewhere (12) he says only that Magellan "had plenty of information about the Moluccas" (tenia mucha notícia . . . de las islas del Maluco).

68. Garcia, *Viagem*, 22–26, 266–67.

69. Cortesão, ed., *Suma Oriental*, lxxxi–ii.

70. Antonio Galvão, *Tratado dos descobrimentos antigos, e modernos* (Lisbon: Ferreiriana, 1731), 2:35–36; Cortesão, *Suma Oriental*, lxxxi.

71. "Levantou as manos dando louvores a Deos, pois lhe mostrára ante da sua morte os homens de fero, em cujas forças estava a segiuridade de sue Reyno." Barros, *Decadas da Asia*, 5:593 (Dec. III, bk. 5, chap. 6).

72. Artur Basílio de Sá, ed., *Documentação para a história das missões do padroado português do Oriente*, 1 (Lisbon: Agência Geral do Ultramar, 1954), 80. Professor Sanjay Subrahmanyam kindly gave me a copy of this work.

73. "Sabemos per nossas navegacoes per mar e terra retalhada em muitas mil ilhas, que juntamente elle e ellas contem en si grande parte da redndeza da terra . . . e no meio deste gran numero de ilhas estam as chamadas Maluco." Barros, *Decadas da Asia*, 5:564–65 (Dec. III, bk. 5, chap. 6).

74. "Ampliando isto com tantas palavras, e mysterios, fazendo tanta distancia donde estava a Malaca, por fazer de seu galardão ante El Rey D. Manuel, que parecia virem aquellas cartas de mais longe que dos Antipodas, e de outro novo mundo." Barros, *Decadas da Asia*, 5:599–600 (Dec. III, bk. 5, chap. 6), 622–23 (Dec. III, bk. 5, chap. 8).

75. "Una carta de francisco serrano, . . . escrita en los Malucos, en la qual le rogava que se fuesse alli si queria ser presto rico." F. López de Gómara, *Historia general de las Indias* (Zaragoza: Miguel de Papila, 1555), fol. 45v, facs. ed., ed. F. Pease (Lima: Comisión Nacional del Quinto Centenario del Descubrimiento de América, 1993).

76. "Das quaes cartas começou este Fernão de Magalhães tomar huns novos conceitos, que lhe causáram morte, e metteo este Reyno en algum desgosto." Barros, *Decadas da Asia*, 5:600 (Dec. III, bk. 5, chap. 6).

77. A. Pigafetta, *Il primo viaggio intorno al globo di Antonio Pigafetta, e le sue Regole sull'arte del navigare*, ed. A. Da Mosto, Raccolta Colombiana, pt. 5, vol. 3 (Rome: Ministero della pubblica istruzione, 1894), 81, and *The First Voyage around the World (1519–22): An Account of Magellan's Expedition by Antonio Pigafetta*, ed. T. Cachey (New York: Marsiglio, 1995), 35. Cf. above, 99.

78. "Testimonio del testamento que otorgó Fernando de Magallanes en los Alcázares Reales de Sevilla," August 24, 1519, in "Documentos para el quinto centenario de la primera vuelta al mundo: La huella archivada del viaje y sus protagonistas," transcribed by Cristóbal Bernal, 2019–22, http://sevilla.2019–2022.org/wp-content/uploads/2016/03/10.ICSevilla2019_Testamento-de-Magallanes-g15.pdf.

79. José Manuel Garcia, "Documentos existentes em Portugal sobre Fernão de Magalhães e as suas viagens," *Abriu*, 8 (2019): 24–5. Professor Sanjay Subrahmanyam kindly gave me a copy of this article.

80. *Documentação para a história das missões do Padroado portugues do oriente: Insulindia*, i (Lisbon, 1954), 79–80, quoted in Garcia, "Fernando de Magallanes," 104.

81. Las Casas, *Historia de las Indias*, 1:221–22, 266 (bk. 1, chap. 38).

82. Fernández de Oviedo, *Historia natural y general*, 1:46–51.

83. L. A. Brown, *The Story of Maps* (New York: Little, Brown, 1949), 29–33.

84. M. Fernández de Navarrete, *Obras*, ed. C. Seco Serrano, 3 vols. (Madrid, 1954–55), 2:612.

85. P. de Medina, *L'art de naviguer* (1554), facs. ed., ed. C. Rahn Phillips (New York: Scholars' Facsimiles, 1992), fol. 3.

86. Brown, *Story of Maps*, 28–32.

87. C. Varela, ed., *Cristóbal Colón: Textos y documentos completos* (Madrid: Alianza, 1984), 217; G. E. Nunn, *The Geographical Conceptions of Columbus* (New York: American Geological Society, 1924), 1–30.

88. P. E. Taviani, *Christopher Columbus: The Grand Design* (London: Orbis, 1985), 413–27; H. Wagner, "Die Rekonstruction der Toscanelli Karte von Jahre 1474," in *Nachrichten von der Königliche Gesellschaft der Wissenschaften zu Göttingen* (Göttingen: Commissionsverlag der Dieterich'schen Verlagsbuchhandlung, 1894), 208–312; J. K. Willers, ed., *Focus Behaim Globus*, 2 vols. (Nuremberg: Verlag des Germanischen Nationalmuseums, 1992), 1:143–66, 217–22, 239–72.

89. J. Gil and C. Varela, eds, *Cartas de particulares a Colón y relaciones coetáneas* (Madrid: Alianza, 1984), 145.

90. F. Fernández-Armesto, *Amerigo: The Man Who Gave His Name to America* (New York: Random House, 2007), 68–73.

91. *Colección de documentos inéditos para la historia de España*, vol. 16 (Madrid, 1850), 382–420; M. Fernández de Navarrete, *Colección de los viajes y descubrimientos que hicieron por mar los españoles* (Madrid: Imprenta Nacional, 1837), 2:612.

92. Navarrete, *Colección de los viajes*, 2:611.

93. G. Symcox, G. Rabitti, and P. D. Diehl, eds., *Italian Reports on America, 1493–1520: Letters, Dispatches, and Papal Bulls* (Turnhout: Brepols, 2001), 31–34, 95–97.

94. J. F. Pacheco, F. de Cárdenas, and L. Torres de Mendoza, eds., *Colección de documentos inéditos relativos al descubrimiento, conquista y organización de las posesiones españolas en América y Oceanía*, 42 vols. (Madrid: Quirós, 1864–84), 1:296.

95. The effect was to endorse a division of lands and rights already stipulated in a series of papal donations, confirmed, with specific reference to the treaty of 1479, in the bull *Aeterni Regis* of 1481.

96. A. Szászdi Nágy, *La legua y la milla de Colón* (Valladolid: Seminario Americanista de la Universidad de Valladolid, 1991), 35; L. Mendonça de Alburquerque, "O tratado de Tordesillas e as dificuldades técnicas da sua aplicacão rigorosa," in *El tratado de Tordesillas y su proyección*, 2 vols. (Valladolid: Universidad de Valladolid, 1973), 1:119–36; J. Millas Vallicrosa, *Estudios sobre la historia de la ciencia española* (Madrid: CSIC, 1987), 545–78.

97. Navarrete, *Colección de los viajes*, 1:362.

98. D. Goodman, *Power and Penury: Government, Technology and Science in Philip II's Spain* (Cambridge: Cambridge University Press, 1988), 53–54.

99. J. Juan and A. de Ulloa, *Disertación histórica y geográfica sobre el meridian de demarcación entre los dominios de España y Portugal* (Madrid, 1749), 50–79.

100. Report of Fray Tomás Durán, Sebastiano Caboto, and Juan Vespucio: "Tenemos de graduar las leguas y darle menos legua que pudiéremos al grado del cielo, porque dando menos leguas, menos habrán en toda la tierra, lo cual mucho cumple al servicio de sus majestades." Navarrete, *Colección de los viajes*, 2:614.

101. R. Ezquerra, "Las juntas de Toro y Burgos," in *Tratado de Tordesillas*, 149–70; U. Lamb, "The Spanish Cosmographic Juntas of the Sixteenth Century," *Terrae Incognitae* 6 (1974): 51–62; A. Sánchez Martínez, "De la 'cartografía oficial' a la 'cartografía jurídica': La querella de las Molucas reconsiderada, 1479–1529," *Nuevo Mundo, Mundos Nuevos, Debates*, September 8, 2009, http://nuevomundo.revues.org/index56899.html.

CHAPTER 3. THE TRAJECTORY OF A TRAITOR

1. P. E. Russell, *Prince Henry "the Navigator": A Life* (New Haven, CT: Yale University Press, 2000), 291–315.

2. J. Ramos Coelho, ed., *Alguns documentos da Torre do Arquivo Nacional da Torre do Tumbo ácerca das navegações e conquistas portuguezas publicados por ordem do governo de sua majestade fidelissima ao celebrar-se a commemoração quadricentenaria do descobrimento da America* (Lisbon: Impresa Nacional, 1892), 262–63; R. A. Laguarda Trías, "Pilotos portugueses en el Río de la Plata en el siglo XVI," *Revista da Universidade de Coimbra* 34 (1988): 61.

3. S. E. Morison, *The Northern Voyages*, vol. 1 of *The European Discovery of America* (Oxford: Oxford University Press, 1970), 326–36.

4. A. Teixeira da Mota, ed., *A viagem de Fernão de Magalhães e a questão das Molucas: Actas do II Congreso luso-espanhol de História Ultramarina* (Lisbon: Junta de Investigacões Científicas, 1975), 144–46.

5. "Sentimentos de honra e direitos de justiça," in Teixeira da Mota, *Viagem*, 455.

6. J. Manoel Noronha, "Algumas observações sobre a naturalidade e a família de Fernão de Magalhães," *O Instituto* 77 (1921): 41–45.

7. "Y le recibiesse el juramento y pleyto omenage, según fuero y costumbre de Castilla, que haría el viaje con toda fidelidad como buen vasallo de Su Magestad." A. de Herrera, *Historia general de los hechos de los castellanos en las islas y tierra-firme de el mar occeano* [*sic*], ed. J. Natalicio Gómez, 10 vols. (Asunción: Guarania, 1945), 3:78 (Dec. II, bk. 3, chap. 9).

8. "Es mi voluntad que aya todo lo susodicho por título de mayorazgo Diego de Sosa, mi hermano, que agora bive con el serenísimo señor rey de Portugal, viniéndose a vebir a esos reynos de Castilla e casándose en ellos, e con tanto que se llame de Magallaes e tenga las armas de Magallaes, segund e de la maner que las yo traygo, que son de Magallaes e Sosa." *Colección de documentos relativos a las Islas Filipinas en el Archivo General de Indias* (Barcelona: Co. de Tabacos de Filipinas), 2:319–20; S. Bernabéu Albert, "Magallanes: Del héroe al hombre," in *Magallanes y Sevilla*, ed. E. Vila Vilar (Seville: Universidad de Sevilla, 2019), 118.

9. "O no feito, com verdade, Portugués, poém, não na lealdade." *Os lusíadas,* canto X, my translation.

10. "Nas quaes dizia, que prazendo a Deos, cedo se veria com elle; e que quando não fosse per via de Portugal, sería per via de Castella, porque en tal estado andavam suas cousas; per tanto que o esperasse lá. . . . E como o demonio sempre no anmio dos homens move cousas pera algum máo feito, e os acabar nelle, ordenou caso pera que este Fernão de Magalhães se descontentasse de seu Rey, e do Reyno." J. de Barros, *Decadas da Asia,* 24 vols. (Lisbon: Regia Officina, 1777–88), 5:623 (Dec. III, bk. 5, chap. 8).

11. "Poz elle em obra o que tinha escrito a Francisco Serrão seu amigo, que estava em Maluco; donde parece que sua ida pera Castella andava no seu animo de mais dias, que movida de accidente do despacho." Barros, *Decadas da Asia,* 5:623 (Dec. III, bk. 5, chap. 8).

12. A. Pigafetta, *The First Voyage around the World (1519–22): An Account of Magellan's Expedition by Antonio Pigafetta,* ed. T. Cachey (New York: Marsiglio, 1995), 86.

13. J. M. Garcia, *A viagem de Fernão de Magalhães e os portugueses* (Lisbon: Editorial Presença, 2007)145–47.

14. Luís Adão da Fonseca, *D. João II* (Lisbon: Temas e Debates, 2007), 93–106; Juan Gil, *El exilio portugués en Sevilla: de los Braganza a Magallanes* (Sevilla: Fundación Cajasol, 2009).

15. Barros, *Decadas da Asia,* 5:624 (Dec. III, bk. 5, chap. 8).

16. J. Denucé, *Magellan, la question des Moluques et la première circumnavigation du monde* (Brussells: Hayez, 1911), 133.

17. "Pois mo mataram porvosso serviço e em lugar honrado e com grande perigo de minha pessoa, onde a pé me salvei." Garcia, *Viagem,* 27.

18. Barros, *Decadas da Asia*, 5:625 (Dec. III, bk. 5, chap. 8).

19. "Dizem que foi acrescentamento de sua moradia: cousa que tem dado aos homens nobres deste Reyno muyto trabalho; e parece que he huma especie de martyrio entre os portuguezes, e acerca dos Reys causa de escandalo. . . . Mas quando vem exemplo em seu igual, principalmente naquelles a que apreoveitou mais artificios e amigos, que meritos propios, aqui se perde toda paciencia, daqu nasce a indignação, e della odio, e finalmente toda desesperação, té que vem commetter com que damnam a si, e a outrem." Barros, *Decadas da Asia*, 5:626 (Dec. III, bk. 5, chap. 8).

20. "Ou porque ell sería limpio desta culpa, ou (segundo se mais affirma) os fronteiros de Aazamor polo não avexar; pero sempre lhe ElRey teve hum entejo." Garcia, *Viagem*, 106.

21. S. E. Morison, *The Southern Voyages*, vol. 2 of *The European Discovery of America* (Oxford: Oxford University Press, 1974), 318.

22. Garcia, *Viagem*, 29–36; F. M. de Sousa Viterbo, *Trabalhos náuticos dos portugueses nos séculos XVI e XVII*, ed. J. M. Garcia (Lisbon: Casa da Moneda, 1968), 227, cited in R. M. Loureiro, "Los años portugueses de Magallanes," in Vila Vilar, *Magallanes y Sevilla*, 33.

23. Garcia, *Viagem*, 31–37.

24. "Que lhe vieram nesta nau ao partido do meio." José Manuel Garcia, "Documentos existentes em Portugal sore Fernão de Magalhães e as suas viagens," *Abriu*, 8 (2014), 15-33, p. 18. I am grateful to Professor Sanjay Subrahmanyam for a copy of this article.

25. I. Soler, "Magallanes y el dibujo del mundo," *Anais da história de alemmar* 20 (2019): 34.

26. Garcia, *A viagem de Fernão de Magalhães e os portugueses*, 37.

27. Garcia, "Documentos existentes," 18.

28. Loureiro, "Años portugueses de Magallanes," 34.

29. "Lyçemça pera hyr buscar vyda omde lhe fyzessem merçe, ao que elrrey respomdeo secamente que nynguem lho nom tolhya." Magellan "se alevamtou a sahyo da casa omde elrrey estava logo rompenmdo o seu alvara de fylhamento e os pedaços deytou da mao." Garcia, *Viagem*, 106; G. Correia, *Crónicas de D. Manuel e de D. João III*, ed. J. Pereira da Costa (Lisbon: Academia das Ciencias, 1992), 200; Loureiro, "Años portugueses de Magallanes," 34.

30. C. A. Wilkens, *Spanish Protestants in the Sixteenth Century* (London: Heinemann, 1897), 66.

31. R. Jiménez Fraile, "Cristóbal de Haro, el imprescindible agente en la sombra," *Sociedad Geográfica Española* 64 (September-December 2019): 40–52.

32. A. Sagarra Gamazo, "La empresa del Pacífico o el sueño pimentero burgalés," *Revista de estudios colombinos* 9 (2013): 23.

33. "La isla de Maluque, que es en los límites de nuestra demarcación y tomaréis la posesión della." Quoted in R. A. Laguarda Trías, "Las longitudes geográficas de la membranza de Magallanes y del primer viaje de cirunnavegación," in Teixeira da Mota, *Viagem*, 144.

34. "Debeis mucho mirar en ello, para que en Dios y en vuestra conciencia hagais la demarcación la más justamente que pudiérdes." Quoted in Laguarda Trías, "Longitudes geográficas," 144.

35. D. Ramos Pérez, "Magallanes en Valladolid: La capitulación," in Teixeira da Mota, *Viagem*, 194; H. Kellenbenz, *Los Fugger en España y Portuga hasta 1560* (Valladolid: Junta de Castilla y León, 2000), 213–14; P. Gallez, *Cristóbal de Haro: Banqueros y pimenteros en busca del Estrecho Magallánico* (Bahía Blanca: Instituto Patagónico, 1991), 52–67.

36. "Als sie bey Sechzig meillen umb den Capo kommen sein, zu gerleicher weyss als wenn ainer in Levanten fert, und die stritta de gibraltearra passiert, das ist, furfert, oder hyndurch einsam, und das lande von Barbaria sicht. Und als sie umb den Capo kunnen sien, wie gemelt ist, und gegen uns Nord westwerz geseylet oder gefaren haben. Do ist ungewitter so gross worden, auch windt gewesen, das sie nicht weyter haben kunnen saylen, oder faren. Do haben sie durch Tramontana, das ist Nort, oder mitternacht, wider her umb auss die annder seyten und Costa, das ist landt von Bresill müssen faren. . . . Von sollichem Cabo . . . uber Sechs hundert meyl gen Malaqua nit sey." *Copia der newen Zeytung aus Presillg Landt*, printed pamphlet with no indication of place, publisher, or date, John Carter Brown Library, https://archive.org/details/copiadernewenzeyoounkn/page/n5/mode/2up.

37. N. Holst, *Mundus, Mirabilia, Mentalität: Weltbild und Quellen des Kartographen Johannes Schöner: eine Spurensuche* (Frankfurt: Scripvaz, 1999); C. van Duzer, *Johann Schöner's Globe of 1515: Transcription and Study* (Philadelphia: American Philosophical Society, 2010); J. Schöner, *Luculentissima quaedam terrae totius descriptio: Cum multis utilissimis Cosmographiae iniciis* (Nuremberg: Stuchssen, 1515), 22; C. Sanz, *Bibliotheca Americana Vetustíssima: Últimas adiciones*, 2 vols. (Madrid: Victoriano Suárez, 1960), 2:734–35.

38. J. Incer, ed., *Crónicas de viajeros: Nicaragua*, vol. 1 (San José: Libro Libre, 1990), 102–4, 137–38.

39. A. Sagarra Gamazo, "Cristóbal de Haro," *Sociedad Geográfica Española* 64 (2014): 34–45.

40. "Esta suma constutuye el beneficio y los intereses de las armadas proyectadas y enviadas a las islas de las especias por España, por lo que suplico a su Majestad me considere heredero, al igual de su majestad y otros armadores, ya que hemos sido la causa del descubrimiento de las Molucas. . . . He sufrido grandes pérdidas del lado de Portugal por haber servido de instrumento para el diho descubrimiento y por haberlo organizado." Jiménez Fraile, "Cristóbal de Haro," 52. Haro's and the crown's final agreement, in 1538, is reproduced in facsimile at "Autos de Cristóbal de Haro: Abono de los gastos armada de Magallanes," PARES (Portal de Archivos Españoles), http://pares.mcu.es/ParesBusquedas20/catalogo/show/122249.

41. Navarrete, *Colección de los viajes*, 4:lxxxv.

42. J. Gil, "Magallanes en Sevilla," in Vila Vilar, *Magallanes y Sevilla*, 48.

43. "En joyas e en axuar e alhajas e preseas de casa." Gil, "Magallanes en Sevilla," 43.

44. Gil, "Magallanes en Sevilla," 46–47.

45. F. López de Gómara, *Historia general de las Indias* (Zaragoza: Miguel de Papila, 1555), fol. 45, facs. ed. F. Pease (Lima: Comisión Nacional del Quinto Centenario del Descubrimiento de América, 1993).

46. Gómara, *Historia general*, fols. 48–49. See above, 258.

47. Barros, *Decadas da Asia,* 3:628.

48. A. Pigafetta, *The First Voyage round the World, by Magellan,* ed. H.E.J. Stanley, Lord Stanley of Alderley (London: Hakluyt Society, 1874), Appendix, x.

49. Russell, *Prince Henry "the Navigator,"* 15.

50. "Que lo que supiese el uno así de Portugal como de Castilla tocante a su negociación lo comunicase con el otro dende a seys horas e que, si alguno de éllos con su honra se quisiese bolver a Portugal, que lo pudiese fazer." D. Ramos Pérez, "Magallanes en Valladoid: La capitulación," in Teixeira da Mota, *Viagem,* 181–241.

51. "Ajuntou-se com hum Ruy Faleiro Portuguez de nação Astrologo judiciario, tambem aggravado d'ElRey, porque o não quiz tomar por este officio, como se fora cousa de que ElRey tinha muita necesidade." Barros, *Decadas da Asia,* 5:628 (Dec. III, bk. 5, chap. 8).

52. J.T. Medina, ed., *El descubrimiento del Océano Pacífico: Hernando de Magallanes y sus compañeros* (Santiago: Elzeveriana, 1920), 278.

53. "Se ofreció á descubrir muy grande especiería y otras riquezas en el mar Océano dentro de los límites de Castilla." A. de Santa Cruz, *Crónica del emperador Carlos V,* ed. R. Beltrán y Rózpide and A. Blázquez y Delgado-Aguilera, 5 vols. (Madrid: Patronato de Huérfanos, 1920–25), 1:17.

54. Gil, "Magallanes en Sevilla," 50.

55. A. Teixeira da Mota, *O Regimento da altura de leste-oeste de Rui Faleiro* (Lisbon: Agência Geral do Ultramar, 1953), 9.

56. Gil, "Magallanes en Sevilla," 50.

57. "Enloqueció Ruy Faleiro de pensamieno de no poder cumplir con lo prometido." Gómara, *Historia general,* fol. 411.

58. W. G. Randles, "Spanish and Portuguese Attempts to Measure Longitude in the Sixteenth Century," *Boletim da Biblioteca da Universidade de Coimbra* 39 (1984): 143–60.

59. "Muy Poderoso Señor, Porque podría ser que el Rey de Portugal quisiese en algund tiempo decir que las Islas de Maluco estan dentro de su demarcacion, . . . sin que nadie ge lo entendiese, ansi como yo lo entiendo, y sé como se podria hacer, quise por servicio á V.A. dejarle declarado las alturas de las tierras y cabos principales, y de las alturas en que estan ansi de latitud como de longitud." Navarrete, *Colección de los viajes,* 4:188–89; Laguarda Trías, "Longitudes geográficas," 177–78.

60. "Y esta membranza que á V.A. doy mande muy bien guardar que podrá venir tiempo que sea necesaria, y excusará diferencias; y esto digo con sana conciencia, not teniendo respeto a otra cosa sino a decir verdad." Navarrete, *Colección de los viajes,* 4:189.

61. Laguarda Trías, "Longitudes geográficas," 151–54, 176–78.

62. G.E. Nunn, *The Columbus and Magellan Concepts of South American Geography* (Glennside, PA:, private printing, 1932).

63. "ElRey de Castella como estava namorado das cartas e pomas de marear, que Fernã de Magalhães lhe tinha mostrado, e principalmente da Carta que Francisco Serrão escreveo a elle Fernão de Magalhães de Maluco, em que elle mais escorava, e assi das razões dell, e do Faleiro Astrologo." Barros, *Decadas da Asia,* 629 (Dec. III, bk. 5, chap. 8).

64. C. Manso Porto, "La cartografía de la expedición Magallanes-Elcano," in *Congreso Internacional de Historia: "Primus Circumdisti Me,"* ed. C. Martínez Shaw (Madrid: Ministerio de Defensa, 2018), 271–301.

65. A. Pigafetta, *Il primo viaggio intorno al globo di Antonio Pigafetta, e le sue Regole sull'arte del navigare,* ed. A. Da Mosto, Raccolta Colombiana, pt. 5, vol. 3 (Rome: Ministero della pubblica istruzione, 1894), 61, and *First Voyage,* ed. Cachey, 21.

66. "Una carta de francisco serrano, . . . escrita en los Malucos, en la qual le rogava que se fuesse alli si queria ser presto rico." Gómara, *Historia general,* fol. 45v.

67. E. Casamassima, "Ludovico degli Arrighi detto Vicentino copista dell' 'Itinerario' di Varthema," *La Bibliofila* 64 (1962): 123.

68. G. P. Badger, ed., *The Travels of Ludovico di Varthema* (London: Hakluyt Society, 1863), 104.

69. R. C. Temple, ed., *The Itinerary of Ludovico di Varthema in Southern Asia* (London: Argonaut, 1928), xxx; cf. J. Aubin, "'L'Itinerario' de Ludovico di Varthema," in *Le Latin et l' astrolabe,* 2 vols. (Paris: Centre Gulbenkian, 2000), 485.

70. Aubin, "'Itinerario,'" 489.

71. Gómara, *Historia general,* fol. 4v.

72. Gil, "Magallanes en Sevilla," 55–56.

73. Badger, *Travels,* 243–45.

74. B. de Las Casas, *Historia de las Indias,* ed. el Marqués de la Fuensanta del Valle and D. José Sancho Rayón, 5 vols. (Madrid: Aguilar, 1927), 4:378 (bk. 3, chap. 1); H. Thomas, *Rivers of Gold* (New York: Random House, 2003), 437.

75. "Tantos etorbos y embarazos para que no se hiciese, cuantas malas voluntades para ello algunos mostraron." Gil, "Magallanes y Sevilla," 56–57.

76. "Por no estar acabadas de pintar y yo con el trabajo de sacar la nao no lo miré." Navarrete, *Colección de los viajes* 4:125.

77. "Como aquellas armas no eran del Rey de Portugal, antes eran mias é yo vasallo de V.A." Navarrete, *Colección de los viajes* 4:125.

78. "Llamando gente para prender al Capitán portugués que levantaba banderas del Rey de Portugal." Navarrete, *Colección de los viajes* 4:125.

79. "Me parece cosa muy agena de V.A. ser maltratados los hombres que dejan su reino y naturaleza por le servir en cosa tan señalada." Navarrete, *Colección de los viajes* 4:126.

80. "Me placia puesto que me era afrenta hacerlo por estar alli presente un caballero del Rey de Portugal, que por su mandado vino á esta ciudad á contratar conmigo que volviese á Portugal." Navarrete, *Colección de los viajes* 4:125.

81. "Cuan feo era receber hum Rei os vasalos de outro . . . que era cousa que entre caballeiros se nam acostumbraba." Pigafetta, *First Voyage,* ed. Stanley, Appendix, i.

82. "Mais em cousa que lhe tam pouco inportaua e tam incerta e que muitos uasalos e omens tinha pera fazer seuos descobrimentos quando fore tempo e nam e os que de uosalteza uinham descontentes." Pigafetta, *First Voyage,* ed. Stanley, Appendix, i.

83. "Fazer crer a el Rei que ele nom eraua nisto a uostalteza porque nom mandaua descobrir sénam dentro no seu lemite." Pigafetta, *First Voyage*, ed. Stanley, Appendix, i.

84. "Del bachiller no se haga caso; duerme poco, y anda casi fuera de seso." Navarrete, *Colección de los viajes*, 4:123.

85. "He sabido que vos teneys alguna sospecha, que del armada que mandamos hazer para yr a las Indias, de que van por capitanes Hernando de Magallanes y Ruy Faleiro, podria venir algun prrejuizio a lo que vos pertenece de aquellas partes . . . el primer captulo y mandamiento nuestro, que llevan los dichos capitanes, es que guarden la demarcacion, y que no toquen en ninguna manera, y so graves penas, en las partes y tierras y mares qie por la demarcacion a vos estan senaladas." Ramos Coelho, *Alguns documentos*, 422–23.

86. Gil, "Magallanes y Sevilla," 52; Garcia, *Viagem*, 155–58.

87. "Viendo ocasion oportuna para hacer lo que mandó V.A. fuime á la posada de Magalheas [*sic*], y halléle componiendo vituallas, conservas, etc., y le dije que aquello me parecia conclusion de su mal propósito, é porque esta seria la última vez que como su amigo y buen portugés le hablaria, pensase bien el yerro que iba a hacer."

88. "Dijome que era punto suyo seguir lo empezado. Acudí que no era honra lo que se ganaba indebidamente; que hasta los castellanos le miraban como ruin y traidor contra su patria. Respondió que él pensaba en su viage hacer servicio á V.A. y no tocar en cosa suya. Dígele que bastaba descubrir la riqueza que ofrecia en demarcacion de Castilla para hacer un gran daño a Portugal. . . . No sabia causa por dejar al Rey de España que tanta merced le habia hecho. Dígele que por hacer lo que debia y no perder su honra, y que pesase su venida de Portugal que fue por cien reis mas al año d moradia que V.A. dejó de darle por no quebrantar su ordenanza."

89. "Que mais queria veer que os regimentos e Ruy Faleiro que dezia abiertamente que no avia de navegar ao sull ou non hira na armada, e que ele cuidara que hia por capitao mor e qu eu sabia que avia outros mandados en contraio os quaes ello no saberia senao a tempo que non rremedear su onrra, e que non curasse do mell que lhe punha pellos beiços do bispo de Burgos."

90. "Qué no dejaria la empresa sino en caso de faltarle á algunas de las cosas capituladas; y ntonces queria saber qué mercedes le prometía V.A."

91. "E que me desse carta para vossa alteza e que eu por amoor dele yria a vossa alteza a fazer seu partido, porque eu non tinha nenhum rrecado de vosa alteza para en tall entender somente falava o que me pareçia como outras vezes lhe avia falado. Dyseme que non me dezia nada ate veer o rrecado que o correo trazia."

92. Navarrete, *Colección de los viajes*, 4:153–54: Pigafetta, *First Voyage*, ed. Stanley, Appendix, iii–xi.

93. G. Fernández de Oviedo, *Historia natural y general de las Indias occidentales*, ed. J. Amador de los Ríos, 14 vols. (1851–55; repr., Asunción: Guarania, 1944–45), 2:9.

94. "Todo o que fingia que sabia era por hum spirito familiar." F. Lopes de Castanheda, *História do descobrimento e conquista da India pelos portugueses*, 7 vols. (Lisbon: Rollandiana, 1833), 6:9 (bk. 6, chap. 6).

95. "Por quanto yo tengo por çcierto, segund la mucha informaçcion que he havido de personas, qie por esperiençia lo han visto, que en las islas de Maluco ay la espeçiairia, que principalmente ys a buscar com esa dicha armada, e mi voluntad es que derechamente sigais el viage a las dichas islas . . . para que, antes e primero que a otra parte alguna, vais a las dichas islas de Maluco, sin que aya ninguna falta." Ramos Coelho, *Alguns documentos,* 430.

96. Luís Filipe F.R. Thomaz, "As cartas malaias de Abu Hayt, Sultão de Ternate, a El-Rei de Portugal," *Anais de história de Além-mar,* 4 (2003), 416–21; Artur Basílio de Sá, ed., *Documentação para a históra das missões do padroado português do Oriente,* 1 (Lisbon: Agência Geral do Utramar, 1954), 85–7, 113–15. Professor Sanjay Subrahmanyam kindly gave me a copy of this work.

97. Bernabéu Albert, "Magellanes," 129–33.

98. "En una isla grande llamada Seilani, la cual es habitada y tiene oro en ella, y la costeamos, y fuimos al Oessudoeste a dar en una isla pequeña, es habitada y llamse Mazava, y la gente es muy buena. Allí pusimos una cruz encima de un monte, y de allí nos mostraron 3 islas a la parte del Oessudoeste, y dicen que hay mucho oro, y nos mostraron cómo lo cogían y hallan pedacicos como garbanzos y como lentejas. Esta isla está en nueve grados y dos tercios de la parte del Norte." Francisco Albo, "Derrotero del viaje al Maluco, formado por Francisco Albo, piloto de la nao Trinidad y, posteriormente, de la nao Victoria," 1519, in "Documentos para el quinto centenario de la primera vuelta al mundo: La huella archivada del viaje y sus protagonistas," transcribed by Cristóbal Bernal, 2019–22, http://sevilla.2019-2022.org/el-derrotero-de-francisco-albo-en-la-octava-entrega-de-los-documentos-para-el-vo-centenario/, 14.

99. "Que ninguno fuere osado, so pena de muerte, de rescatar oro, ni tomar oro, porque quería despreciar el oro . . . que los rescates se asentaban todos en el libro del contador o del tesorero, después que este testigo fue Capitán e tesorero." V.M. de Sola, *Juan Sebastián de Elcano* (Bilbao: Caja de Ahorros, 1962), 608–9.

100. A. Blázquez y Delgado-Aguilera, ed., *Descripción de los reinos, puertos e islas que hay desde el Cabo de Buena Esperanza hasta los Leyquios* (Madrid: Torrent, 1921), 192.

CHAPTER 4. THE MAKING AND MARRING OF A FLEET

1. E. Calderón, *El rey ha muerto* (Madrid: Cirene, 1991), 11–17.

2. I. Soler, "Magallanes y el dibujo del mundo," *Anais da história de alem-mar* 20 (2019): 21.

3. "No faltó ninguno tan aparejado para se informar de lo que Magallanes y Faleiro querían como a Juan de Aranda." J. Gil Fernández, "Magallanes, de Sevilla a Valladolid," in *Congreso Internacional de historia: "Primus Circumdedisti Me,"* ed. C. Martínez Shaw (Madrid: Ministerio de Defensa, 2018), 79–94.

4. "Yo les ayudé en hacer relación de lo que yo sabía de la abilidad de sus personas e del sentimiento que de su venida abía en Portugal." J.T. Medina, ed., *Colección de documentos inéditos para la historia de Chile,* 14 vols. (Santiago: Ercilla, 1888–1902), 1:21–53.

5. D. Ramos Pérez, "Magallanes en Valladoid: La capitulación," in *A viagem de Fernão de Magalhães e a questão das Molucas: Actas do II Congreso luso-espanhol de História Ultramarina*, ed. A. Teixeira da Mota (Lisbon: Junta de Investigacões Científicas, 1975), 179–241.

6. F. López de Gómara, *Historia general de las Indias* (Zaragoza: Miguel de Papila, 1555), fol. 40v, facs. ed. F. Pease (Lima: Comisión Nacional del Quinto Centenario del Descubrimiento de América, 1993).

7. Gómara, *Historia general*, fol. 40v; Navarrete, *Colección de los viajes*, 4:lxxxv.

8. "Ya no estaréis quexosos de lo que tengo escripto al gran chanciller; antes por ello e por lo que yo faré en dezir a Su Alteza lainformación que de vos tengo de Portugal, me devriades de dar parte del bien que Dios vos fiziesse." Ramos Pérez, "Magallanes en Valladolid," 209.

9. Ramos Pérez, "Magallanes en Valladolid," 210–11.

10. "Si le querían satisfacer algo por el trabajo que avía rescebido e ayuda que les avía dado, que se lo ternía en merced." Ramos Pérez, "Magallanes en Valladolid," 211.

11. "De todo el provecho e interese que oviéremos del descubrimiento de las tierras e islas que, plaziendo a Nro Sñr, hemos de descubrir e de hallar en las tierras e límites e comarcaciones del rey don Carlos, nro sñr, que vos ayais la octava parte." Navarrete, *Colección de los viajes*, 4:208.

12. *Opus epistolarum*, February 4, 1518, in Petrus Martyr Anglerius, *Opus epistolarum*, ed. C. Patin (Amsterdam: Elsevir, 1670), 335.

13. Ramos Pérez, "Magallanes en Valladolid," 221–22.

14. "Había por ventura dejado cerradas y distinguidas las aprtes orientales de los occidentales, en tal manera que no se pudiese navegar ni pasar de las unas a las otras partes. O que por ventura, aquella gran tierra firme . . . era tan per-petua y sinfin que apartaba, determinaba y distinguía los mares occidentales de los orientales, de forma que de ninguna manera se pudiese pasar ni navegar por allí hacia el Oriente." Ramos Pérez, "Magallanes en Valladolid," 229.

15. Peter Martyr, *Cartas sobre el Nuevo Mundo*, trans. J. Bauzano (Madrid: Polifemo, 1990), 98 (ep. 634).

16. J. Gil, "Magallanes en Sevilla," in *Magallanes y Sevilla*, ed. E. Vila Vilar (Seville: Universidad de Sevilla, 2019), 88.

17. J. Denucé, *Magellan, la question des Moluques et la première circum-navigation du monde* (Brussels: Hayez, 1911), 210.

18. Medina, *Colección de documentos inéditos*, 2:38.

19. "Dezia sienpre Juan de Aranda que demandaban mucho, y por su causa bolvieron a demandar mucho menos de lo que tenían determinado de demandar." Medina, *Colección de documentos inéditos*, 1:55.

20. Medina, *Colección de documentos inéditos*, 1:55.

21. "Escogendo primer V.A. las seis, y que despues ente todas las otras nosotros podamos tomar las dos mejores que nos pareciere, de las cuales V.A. nos dará el Señorío con todo lo que al presente y adelante rentasen, y con todo el trato, i que V.A. no haya mas derechos de diez por ciento de lo que nos rent-are." Navarrete, *Colección de los viajes*, 4:114.

22. Navarrete, *Colección de los viajes*, 4:118.

23. "En las dichas tierras é islas que descubrierdes . . . que . . . os acaten é cumplan vuestros mandamientos, so la pena ó penas que vosotros de nuestra parte les pusierdes é mandardes poner." Navarrete, *Colección de los viajes*, 4:121–22.

24. "Nombremos un factor é tesorero é contador y escribanos de las dichas naos que lleven é tengan cuenta é razon de todo é ante quien pase é se asiente todo lo que de la dicha Armada se hobiere." Navarrete, *Colección de los viajes*, 4:127.

25. Navarrete, *Colección de los viajes*, 4:127–28.

26. Denucé, *Magellan, la question*, 220.

27. Navarrete, *Colección de los viajes*, 4:lxxxi.

28. "El cual descubrimiento habeis de hacer, con tanto que no descubrais ni hagais cosa en la demarcacion é límites del serenísimo Rey de Portugal, mi muy caro y muy amado tio é hermano, ni en perjuicio suyo, salvo dentro de los límites de nuestra demarcación." Navarrete, *Colección de los viajes*, 4:117–19.

29. "Tan poca antidad no bastará para cargar las naves de especerías." Navarrete, *Colección de los viajes*, 4:124–27. For the allocation, see Gil, "Magallanes en Sevilla," 56.

30. Sergio Sardone, "El 'Maluco': La financiación de las expediciones, 1518–29," in Martínez Shaw, *Congreso Internacional de historia*, 225–57, makes some modifications to the figures provided in L. Díaz-Trechuelo, "La organización del viaje magallánico: Financiación, engaches, acopios y preparativos," in Teixeira da Mota, *Viagem*, 267–314.

31. Navarrete, *Colección de los viajes*, 4:9.

32. Navarrete, *Colección de los viajes*, 4:179; Denucé, *Magellan, la question*, 207.

33. Díaz-Trechuelo, "Organización del viaje magallánico," 274.

34. "Por cuenta del capitán Juan Sebastián Elcano, cuyo solar es en Aya." I. Fernández Vial, "La nao Victoria," *Sociedad Geográfica Española* 64 (September–December 2019): 56.

35. I. Fernández Vial, "Nao Victoria," 54–65; Francisco Fernández González, "Los barcos de la armada del Maluco," in Martínez Shaw, *Congreso Internacional de historia*, 177–82.

36. Navarrete, *Colección de los viajes*, 4:162–79.

37. Navarrete, *Colección de los viajes*, 4:130–52.

38. Navarrete, *Colección de los viajes*, 4:182–88.

39. F. López-Ríos Fernández, *Medicina naval española: En la época de los descubrimientos* (Barcelona: Labor, 1993).

40. López-Ríos Fernández, 4:8.

41. M. Destombes, "The Chart of Magellan," *Imago Mundi* 12 (1955): 65–88; R. L. Kagan and B. Schmidt, "Maps and the Early Modern State: Official Cartography," in *The History of Cartography*, vol. 3, *Cartography in the European Renaissance*, ed. D. Woodward (Chicago: University of Chicago Press, 2007), 669; J. H. F. Sollewijn Gelpke, "Afonso de Albuquerque's Pre-Portuguese 'Javanese' Map, Partially Reconstructed from Francisco Rodrigues' Book," *Bijdragen tot de Taal-, Land- en Volkenkunde* 151 (1995): 76–99.

42. S. E. Morison, *The Southern Voyages*, vol. 2 of *The European Discovery of America* (Oxford: Oxford University Press, 1974), 343.

43. Alvares in A. Pigafetta, *The First Voyage round the World, by Magellan*, ed. H. E. J. Stanley, Lord Stanley of Alderley (London: Hakluyt Society, 1874), Appendix, x.

44. Navarrete, *Colección de los viajes*, 4:179–82.

45. E. Vila Vilar, "Los vínculos de Magallanes con Sevilla: Amigos, enemigos y devociones," in Vila Vilar, *Magallanes y Sevilla*, 146.

46. Carta de [Juan Rodríguez] Serrano, Andrés de San Martín, Juan Rodríguez Mafra y Vasco Gallego al rey Carlos I, solicitando un aumento de sueldo durante el tiempo que durara la expedición a la Especiería, June 30, 1519, Portale de Archivos Españoles (PARES), http://pares.mcu.es/ParesBusquedas20/catalogo /description/122218.

47. J. T. Medina, *Colección de documentos inéditos para la historia de Chile*, 14 vols. (Santiago: Ercilla, 1888–1902), 2:198–217, prints two examples of the documents in question; others are reproduced at "Autos de herederos de Martín Méndez," 1533, PARES, http://pares.mcu.es/ParesBusquedas20/catalogo/show /122246.

48. Díaz-Trechuelo, "Organización del viaje magallánico," 286–87.

49. Morison, *Southern Voyages,* 339, who represents Magellan's point of view, rather than the objective facts, accurately.

50. "Por por servir a su Alteza . . . como su Alteza lo manda por su carta . . . e como de antes su Alteza lo tenia mandado por las provisiones e instrucciones quel dicho Juan de Cartagena tiene de su Alteza." Navarrete, *Colección de los viajes*, 4:156.

51. Navarrete, *Colección de los viajes*, 4:156–62.

52. A. Teixeira da Mota, "O 'Regimento de altura de Leste oeste' de Rui Faleiro," *Boletim geral do Ultramar* 28 (1953): 163–71.

53. F. Lopes de Castanheda, *História do descobrimento e conquista da India pelos portugueses*, 7 vols. (Lisbon: Rollandiana, 1833), 6:13–14 (bk. 6, chap. 7).

54. W. G. Randles, "Spanish and Portuguese Attempts to Measure Longitude in the Sixteenth Century," *Boletim da Biblioteca da Universidade de Coimbra* 39 (1984): 143–60.

55. Medina, *Colección de documentos inéditos*, 1:327.

56. Alvares in Stanley, *First Voyage*, Appendix, xliii.

57. B. Vázquez Campos, C. Bernal, and T. Mazón, "Traducción al castellano y comentarios del documento 'Auto das perguntas que se fizeram a dois Espanhois que chegaram à fortaleza de Malaca vindos de Timor na companhia de Álvaro Juzarte, capitão de um junco,'" June 1, 1522, Archivo Nacional de Torre do Tombo (Lisboa) Código de referencia: PT/TT/CC/2/101/87, no translation date, 3d34c7df-4aea-4cd8-8436-9739d0131801.filesusr.com/ugd/9a00c 3_7a2f3072785a4510881dc0f746672476.pdf, n11.

58. "Tienen muchos portugueses para llevar consigo cada uno, y porque paresce que seria inconveniente esto, yo vos mando que luego por la mejor manera que os paresciere hableis á los dichos maestres capitanes que no lleve cada uno mas de hasta cuatro ó cinco personase, é los demas que tomaren para llevar los dejen é despidan, é vosotros proveed como en ninguna manera otra cosa se haga; pero esto se ha de hacer con toda la mejor disimulacion que ser pueda." Navarrete, *Colección de los viajes*, 4:159.

59. Navarrete, *Colección de los viajes*, 4:285.

60. Navarerete, *Colección de los viajes*, 4:153.

61. "Asimismo he sabido que en la dicha Armada estaban tomados diez é seis ó diez é siete portugueses, que son todos grumetes, y que al tiempo que se tomaron fue con necesidad que habia de gente; é que agora hallan hartos grumetes é gente: yo vos mando que hagais que se tomen otros grumetes . . . ni otra gente extrangeira mas de la que os tengo escrito, que vaya para acompañar los capitanes . . . porque mi intincion es que se guarde lo que los católicos Reyes mis señores, que hayan gloria, tienen mandado." Navarrete, *Colección de los viajes*, 4:129.62. "E que si otra cosa S.A. o los del su consejo mandasen en contrario de dicho regimiento é captulacion que é no lo guardaria." Navarrete, *Colección de los viajes*, 4:158.

63. Vila Vilar, "Vínculos de Magallanes," 147.

64. Denucé, *Magellan, la question*, 221.

65. F. de B. Aguinagalde, "¿Qué sabemos realmente de Juan Sebastián de Elcano?," in *In medio orbe: Sanlúcar de Barrameda y la I vuelta al mundo*, ed. M.J. Parodi, 2 vols. (Seville: Junta de Andalucía, 2016–17), 1:25–37; "El archivo personal de Juan Sebastián de Elcano," in Parodi, *In medio orbe*, 2:79–93.

66. "Por los servicios que me ha hecho y porque ruegue a Dios por mi ánima." "Testimonio del testamento que otorgó Fernando de Magallanes en los Alcázares Reales de Sevilla," August 24, 1519, in "Documentos para el quinto centenario de la primera vuelta al mundo: La huella archivada del viaje y sus protagonistas," transcribed by Cristóbal Bernal, 2019–22, http://sevilla.2019–2022.org/wp-content/uploads/2016/03/10.ICSevilla2019_Testamento-de-Magallanes-g15.pdf, 5.

67. Navarrete, *Colección de los viajes*, 4:130; F. Morales Padrón, "Las instrucciones a Magallanes," in Teixeira da Mota, *Viagem*, 243–63.

68. Pigafetta, *First Voyage*, ed. Stanley, Appendix, xii.

69. "Hareis todos juntamente vuestro camino, con la buena ventura, á la tierra que nombraréis á los otros Capitanes é Pilotos." Navarrete, *Colección de los viajes*, 132–33.

70. "Primero que salgais del rio de . . . Sevilla, o despues de salidos dél, llamareis los Capitanes, Pilotos, é Maestres; é darles heis las cartas que tenéis hechas para hacer el dicho viage, é mostrarles la primera tierra que esperáis ir a demandar, porque sepan en que derrota está para la ir a demandar." Navarrete, *Colección de los viajes*, iv, 4:131.1.71. "Asimismo ya sabeis como los dichos Capitanes han de declarar la derrota que han de llevar en el dicho viage: Yo vos mando que la recibais dellos por escrito, é conforme á ella hagais vosotros é los dichos Capitanes una instruccion en que se declare la dicha derrota con todos los regimientos de altura que los dichos Capitanes saben para el dicho viage, é lo mostreis todo á los Pilotos que han de ir en la dicha Armada, é deis á cada uno treslado de la dicha instrucción." Navarrete, *Colección de los viajes*, 4:131–32.

72. "No consintais que se toque, ni descubra tierra, ni otra cosa entro en los límites del serenísimo Rey de Portugal." Navarrrete, *Colección de los viajes*, 4:129–30.

73. "É cuando llegardes á ella saldrieis en tierra é porneis un padrón de nuestras armas." Navarrrete, *Colección de los viajes*, 4:132.

74. "De la demarcacion de entre estos Reinos é los de Portugal." Navarrrete, *Colección de los viajes*, 4:132.

75. E. H.. Blair and J. A. Robertson, eds, *The Philippine Islands, 1493–1898*, 55 vols. (Cleveland, OH: Clark, 1903–9), 1:279; Navarrete, *Colección de los viajes*, 4:132–33.

76. "E cuando con la buena ventura llegardes á las tierras é islas adonde hay las especerias, hareis asiento de paz é trato con el rey o señor de la tierra, como vierdes ques mas nuestro servico é provecho; y porque en esto Yo creo que hareis todo lo que cumple á nuestro servicio, no vos limitamos cosa ninguna, porque bien creemos que terneis habilidad para lo hacer por la espiriencia que ya teneis de las semejantes cosas." Navarrete, *Colección de los viajes*, 4:134.

77. "É seyendo la tal tierra poblada, procuraréis de hablar con la gente della, . . . é terniendo vos habla, procurareis de saber que manera esla que tiene é si en la tierra hay cosa de que nos podamos aprovechar." Navarrete, *Colección de los viajes*, 4:133.

78. "Y en señal de paz é seguridad della le diréis, como tenemos por costumbre mandar poner un patron de nuestrs armas en la tierra, en señal de seguridad, é encuanto por él é por los suyos fuere guardado el dicho patron." Navarrete, *Colección de los viajes*, 4:134.

79. "É que haciéndolo les tomarán sus naos é haciendas é captivaran sus personas." Navarrete, *Colección de los viajes*, 4:136.

80. "Es bien que sean de vos bien tratados, declarándoles la razon por qué tomais las naos, ques por ser de gentes con quien no queremos tener paz ni trato." Navarrete, *Colección de los viajes*, 4:135.

81. "Por donde podrán conocer que nuestra voluntad no es hacer mal á los que con nos quisieren tomar asiento de paz é trato de mercaderías." Navarrete, *Colección de los viajes*, 4:135–6.

82. "Serán tomados de buena guerra . . . é si necesario fuese usar con ellos alguna crueldad, lo podreis hacer moderadamente por dar ejemplo é castigo á otros." Navarrete, *Colección de los viajes*, 4:136.

83. "É puesto que dellos por alguna manera alguna persona de los vuestros resciban algund desaguisado, no sean de vosotros maltratados." Navarrete, *Colección de los viajes*, 4:136.

84. "E lo principal que vos encomendamos es que en cualquier cosa que con los indios contratardes se les mantenga é guarde toda verdad, é por vos no sea quebrado, é aunque no lo hayan seido trabajad por venir en concordia; é no habeis de consentir en ninguna manera que se les haga mal ni daño . . . é mas se gana en convertir ciento por esta manera que mil por otra." Navarrete, *Colección de los viajes*, 4:139.

85. "E porque algunas de las personas que van en la dicha armada les parecerá ser mucho el tiempo que habeis andado sin hallar nada, notificáldes . . . que mientras el mantenimiento tovieren en abundancia, ninguno sea osado á hablar . . . en el dicho viaje ni descubrimiento, estar mucho tiempo ni poco, sino que dejen hacer á los que llevan cargo dél." Navarrete, *Colección de los viajes*, 4:142.

86. Denucé, *Magellan, la question*, 215.

87. J. Manoel Noronha, "Algumas observações sobre a naturalidade e a família de Fernão de Magalhães," *O Instituto* 77 (1921): 41–45; J. Gil, *El exilio portugués en Sevilla de los Braganza a Magallanes* (Seville: Cajasol, 2009), 180–82.

88. "Si fallesciere en esta cibdad de Sevilla que mi cuerpo sea enterrado en el monasterio de Santa María de la Victoria, Triana." "Testimonio del testamento que otorgó Fernando de Magallanes en los Alcázares Reales de Sevilla," August 24, 1519, in "Documentos para el quinto centenario de la primera vuelta al mundo: La huella archivada del viaje y sus protagonistas," transcribed by Cristóbal Bernal, 2019–22, http://civiliter.es/biblioteca/ICSevilla2019_Testamento%20de%20Magallanes%20(g15).pdf, 3.

89. Gil, *Exilio portugués*, 426–30.

90. Morison, *Southern Voyages*, 317, 351–52.

91. "Para que antes e primero que otra parte alguna vais a las dichas islas de maluco . . . e despues de fecho esto se podra buscar lo demas que convenga conforme a lo que llevais mandado." Pigafetta, *First Voyage*, ed. Stanley, Appendix, xii.

92. "Desde este cabo frio ate as ilhas de maluco per esta navegaçao non ha nenhumas terras asentadas nas cartas que levan prazera a dios todo poderoso que tall viajem façan como os conterraneos, e vosa alteza fique descansado." Pigafetta, *First Voyage*, ed. Stanley, Appendix, ix–x.

CHAPTER 5. THE CRUEL SEA

1. *The Rule of the Blessed Raymond du Puy* (London: Grand Priory of England, 2020), 13.

2. F. Borja Aguinagalde, "Las dos 'cartas' que escribió el capitán Juan Sebastián de Elcano a su regreso," in *La primera vuelta al mundo: Edición conmemorativa del V Centenario del viaje de Magallanes y Elcano, 1519–1522* (Madrid: Taverna, 2019), 171.

3. A. Pigafetta, *Il primo viaggio intorno al globo di Antonio Pigafetta, e le sue Regole sull'arte del navigare*, ed. A. Da Mosto, Raccolta Colombiana, pt. 5, vol. 3 (Rome: Ministero della pubblica istruzione, 1894), 28.

4. A. Pigafetta, *Relazione del primo viaggio attorno al mondo*, ed. A. Canova (Padua: Antenore, 1999), 30.

5. A. Pigafetta, *The First Voyage around the World (1519–22): An Account of Magellan's Expedition by Antonio Pigafetta*, ed. T. Cachey (New York: Marsiglio, 1995), xix–xxiv, xliv–xlviii.

6. Pigafetta, *First Voyage*, ed. Cachey, xlv.

7. "Le grande et stupende cose del mare Oceano . . . far experientia di me et andarea vedere quelle cose . . . et potessero parturirmi qualche nome apresso la posterità." Pigafetta, *Primo viaggio*, ed. Da Mosto, 51, and *First Voyage*, ed. Cachey, 4.

8. F. Fernández-Armesto, introduction to Alvise da Mosto, *Questa e una opera necessaria tutti li naviganti* (New York: Scholars' Facsimiles, 1992).

9. Pigafetta, *First Voyage*, ed. Cachey, 134–35.

10. "Non oro nè argento, ma cose da essere assay apreciate da un simil signore." Pigafetta, *Primo viaggio*, ed. Da Mosto, 112, and *First Voyage*, ed. Cachey, 123.

11. Nunziatella Alessandrini, "Antonio Pigafetta, cavaleiro do mar oceano: uma resonstrução biografica," *Anais de história de além-mar*, 20 (2019), 72. I am indebted to Professor Sanjay Surahmanyam for a copy of this paper.

12. "Metendo li piedi sopra una antena per descendere ne la mesa di garnitione, me slizegarono li piedi, perchè era piovesto, et cosi cascai nel mare, che ninguno me viste, et, essendo quasi sumerso, me venne ne la mano sinistra la scota de la vella magiore, che era ascosa ne l'acqua; me teni forte e comensai a gridare, tanto che fui aiuto con lo batelo. non credo ià per mey meriti, ma per la misericordia di quella fonte de pietà, fosse aiutato." Pigafetta, *Primo viaggio*, ed. Da Mosto, 71, and *First Voyage*, ed. Cachey, xviii. All translations from Pigafetta in the present work are Cachey's, unless otherwise stated. Except as indicated in further notes, I have relied on the transcription of the manuscript from which Cachey worked in the Biblioteca Ambrosiana, which purports to be a copy of one by the author himself and is, by scholarly consensus, closest to Pigafetta's intentions, in Pigafetta, *Primo viaggio*, ed. Da Mosto. At times when I have not had this work to hand, I have relied on Cachey's translation, checking it subsequently. Cachey modestly represented his translation as a modification of that of J. A. Robertson, ed., *Magellan's Voyage round the World* (Cleveland, OH: Clark, 1906), which I have not consulted except for Robertson's notes. Da Mosto's transcription is in small details different from a more recent version by A. Canova (Pigafetta, *Relazioni del primo viaggio*), to which I had only intermittent access. A project to produce a new edition has been announced by Dr. C. Jostmann on his website. The history of the text and of attempts to edit and translate it is reviewed in C. McCarl, "The Transmission and Bibliographic Study of the Pigafetta Account: Synthesis and Update," *Abriu* 8 (2019): 85–98.

13. Pigafetta, *First Voyage*, ed. Cachey, 4, and *Primo viaggio*, ed. Da Mosto, 51.

14. M. Fernández de Navarrete, *Colección de los viajes y descubrimientos que hicieron por mar los españoles* (Madrid: Imprenta Nacional, 1837), 4:209–47. Doubts cast on the authenticity of the text (G. E. Nunn, "Magellan's Route in the Pacific," *Geographical Review* 24 [1934]: 615–33) seem forced or fantastical. Unless otherwise stated in the notes that follow, I rely on Francisco Albo, "Derrotero del viaje al Maluco, formado por Francisco Albo, piloto de la nao Trinidad y, posteriormente, de la nao Victoria," 1519, in "Documentos para el quinto centenario de la primera vuelta al mundo: La huella archivada del viaje y sus protagonistas," transcribed by Cristóbal Bernal, 2019–22, http://sevilla.2019–2022.org/wp-content/uploads/2016/03/8.ICSevilla2019_Derrotero-de-Francisco-Albo-f15.pdf.

15. J. T. Medina, *El descubrimiento del Océano Pacífico: Hernando de Magallanes y sus compañeros* (Santiago: Elzeviriana, 1920), 60.

16. C. Varela, "Los cronistas del viaje de Magallanes y Elcan," in *Congreso Internacional de Historia: "Primus Circumdedisti Me,"* ed. C. Martínez Shaw (Madrid: Ministerio de Defensa, 2018), 257 ff.; Prospero Peragallo, *Sussidi*

documentari per una monografia su Leone Pancaldo, Raccolta di documenti e studi pubblicati dalla R. Commissione Colombiana (Rome, 1894), 284–89; Massimo Donattini, *Dizionario biografico degli Italiani* 80 (2014): s.v. Pancaldo.

17. G. B. Ramusio, ed., *Navigationi et viaggi*, 3 vols. (Venice: Giunti, 1563–83), 1:222–23.

18. "Mettemmo nome stretto della Vittoria perche la nave Vittoria fu la prima che lo vidde; alcuni le dissero il stretto de Magiglianes perch'il nostro capitan o si chiamava Fernando di Magiglianes." Ramusio, *Navigationi et viaggi*, 1:222–23.

19. M. J. Benites, "Palabras pulverizadas: El relato de Ginés de Mafra sobre la travesía de Magallanes-Elcano," *Bibliographia americana* 15 (2019): 21–29.

20. "Este Ginés de Mafra conservaba, escrito de su mano, una relación de todo lo que había pasado en el viaje de Magallanes, del que había sido testigo. Lo remitió al autor conociendo que este último quería hacer un libro de todo esto." Benites, "Palabras pulverizadas."

21. A. Blázquez y Delgado-Aguilera, *Libro que trata del descubrimento y principio del estrecho que se llama de Magallanes por Ginés de Mafra que se halló en todo y lo vio por vista de los ojos* (Madrid: Torrent, 1920).

22. P. G. H. Schreurs, ed., *The Voyage of Ferdinand Magellan: From the Original Portuguese Manuscript in the University Library of Leiden, Netherlands* (Tilburg: Schreuers, 1999); J. M. Garcia, *A viagem de Fernão de Magalhães e os portugueses* (Lisbon: Editorial Presença, 2007) 195–209.

23. cf. J. Aubin, "Études magallaniennes," in *Le Latin et l'astrolabe*, 2 vols. (Paris: Centre Gulbenkian, 2000), 2:584–85.

24. *Colección general de documentos relativos a las Islas Filipinas existentes en el Archivo de Indias de Sevilla, publicada por la Compania General de Tabacos de Filipinas*, 5 vols. (Barcelona: Viuda de Tasso, 1918–23), 5:194–98; F. B. Aguinagalde, "El archivo personal de Juan Sebastián de Elcano (1487–1526), marino de Guetaria," in *In medio orbe: Sanlúcar de Barrameda y la 1 vuelta al mundo: Actas del II Congreso Internaciona sobre la 1a vuelta al mundo*, ed. M. J. Parodi, 2 vols. (Seville: Junta de Andalucía, 2016), 1:76. Medina's transcription of Gómez's surviving narrative is at "Carta de Gonzalo Gómez de Espinosa al Rey, fecha in Cochín a 12 de enero de 1525, dándole cuenta del viaje que hizo desde que salió de Tidori hasta su llegada allí despues de siete meses," January 12, 1525, https://3d34c7df-4aea-4cd8–8436–9739d0131801.filesusr.com/ugd/9a00c 3_2817732d569243a2a31b17df363a574a.pdf, posted by Tomás Mazón.

25. Medina, *Descubrimiento del Océano Pacífico*, 291–94; *As gavetas da Torre do Tombo*, ed. Arquivo Nacional da Torre do Tombo, 12 vols. (Lisbon: Centro de Estudos Históricos Ultramarinos, 1960–77), 4:316–19, 7:381–84.

26. "Mientras fue vivo Fernando de Magallanes este testigo no ha escrito cosa ningun porque no osaba, e despues que a este testigo eligieron por capitán e tesorero lo que pas'ó lo tiene escrito." Navarrete, *Colección de los viajes*, 4:285–95.

27. Navarrete, *Colección de los viajes*, 4:285–95; J. T. Medina, *Colección de documentos inéditos para la historia de Chile*, 14 vols. (Santiago: Ercilla, 1888–1902), 1:299–310.

28. Medina, *Colección*, 1:323–30; Garcia, *Viagem*, 167–83; Artur Basílio de Sá, ed., *Documentaçã para a história das missões do padroado português do*

Oriente, 1 (Lisbon: Agência Geral do Ultramar, 1954), 141–53. Professor Sanjay Subrahmanyam kindly provided a copy of this work.

29. A. Baião, "A viagem de Fernão de Magalhães por uma testemunha presencial," *Arquivo Histórico de Portugal* 1 (1933): 276–81; N. Águas, ed., *Fernão de Magalhães, A primeira viagem à volta do mundo contada pelos que nela participaram* (Lisbon: Europa-America, 1987), 187–97; Garcia, *Viagem*, 184–89. I have relied for direct quotations on the transcription at "Auto das perguntas que se fizeram a dois espanhóis que chegaram à fortaleza de Malaca vindos de Timor na companhia de Álvaro Juzarte, capitão de um junco," June 1, 1522, Arquivo Nacional Torre do Tombo, https://digitarq.arquivos.pt/details?id=3801974.

30. *De Orbe Novo* (Alcalá: Eguía, 1530), fols. 76–80, Dec. V, chap. 7.

31. Maximilianus Transylvanus, *Maximiliani Transyluani Cæsaris a secretis epistola, de admirabili & nouissima Hispanorum in Orientem nauigatione* (Rome: Calvi, 1524); I reproduce Stanley's translations of this document, included in A. Pigafetta, *The First Voyage round the World, by Magellan*, ed. H. E. J. Stanley, Lord Stanley of Alderley (London: Hakluyt Society, 1874), unless otherwise stated. For the pagination of the original work, I rely on that of the Gale virtual reference library, Hesburgh Libraries,University of Notre Dame, accessed January 27, 2021, https://go-gale-com.proxy.library.nd.edu/ps /i.do?action=interpret&id=GALE%7CCY0102237937&v=2.1&u=nd_ref&it= r&p=SABN&sw=w.

32. Aguinagalde, "Dos 'cartas,'"169.

33. F. Leite de Faria, "As primeiras relações impressas sobre a viagem de Fernão de Magalhães," in *A viagem de Fernão de Magalhães e a questão das Molucas: Actas do II Congreso luso-espanhol de História Ultramarina*, ed. A. Teixeira da Mota (Lisbon: Junta de Investigações Científicas, 1975), 479–80.

34. "Asimismo se ha de poner a cuenta della lo que más gastó la persona que fué a Canarias de lo que se le dio para la ida y tornada a Castilla." Medina, *Descubrimiento del Océano Pacífico*, 108–9.

35. "Hareis todos juntamente vuestro camyno con la buena ventura a la tierra que nonbrareis a los otros captanes e pilotos" and "dar les eys las cartas que teneys hechas para hazer el dicho viaje." *Colección general de documentos relativos a a las Islas Filipinas*, 2:243, 245; S. Bernabéu Albert, "Magallanes: Del héroe al hombre," in *Magallanes y Sevilla*, ed. E. Vila Vilar (Seville: Universidad de Sevilla, 2019), 122–23.

36. "Entrambos juntamente habían de proveer en todas las cosas que fuesen necesarias; e que el dicho Fernando de Magallanes le decía, que no se había en aquello proveído bien, ni él lo entendía." V. M. de Sola, *Juan Sebastián de Elcano* (Bilbao: Caja de Ahorros, 1962), 608, 620; Bernabéu Albert, "Magallanes," 123.

37. "Havendo deliberato il capitanio generale di fare cosi longa navigatione per lo mare Occeanno, dove sempre sonno inpetuosi venti et fortune grandi, et non volendo manifestare a niuno de li suoi el viagio, che voleva, fare, açiò non fosse smarito in pensare de fare tanto grande et stupenda cosa." Pigafetta, *Primo viaggio*, ed. Da Mosto, 52, and *First Voyage*, ed. Cachey, 5.

38. "Dar fine a questo, que promise con iuramento a lo imperatore don Carlo re di Spagna." Pigafetta, *Primo viaggio*, ed. Da Mosto, 112.

39. "Per andare a scopriure la speceria ne le ysolle de Maluco." Pigafetta, *Primo viaggio*, ed. Da Mosto, 51.

40. Pigafetta, *First Voyage*, ed. Cachey, 5, and *Primo viaggio*, ed. Da Mosto, 52.

41. "E la otra gente tenían miedo que los tomaría presos por los muchos portugueses e gente de muchas naciones que había en la armada." Navarrete, *Colección de viajes*, 4:287.

42. Garcia, *Viagem*, 198.

43. A. de Herrera, *Historia general de los hechos de los castellanos en las islas y tierra-firme de el mar occeano* [sic], Dec. II, bk. 5, chap. 10, ed. J. Natalicio Gómez, 10 vols. (Asunción: Guarania, 1945), 3:80, confirmed and perhaps informed by the Leiden MS: "e o capitão-mor lhe respondeu que lhe não havia de dar conto de seu caminho e ele se fosse pera sua nau e seguisse seu farol, senão que o mandaria prender e castgar como revel." Garcia, *Viagem*, 198.

44. Garcia, *Viagem*, 108–26.

45. N. Wey-Gómez, *The Tropics of Empire: Why Columbus Sailed South to the Indies* (Cambridge, MA: MIT Press, 2008).

46. "Molte gropade de venti impetuosi et correnti de acqua ne asaltaronno contra el viagio. . . . et de queta sorte andavamo de mare in trrraverso fin che passava la grupada, perchè veniva molto furioso." Pigafetta, *Primo viaggio*, ed. Da Mosto, 54, and *First Voyage*, ed. Cachey, 8.

47. "Benchè non sonno bonni da mangiare, se non li piccoli, et anche loro mal bonny." Pigafetta, *Primo viaggio*, ed. Da Mosto, 54 and *First Voyage*, ed. Cachey, 8.

48. "E le privó de la capitanía e veeduría, e quísole echar desterrado en la costa de Brasil, y por ruego de los otros Capitanes no le echó entonces, e dióle preso a Gaspar de Quesada sobre su pleito homenaje para que le tuviese preso." Sola, *Juan Sebastián de Elcano*, 607; Bernabeu Albert, "Magallanes," 124.

49. Pigafetta, *First Voyage*, ed. Cachey, 9, and *Primo viaggio*, ed. Da Mosto, 59.

50. "Hay buena gente y mucha, y van desnudos, y contratan con anzuelos, y espejos y cascabeles por cosas de comer, y hay mucho brasil." Stanley, *First Voyage*, 193.

51. F. Fernández-Armesto, "The Stranger-Effect in Early Modern Asia," *Itinerario* 24 (2000): 80–103.

52. M. W. Helms, *Ulysses' Sail: An Ethnographic Odyssey of Power, Knowledge, and Geographical Distance* (Princeton, NJ: Princeton University Press, 2008), and *Craft and the Kingly Ideal: Art, Trade, and Power* (Austin: Texas University Press, 1993).

53. M. Restall, L. Sousa, and K. Terraciano, eds., *Mesoamerican Voices: Native-Language Writings from Colonial Mexico, Oaxaca, Yucatan, and Guatemala* (Cambridge: Cambridge University Press, 2005), 47–54, 64–96, 101–13.

54. Pigafetta, *First Voyage*, ed. Cachey, 9.

55. Pigafetta, *Primo viaggio*, ed. Da Mosto, 55, and *First Voyage*, ed. Cachey, 11; Ayamonte's testimony at "Auto das perguntas," https://digitarq.arquivos .pt/details?id=3801974.

56. Pigafetta, *Primo viaggio*, ed. Da Mosto, 76.

57. Navarrete, *Colección de los viajes*, 4:192.

58. M. Justo Guedes, "A Armada de Fernão de Magalhães e o Brasil," in Teixeira, *Viagem,* 361–77.

59. Medina, *Descubrimiento del Océano Pacífico,* 225.

60. "Ne davano per una acceta ho cortello grande una ho due de le sue figliole per schiave. . . . Una iovene bella vene un dì ne la nave capitania, . . . non per altro se non per trovare alguno recapito." Pigafetta, *Primo viaggio,* ed. Da Mosto, 56.

61. "Stando cossì et aspettando, butò lo ochio supra la camera del maistro et victe uno quiodo longo più de un dito, il que pigliando, con grande gentilessa et galantaria, se lo ficò a parte de li labri della sua natura, et subito bassa se partite." Pigafetta, *Primo viaggio,* ed. Da Mosto, 56, and *First Voyage,* ed. Cachey, 11–12, 138.

62. Pigafetta, *Primo viaggio,* ed. Da Mosto, 55–56, and *First Voyage,* ed. Cachey, 10–11.

63. Pigafetta, *Primo viaggio,* ed. Da Mosto, 55. A. Gerbi, *Nature in the New World: From Christopher Columbus to Gonzalo Fernández de Oviedo* (Pittsburgh, PA: University of Pittsburgh Press, 1985), 106, quoted in Pigafetta, *First Voyage,* ed. Cachey, 135.

64. F. Fernández-Armesto, *Amerigo: The Man Who Gave His Name to America* (New York: Random House, 2007), 100–102, 137–38.

65. Pigafetta, *Primo viaggio,* ed. Da Mosto, 55; David Bindman and Henry Louis Gates Jr., eds., *The Image of the Black in Western Art,* vol. 3, *From the "Age of Discovery" to the Age of Abolition, Part 1: Artists of the Renaissance and Baroque* (Cambridge, MA: Harvard University Press, 2010).

66. "Questi stavano con tanta contritione in genoquioni alsando la mane giunte che era grandissimo piacere vederli . . . Questi populi facilmente se converterebonno a la fede de Iesu Cristo." Pigafetta, *Primo viaggio,* ed. Da Mosto, 56, and *First Voyage,* ed. Cachey, 11–12.

67. St. Augustine, *City of God* 16.8, "Whether Certain Monstrous Races of Men Are Derived from the Stock of Adam or Noah's Sons." In *A Select Library of the Nicene and Post-Nicene Fathers,* ed. Philip Schaff (Buffalo, NY: Christian Literature Publishing Co., 1887), vol. 2, https://oll.libertyfund.org/title/schaff-a-select-library-of-the-nicene-and-post-nicene-fathers-of-the-christian-church-vol-2#lf1330–02_label_1353.

68. "Retulerunt autem . . . ut non modo nihil fabulosi afferre, sed fabulosa omnia alia veteribus authoribus prodita, refellere t reprobare, narratione sua viderentur." *Maximilanus Transylvanus Cæsaris,* 8; Gale virtual reference library pagination.

69. Evidently not Pentecost, as proposed by Abel Fabre, 'L'Iconographie de la Pentecôte," *Gazette des Beaux-Arts* 2 (1923): 33–42. Detailed exposition will be found in F. Fernández-Armesto and J. Peterson, *Under the Angels,* in progress.

70. M.L. Price, *Consuming Passions: The Uses of Cannibalism in Late Medieval and Early Modern Europe* (London: Routledge, 2003).

71. "Non se mangiano subito, ma ogni uno taglia uno pezo et lo porta in casa metendolo al fumo; per ogni 8 iorni taglia uno pezeto, mangiandolobrutolado con le altre cose per memoria degli sui nemici." Pigafetta, *Primo viaggio,* ed. Da Mosto, 55, and *First Voyage,* ed. Cachey, 11.

72. G. Fernández de Oviedo, *Historia natural y general de las Indias occidentales,* ed. J. Amador de los Ríos, 14 vols. (1851–55; repr., Asunción: Guarania, 1944–45), 2:37–40.

73. A. A. Duker, "The Protestant Israelites of Sancerre: Jean de Léry and the Confessional Demarcation of Cannibalism," *Journal of Early Modern History* 18 (2014): 255–86; Anne B. McGinness, "Christianity and Cannibalism: Three European Views of the Tupi in the Spiritual Conquest of Brazil, 1557–1563," *World History Connected* 7, no. 3 (October 2010), https://worldhistoryconnected.press.uillinois.edu/7.3/mcginness.html.

74. "Mangiano carnbe humana de li sui nemici, non per bonna, ma per una certa usanza." Pigafetta, *Primo viaggio,* ed. Da Mosto, 55.

75. "Ricordandosi del suo figliolo, como cagnia rabiata li corse adosso et lo mordete in una spala." Pigafetta, *Primo viaggio,* ed. Da Mosto, 55.

76. P. R. Sanday, *Divine Hunger: Cannibalism as a Cultural System* (Cambridge: Cambridge University Press, 1986), 21–22, 69.

77. Felipe Fernández-Armesto, *Food: A History* (London: Macmillan, 2004), 25–34.

78. D. Dalby and P. E. H. Hair, "'Le Langaige du Brésil': A Tupi Vocabulary," *Transactions of the Philological Society* 65 (1966): 61.

79. Jan de Léry, *History of a Voyage to the Land of Brazil,* ed, J. Waley (Berkeley: University of California Press, 1993), 101.

80. Pigafetta, *Primo viaggio,* ed. Da Mosto., 55–56, and *First Voyage,* ed. Cachey, 10–11.

81. Albo, "Derrotero," 4n223.

82. J. Denucé, *Magellan, la question des Moluques et la première circumnavigation du monde* (Brussells: Hayez, 1911), 263.

83. "Già se pensava che de qui se pasasse al mare de Sur." Pigafetta, *Primo viaggio,* ed. Da Mosto, 56, and *First Voyage,* ed. Cachey, 13.

84. "Declaración de las personas fallecidas en el viaje al Maluco (del 20-XII-1519 al 29-VII-1522)," in "Documentos para el quinto centenario de la primera vuelta al mundo: La huella archivada del viaje y sus protagonistas," December 20, 1519–July 29, 1522, transcribed by Cristóbal Bernal, 2019–22, http://sevilla.2019–2022.org/wp-content/uploads/2016/03/4.ICSevilla2019_Declaracion-de-fallecidos-en-el-viaje-s14.pdf.

85. Pigafetta, *Primo viaggio,* ed. Da Mosto, 57, and *First Voyage,* ed. Cachey, 14; Albo, "Derrotero," 7.

86. F. P. Moreno, *Viaje a la Patagonia austral* (Buenos Aires: Sociedad de Abogados, 1877), 23.

87. "In una hora cargasimo le cinque nave. . . . Eranno tanti grossi che non bisogniva pelarli ma scortigarli." Pigafetta, *Primo viaggio,* ed. Da Mosto, 57. Quoted in Albo, "Derrotero," 14n362.

88. R. Laguarda Trías, *El enigma de las latitudes de Colón* (Valladolid: Universidad de Valladolid, 1974); F. Fernández-Armesto, *Columbus* (London: Duckworth, 1996), 74–78, and *Amerigo,* 74–83.

89. E. G. R. Taylor, "A Log-Book of Magellan's Voyage," *Journal of Navigation* 17 (1964): 83–87.

90. "Los otros Capitanes, juntamente con el dicho Cartagena, requerían al dicho Magallanes que tomase consejo con sus oficiales, e que diese la derrota a donde quería ir, e que no anduviese ansí perdido." Navarrete, *Colección de viajes*, 4:287.

91. "E que no tomase puerto donde invernasen e comisen los bastimentos, e que caminasen hasta donde podiesen sufrir el frío para que si hobiese lugar pasasen adelante." Sola, *Juan Sebastián de Elcano*, 607; Bernabeu Albert, "Magallanes," 124–25.

92. "Les diese favor y ayuda para que hiciesen cumplir los mandamientos del Rey, como en sus instrucciones lo mandaba." Navarrete, *Colección de viajes*, 4:287.

93. "Y este testigo, que obedecía, e que está preso para facerle cumplir e requerir con aquello al dicho Fernando de Magallanes." Bernabeu Albert, "Magallanes," 127.

CHAPTER 6. THE GIBBET AT SAN JULIÁN

1. W. S. W. Vaux, ed., *The World Encompassed* (London: Hakluyt Society, 1854), 56, 68–70.

2. Vaux, *World Encompassed*, 5.

3. Maximilianus Transylvanus, *Maximiliani Transyluani Cæsaris a secretis epistola, de admirabili & nouissima Hispanorum in Orientem nauigatione* (Rome: Calvi, 1524), 13.

4. "Essendo l'inverno le navi intrarono in uno bon porto per invernarse." A. Pigafetta, *Il primo viaggio intorno al globo di Antonio Pigafetta, e le sue Regole sull'arte del navigare*, ed. A. Da Mosto, Raccolta Colombiana, pt. 5, vol. 3 (Rome: Ministero della pubblica istruzione, 1894), 57, *The First Voyage around the World (1519–22): An Account of Magellan's Expedition by Antonio Pigafetta*, ed. T. Cachey (New York: Marsiglio, 1995), 14, and *Primo viaggio intorno al globo terracqueo*, ed. C. Amoretti (Milan: Galeazzi, 1800), 24. In this chapter, the translations are my own and are based on Amoretti's, the text I had to hand while writing. The transcriptions are—to choose a positive term— creative; so I have checked them against Pigafetto, *Primo viaggio*, ed. Da Mosto, and my translations against Cachey's.

5. F. Albo, "Derrotero del viaje al Maluco, formado por Francisco Albo, piloto de la nao Trinidad y, posteriormente, de la nao Victoria," 1519, in "Documentos para el quinto centenario de la primera vuelta al mundo: La huella archivada del viaje y sus protagonistas," transcribed by Cristóbal Bernal, 2019– 22, http://sevilla.2019–2022.org/wp-content/uploads/2016/03/8.ICSevilla2019_ Derrotero-de-Francisco-Albo-f15.pdf, 7.

6. F. López de Gómara, *Historia general de las Indias* (Zaragoza: Miguel de Papila, 1555), fol. 41v, facs. ed., ed. F. Pease (Lima: Comisión Nacional del Quinto Centenario del Descubrimiento de América, 1993).

7. "Yo he seguido la relaçion que Johan Sebastian del Cano me dió . . . é quasi la misma relaçion escribió el bien enseñado secretario de César llamado Maximiliano Transilvano." G. Fernández de Oviedo, *Historia natural y general*

de las Indias occidentales, ed. J. Amador de los Ríos, 14 vols. (1851–55; repr., Asunción: Guarania, 1944–45), 2:15. Oviedo also referred to conversations with Elcano's fellow survivor Hernando de Bustamante (18, 23) and with Gómez de Espinosa (32).

8. "De vetere atque aeterno Portugallensium atque Castellanorum odio Magellanum Portugalensem esse." Transylvanus, *Maximiliani Transyluani Cæsaris a secretis epistola,* 17; A. Pigafetta, *The First Voyage round the World, by Magellan,* ed. H.E.J. Stanley, Lord Stanley of Alderley (London: Hakluyt Society, 1874), 194.

9. Pigafetta, *Primo viaggio,* ed. Amoretti, 33, and *First Voyage,* ed. Cachey, 19.

10. A. Blázquez y Delgado-Aguilera, ed., *Descripción de los reinos, costas, puertos e islas* (Madrid: Torrent, 1921), 187.

11. A. de Herrera, *Historia general de los hechos de los castellanos en las islas y tierra-firme de el mar occeano [sic],* ed. J. Natalicio Gómez, 10 vols. (Asunción: Guarania, 1945), 2:235, 281–95 (Dec. II, bk. 4, chap. 10, and Dec. II, bk. 9, chaps. 10–15).

12. Steve Zimmerman, *Food in the Movies* (Jefferson, NC: McFarland, 2010), 244–64.

13. M. Fernández de Navarrete, *Colección de los viajes y descubrimientos que hicieron por mar los españoles* (Madrid: Imprenta Nacional, 1837), 4:302.

14. Blázquez y Delgado-Aguilera, *Descripción,* 187.

15. Navarrete, *Colección de los viajes,* 4:190.

16. "Hago saber quel dicho Domingo de Ramos, en la noche, primero dia del mes de Abril de este año de 1520 años, estando en mi cámara en la dicha nao, e reposada ya toda la gente, vino Gaspar de Quesada, capitan de la nao Concepcion, e Juan de Cartagena, armados con cerca de de treinta hombres armados todos, e se allegaron a mi cámara con las espadas sacadas, e me tomaron poniéndome las dichas armas en los pechos, e se alzaron con la nao, e me llevaron despues de tomado debajo de la cubierta e me metieron en la cámara de Berónio Guerra, escribano de la dicha nao, e me echaron los grillos e no bastó echar los dichos grillos, sino que me cerraron la puerta de la dicha cámara con un candado, e demas desto pusieron un hombre a la puerta para que la guardase." Navarrete, *Colección de los viajes,* 4:190.

17. "Q: ¿Quién aprueba eso? Priest: El profeta David. Q: No conocemos agora padre el profeta David." Navarrete, *Colección de los viajes,* 4:192.

18. "Requieros de parte de Dios e del Rey Don Carlos que vos vais a vuestr nao, porq no es este tiempo de andar con hombre armados porlas naos, y tambien vos requiero q soltéis nro capitán." Navarrete, *Colección de los viajes,* 4:192.

19. Navarrete, *Colección de los viajes,* 4:136.

20. Navarrete, *Colección de los viajes,* 4:191.

21. "Declaración de las personas fallecidas en el viaje al Maluco," from December 20, 1519, to July 29, 1522, in "Documentos para el quinto centenario de la primera vuelta al mundo: La huella archivada del viaje y sus protagonistas," transcribed by Cristóbal Bernal, 2019–22, http://sevilla.2019–2022 .org/wp-content/uploads/2016/03/4.ICSevilla2019_Declaracion-de-fallecidos-en-el-viaje-s14.pdf.

22. "Grandísimo estrago en los mantenimientos sin haber peso ni medida sino todo abierto a quien lo quería tomar." Navarrete, *Colección de viajes*, 4:191.

23. Navarrete, *Colección de los viajes*, 4:197.

24. "E mataba la gente a palos e no les daba de comer." Navarrete, *Colección de viajes*, 4:191.

25. Navarrete, *Colección de los viajes*, 4:204.

26. Navarrete, *Colección de los viajes*, 4:201–8.

27. A. Baião, ed., "El viaje de Magallanes según un testigo presencial," *Revista chilena de historia y geografía* 79–81 (1936): 31–41.

28. Navarrete, *Colección de los viajes*, 4:36.

29. Navarrete, *Colección de los viajes*, 4:37.

30. Blázquez y Delgado-Aguilera, *Descripción*, 190.

31. Navarrete, *Colección de los viajes*, 4:38.

32. Blázquez y Delgado-Aguilera, *Descripción*, 191.

33. Pigafetta, *First Voyage*, ed. Stanley, 250.

34. Transylvanus, *Maximiliani Transyluani Cæsaris a secretis epistola*, 16–17; Navarrete, *Colección de los viajes*, 4:163.

35. Peter Martyr, *Cartas sobre el Nuevo Mundo,* fol. 77v. (Dec. V, chap. 7), trans. J. Bauzano (Madrid: Polifemo, 1990).

36. The most egregious such case, and apparently the source of many others (and of further errors apparently diffused via the Web), is that of L. Bergreen, *Over the Edge of the World* (London: Harper Perennial, 2003), 150.

37. Navarrete, *Colección de los viajes*, 4:206, 276.

38. Navarrete, *Colección de los viajes*, 4:14.

39. J. de Barros, *Decadas da Asia,* 24 vols. (Lisbon: Regia Officina, 1777–88), 5:640, 657 (Dec. III, bk. 5, chaps. 9 and 10).

40. R. A. Laguarda Trías, "Las longitudes geográficas de la membranza de Magallanes y del primer viaje de circunnavegación," in *A viagem de Fernão de Magalhães e a questão das Molucas: Actas do II Congreso luso-espanhol de História Ultramarina,* ed. A. Teixeira da Mota (Lisbon: Junta de Investigacões Científicas, 1975)158–60.

41. J. Denucé, *Magellan, la question des Moluques et la première circumnavigation du monde* (Brussells: Hayez, 1911), 207, 270; Transylvanus, *Maximiliani Transyluani Cæsaris a secretis epistola,* 13 (also in Pigafetta, *First Voyage,* ed. Stanley, 189).

42. Laguarda Trías, "Longitudes geográficas," 162; R. Cerezo Martínez, "Conjetura y realidad en la primera circunnavegación," in *Congreso de historia del descubrimient: Actas,* ed. C. Seco Serrano et al. (Madrid: Real Academia de la Historia, 1992), 2:137–92.

43. Laguarda Trías, "Longitudes geográficas," 163.

44. On the data available to Herrera, see A. Ballesteros y Beretta, "Proemio," in A. Herrera y Tordesillas, *Historia general de los hechos de los castellanos en las islas y Tierra firme del Mar Océano,* ed. A. Ballesteros y Beretta (Madrid: Real Academia de la Historia, 1936).

45. Herrera, *Historia general,* ed. Gómez, 2:285 (Dec. II, bk. 9, chap. 14); "Annular Eclipse of the Sun: 1520 October 11," Astronomical Data Portal @

UK Hydrographic Office, accessed August 30, 2020, http://astro.ukho.gov.uk /eclipse/0331520/.

46. F. Lopes de Castanheda, *História do descobrimento e conquista da India pelos portugueses*, 7 vols. (Lisbon: Rollandiana, 1833), 6:13 (bk. 6, chap. 7).

47. Navarrete, *Colección de los viajes*, 4:307.

48. Transylvanus, *Maximiliani Transyluani Cæsaris a secretis epistola*, 18; cf. Pigafetta, *First Voyage*, ed. Stanley, 194.

49. Pigafetta, *First Voyage*, ed. Stanley, 192.

50. Pigafetta, *First Voyage*, ed. Stanley, 192–94; Transylvanus, *Maximiliani Transyluani Cæsaris a secretis epistola*, 15–18.

51. Pigafetta, *Primo viaggio*, ed. Da Mosto, 34, and *First Voyage*, ed. Cachey, 20.

52. Transylvanus, *Maximiliani Transyluani Cæsaris a secretis epistola*, 16–17; cf. Pigafetta, *First Voyage*, ed. Stanley, 192–93.

53. "Tantoque fore uberiora quanto maiora laboribus et periculis Caesari novum et incognitum Aromatum atque Auro divitem Orbem apperuissent." Transylvanus, *Maximiliani Transyluani Cæsaris a secretis epistola*, 17.

54. Transylvanus, *Maximiliani Transyluani Cæsaris a secretis epistola*, 17–18; cf. Pigafetta, *First Voyage*, ed. Stanley, 193–94.

55. Barros, *Decadas da Asia*, 5:634 (Dec. III, bk. 5, chap. 9).

56. Barros, *Decadas da Asia*, 633–34 (Dec. III, bk. 5, chap. 9).

57. Herrera, *Historia general*, ed. Gómez, 2:283–84 (Dec. II, bk. 9, chap. 13); Blázquez y Delgado-Aguilera, *Descripción*, 291.

58. "El suelo es arenoso, arcilloso, y cubierto casi completamente de casajo; grandes cantidades de arbustos (inciensos, calafates, etc.) de hojas de distintos colores, armonizan el paisaje, y entre ellos, manchones con pasto amarillento de penaches plateadosle dan apariencia metálica que alegra el suelo. Este está surcado por infinidad de pequeñas sendas de guanacos, que facilitan la marcha a pie, pues los cactus, las espinas de los arbustos y la fabulos cantidad de cuevas de ctenomys cansan y maltratan cruelmente al caminante." F. P. Moreno, *Viaje a la Patagonia austral*, 248.

59. "Haviendo tardado once dias llegaron tan desmejados que no los conocian." Herrera, *Historia general*, ed. Gómez, 2:283 (Dec. II, bk. 9, chap. 13).

60. "Porque la Mar estaba tan alterada que era imposible andar por ella." Herrera, *Historia general*, ed. Gómez, 2:283 (Dec. II, bk. 9, chap. 13).

61. "Dixeron los de la Nao perdida que havia treinta i cinco dias que no lo comian." Herrera, *Historia general*, ed. Gómez, 2:283 (Dec. II, bk. 9, chap. 13).

62. "Le navi quasi persenno per li venti terribili, ma Dio et li Corpi sancti le aiutarono." Pigafetta, *Primo viaggio*, ed. Da Mosto, 60, and *First Voyage*, ed. Cachey, 20.

63. Pigafetta, *First Voyage*, ed. Stanley, 4, 218.

64. Blázquez y Delgado-Aguilera, *Descripción*, 193; Albo, "Derrotero," 7.

65. Blázquez y Delgado-Aguilera, *Descripción*, 191.

66. "Un dì a l'inproviso vedessemo uno homo, di statura de gigante, che stava nudo ne la riva del porto. . . . Questo erra tanto grande che li davamo a la cintura." Pigaftetta, *Primo viaggio*, ed. Da Mosto, 57, and *First Voyage*, ed. Cachey, 14.

67. Pigafetta, *Primo viaggio*, ed. Da Mosto, 56, and *First Voyage*, ed. Cachey, 13.

68. Transylvanus, *Maximiliani Transyluani Cæsaris a secretis epistola*, 14; Pigafetta, *First Voyage*, ed. Stanley, 5, 189–90.

69. G. F. Dille, ed., *Spanish and Portuguese Conflict in the Spice Islands: The Loaisa Expedition to the Moluccas* (London: Hakluyt Society, 2021), 49.

70. Vaux, *World Encompassed*, 58–61.

71. V. Kogut Lessa de Sá, ed., *The Admirable Adventures and Strange Fortunes of Master Anthony Knivet: An English Pirate in Sixteenth-Century Brazil* (Cambridge: Cambridge University Press, 2015), 132–33.

72. J. Hawkesworth, *An Account of the Voyages Undertaken by the Order of His Present Majesty for Making Discoveries in the Southern Hemisphere and Successively Performed by Commodore Byron, Captain Wallis, Captain Carteret, and Captain Cook . . .* (London, 1773), 27–28, 31–32.

73. S. Davies, *Renaissance Ethnography and the Invention of the Human* (Cambridge: Cambridge University Press, 2016), 148–65.

74. Pigafetta, *Primo viaggio*, ed. Da Mosto, 55, *First Voyage*, ed. Cachey, 10, and *First Voyage*, ed. Stanley, 5.

75. Pigafetta, *First Voyage*, ed. Stanley, 5, *Primo viaggio*, ed. Da Mosto, 59, and *First Voyage*, ed. Cachey, 18–19; Transylvanus, *Maximiliani Transyluani Cæsaris a secretis epistola*, 14.

76. Pigafetta, *First Voyage*, ed. Stanley, 190–91, *Primo viaggio*, ed. Da Mosto, 58–59, and *First Voyage*, ed. Cachey, 13–18.

77. Transylvanus, *Maximiliani Transyluani Cæsaris a secretis epistola*, 14; Pigafetta, *First Voyage*, ed. Stanley, 190, *Primo viaggio*, ed. Da Mosto, 58, and *First Voyage*, ed. Cachey, 16.

78. Transylvanus, *Maximiliani Transyluani Cæsaris a secretis epistola* op.cit., 15; Pigafetta, *First Voyage*, ed. Stanley, 191.

79. Pigafetta, *Primo viaggio*, ed. Da Mosto, 58, and *First Voyage*, ed. Cachey, 19.

80. "Tehuelche," Endangered Languages Project, accessed July 4, 2020, www.endangeredlanguages.com/lang/teh.

81. Pigafetta, *Primo viaggio*, ed. Da Mosto, 64, and *First Voyage*, ed. Stanley, 62–65; R. Fernández Rodríguez and M. A. Regúnaga, "Patagonian Lexicography (Sixteenth–Eighteenth Centuries)," in *Missionary Linguistic Studies from Mesoamerica to Patagonia*, ed. A. Alexander-Bakkerus, R. Fernández Rodríguez, L. Zack, and O. Zwartjes (Leiden: Brill, 2020), 236–59.

82. Pigaftetta, *Primo viaggio*, ed. Da Mosto, 59, and *First Voyage*, ed. Cachey, 19.

83. R. A. Spencer, *Harry Potter and the Classical World* (Jefferson, NC: McFarland, 2015).

84. F. Vázquez, *Pantaleón* (Venice: Niccolini di Sabio, 1534), 161.

85. "E cosa tan estraña de mirar: tomole en voluntad de lo llevar preso: e penso si lo pudiesse llevar en sus naos que le seria gran nora porque se señora Gridonia lo viesse." Vázquez, *Pantaleón*, 162.

86. "E sabed que el havia aquella condicion assi como el animal que lo engendro de ser muy ledo contra las mugeres." Vázquez, *Pantaleón*, 63.

87. R. Bernheimer, *Wild Men in the Middle Ages: A Study in Art, Sentiment, and Demonology* (Cambridge, MA: Harvard University Press, 1952).

88. "Drizassemo una croce in segno dequesta terra, che erra del re de Spagna." Pigafetta, *Primo viaggio*, ed. Da Mosto, 60, and *First Voyage*, trans. Cachey, 20.

89. Vaux, *World Encompassed*, xxx–xxxii, 62–63, 65–74.

CHAPTER 7. THE GATES OF FAME

1. J. de Barros, *Decadas da Asia*, 24 vols. (Lisbon: Regia Officina, 1777–88), 5:641 (Dec. III, bk. 5, chap. 9).

2. R. A. Laguarda Trías, "Las longitudes geográficas de la membranza de Magallanes y del primer viaje de circunnavegación," in *A viagem de Fernão de Magalhães e a questão das Molucas: Actas do II Congreso luso-espanhol de História Ultramarina*, ed. A. Teixeira da Mota (Lisbon: Junta de Investigacões Científicas, 1975), 167.

3. "Se non era el capitano gennerale, non trovavamo questo strecto, perchè tuti pensavamo et dicevamo como era serato tuto in torno." A. Pigafetta, *Il primo viaggio intorno al globo di Antonio Pigafetta, e le sue Regole sull'arte del navigare*, ed. A. Da Mosto, Raccolta Colombiana, pt. 5, vol. 3 (Rome: Ministero della pubblica istruzione, 1894), 61, and *The First Voyage around the World (1519–22): An Account of Magellan's Expedition by Antonio Pigafetta*, ed. T. Cachey (New York: Marsiglio, 1995), 21.

4. A. Blázquez y Delgado-Aguilera, ed., *Descripción de los reinos, costas, puertos e islas* (Madrid: Torrent, 1921), 193.

5. Pigafetta, *Primo viaggio*, ed. Da Mosto, 62, and *First Voyage*, ed. Cachey, 23.

6. Barros, *Decadas da Asia*, 5:641 (Dec. III, bk. 5, chap. 9).

7. See above, chapter 6, note 54.

8. "Lo quale hodiava molto lo capitabo generale perchè, inanzi se facesse questa armata, costui era andato da lo imperatore per farse dare alcune caravele per discovrire terra, ma, per la venuta del capitano gennerale, su magestà non le li dete." Pigafetta, *Primo viaggio*, ed. Da Mosto, 62, and *First Voyage*, ed. Cachey, 22.

9. M. Fernández de Navarrete, *Colección de los viajes y descubrimientos que hicieron por mar los españoles* (Madrid: Imprenta Nacional, 1837), 4:lxxxii.

10. A. Herrera, *Historia general de los hechos de los castellanos en las islas y tierra-firme de el mar occeano* [*sic*], ed. J. Natalicio Gómez, 10 vols. (Asunción: Guarania, 1945), 4:283 (Dec. II, bk. 4, chap. 9).

11. Navarrete, *Colección de los viajes*, 4:43.

12. Barros, *Decadas da Asia*, 5:644–45 (Dec. III, bk. 5, chap. 9).

13. Barros, *Decadas da Asia*, 5:640–46 (Dec. III, bk. 5, chap. 9).

14. Navarrete, *Colección de los viajes*, 4:305–11; J. T. Medina, *Colección de documentos inéditos para la historia de Chile*, 14 vols. (Santiago: Ercilla, 1888–1902), 1:323–30; J. Ramos Coelho, ed., *Alguns documentos da Torre do Arquivo Nacional da Torre do Tumbo ácerca das navegações e conquistas por-*

tuguezas publicados por ordem do governo de sua majestade fidelissima ao celebrar-se a commemoração quadricentenaria do descobrimento da America (Lisbon: Impresa Nacional, 1892), 464–78.

15. Pigafetta, *Primo viaggio*, ed. Da Mosto, 62, and *First Voyage*, ed. Cachey, 21.

16. "Et cosi stando suspesi vedemo venire due navi con le velle pienne et con lae bandere spiegate verso de noi. essendo così vicine, subito scaricarono molte bombarde et gridi; poy tuti insieme rengratiando Ydio et la Vergine Maria andasemo a cercare più inanzi." Pigafetta, *Primo viaggio*, ed. Da Mosto, 62, and *First Voyage*, ed. Cachey, 22.

17. "Por albricias cuando saltaron en tierra y se descubrió el estrecho." F. L. Jiménez Abollado, "Ocasio Alonso, un marinero en la primera vuelta del mundo: Incidencias y vicisitudes de un superviviente," *Naveg@mérica* 23 (2019): 7, http://revistas.um.es/navegamerica.

18. J. T. Medina, *El descubrimiento del Océano Pacífico: Hernando de Magallanes y sus compañeros* (Santiago: Elzeveriana, 1920), ccliv.

19. G. Fernández de Oviedo, *Historia natural y general de las Indias occidentales*, ed. J. Amador de los Ríos, 14 vols. (1851–55; repr., Asunción: Guarania, 1944–45). 2:38–39; G. F. Dille, ed., *Spanish and Portuguese Conflict in the Spice Islands: The Loaisa Expedition to the Moluccas* (London: Hakluyt Society, 2021), 41.

20. A. Pigafetta, *The First Voyage round the World, by Magellan*, ed. H. E. J. Stanley, Lord Stanley of Alderley (London: Hakluyt Society, 1874), 7.

21. Barros, *Decadas da Asia*, 5:639 (Dec. III, bk. 5, chap. 9).

22. Navarrete, *Colección de los viajes*, 4:43, 307.

23. "La cercassemo per tutolo streto fin in quella boca dove ella fugite." Pigafetta, *Primo viaggio*, ed. Da Mosto, 62, and *First Voyage*, ed. Cachey, 23.

24. Pigafetta, *Primo viaggio*, ed. Da Mosto, 62.

25. "Una bandera in cima di alguno monticello con una lettera in una piniatella." Pigafetta, *Primo viaggio*, ed. Da Mosto, 62; the translation is Stanley's, *First Voyage*, 60.

26. S. E. Morison, *The Southern Voyages*, vol. 2 of *The European Discovery of America* (Oxford: Oxford University Press, 1974), 392.

27. Navarrete, *Colección de los viajes*, 4:202.

28. Pigafetta, *Primo viaggio*, ed. Da Mosto, 63.

29. F. Cervantes, *Conquistadors* (London: Penguin, 2020), 82–83.

30. Navarrete, *Colección de los viajes*, 4:206.

31. Navarrete, *Colección de los viajes*, 4:202.

32. Pigafetta, *Primo viaggio*, ed. Da Mosto, 63.

33. R. J. Campbell, P. T. Bradley, and J. Lorimer, eds., *The Voyage of Captain John Narbrough to the Strait of Magellan and the South Sea in His Majesty's Ship Sweepstakes, 1669–1671* (London: Hakluyt Society, 2018), 274.

34. F. Albo, "Derrotero del viaje al Maluco, formado por Francisco Albo, piloto de la nao Trinidad y, posteriormente, de la nao Victoria," 1519, in "Documentos para el quinto centenario de la primera vuelta al mundo: La huella archivada del viaje y sus protagonistas," transcribed by Cristóbal Bernal, 2019–

22, http://sevilla.2019–2022.org/wp-content/uploads/2016/03/8.ICSevilla2019_
Derrotero-de-Francisco-Albo-f15.pdf, 8; Pigafetta, *First Voyage*, ed. Stanley,
219, translation modified.

CHAPTER 8. THE UNREMITTING WIND

1. P. V. L. Kirch, *On the Road of the Winds: An Archaeological History of
the Pacific Islands before European Contact* (Berkeley: University of California
Press, 2000).

2. J. Needham, *Science and Civilisation in China,* 7 vols. in 15 parts (Cambridge: Cambridge University Press, 1954–98), 4 (pt. 3): 542; Robert Finlay,
"China, the West, and World History in Joseph Needham's Science and Civilisation in China," *Journal of World History* 11 (2000): 265–303.

3. P. Bellwood, *The Polynesians: The History of an Island People* (London,
1978), 39–44; P. Bellwood, *Man's Conquest of the Pacific: The Prehistory of
Southeast Asia and Oceania* (Auckland: Collins, 1979), 296–303; G. Irwin, *The
Prehistoric Exploration and Colonisation of the Pacific* (Cambridge: Cambridge University Press, 1992), 7–9, 43–63.

4. D. L. Oliver, *Oceania: The Native Cultures of Australia and the Pacific
Islands,* 2 vols. (Honolulu: University of Hawaii Press, 1989), 1:361–422;
P. H. Buck (Te Rangi Hiroa), *Vikings of the Sunrise* (New York: Stokes, 1968),
268–69.

5. Maximilianus Transylvanus, *Maximiliani Transyluani Cæsaris a secretis
epistola, de admirabili & nouissima Hispanorum in Orientem nauigatione*
(Rome: Calvi, 1524), 21.

6. A. Pigafetta, *Il primo viaggio intorno al globo di Antonio Pigafetta, e le
sue Regole sull'arte del navigare,* ed. A. Da Mosto, Raccolta Colombiana, pt. 5,
vol. 3 (Rome: Ministero della pubblica istruzione, 1894), 67, and *The First Voyage around the World (1519–22): An Account of Magellan's Expedition by
Antonio Pigafetta,* ed. T. Cachey (New York: Marsiglio, 1995), 27.

7. "Huc perpetuus cursus fuit neque ab eo unquam discessere nisi quantum
tempestatum et ventorum vis aliquando aliorum declinare coegit." Transylvanus, *Maximiliani Transyluani Cæsaris a secretis epistola,* 21; A. Pigafetta, *The
First Voyage round the World, by Magellan,* ed. H. E. J. Stanley, Lord Stanley of
Alderley (London: Hakluyt Society, 1874)197.

8. A. Blázquez y Delgado-Aguilera, ed., *Descripción de los reinos, costas,
puertos e islas* (Madrid: Torrent, 1921), 196.

9. Pigafetta, *Primo viaggio,* ed. A. Da Mosto, 65, and *First Voyage,* ed.
Cachey, 26; G. B. Ramusio, ed., *Navigationi et viaggi,* 3 vols. (Venice: Giunti,
1563–83), 222–23.

10. G. E. Nunn, *The Columbus and Magellan Concepts of South American
Geography* (Glennside, PA:, private printing, 1932).

11. Pigafetta, *Primo viaggio,* ed. A. Da Mosto, 66, and *First Voyage,* ed.
Cachey, 27.

12. G. E. Nunn, "Magellan's Route in the Pacific," *Geographical Journal* 34
(1934): 615–63.

13. "I vero he bene pacifico, perche in questo tempo non havessemo fortuna." Pigafetta, *Primo viaggio*, ed. A. Da Mosto, 65, and *First Voyage*, ed. Cachey, 27.

14. F. Albo, "Derrotero del viaje al Maluco, formado por Francisco Albo, piloto de la nao Trinidad y, posteriormente, de la nao Victoria," 1519, in "Documentos para el quinto centenario de la primera vuelta al mundo: La huella archivada del viaje y sus protagonistas," transcribed by Cristóbal Bernal, 2019–22, http://sevilla.2019–2022.org/wp-content/uploads/2016/03/8.ICSevilla2019_Derrotero-de-Francisco-Albo-f15.pdf, 8–9.

15. "Auto das perguntas que se fizeram a dois espanhóis que chegaram à fortaleza de Malaca vindos de Timor na companhia de Álvaro Juzarte, capitão de um junco," June 1, 1522, Arquivo Nacional Torre do Tombo, https://digitarq.arquivos.pt/details?id=3801974.

16. A. de Herrera, *Historia general de los hechos de los castellanos en las islas y tierra-firme de el mar occeano* [*sic*], ed. J. Natalicio Gómez, 10 vols. (Asunción: Guarania, 1945), 4:290 (Dec. III, bk. 9, chap. 12).

17. J. Denucé, *Magellan, la question des Moluques et la première circumnavigation du monde* (Brussells: Hayez, 1911), 298.

18. M. Fernández de Navarrete, *Colección de los viajes y descubrimientos que hicieron por mar los españoles* (Madrid: Imprenta Nacional, 1837), 4:218; Albo, "Derrotero," 11.

19. Nunn, "Magellan's Route," 224.

20. Pigafetta, *Primo viaggio*, ed. A. Da Mosto, 64, and *First Voyage*, ed. Cachey, 26.

21. Pigafetta, *First Voyage*, ed. Stanley, 9.

22. Pigafetta, *Primo viaggio*, ed. A. Da Mosto, 66–67, and *First Voyage*, ed. Cachey, 28.

23. Navarrete, *Colección de los viajes*, 4:218: Albo, "Derrotero," p. 9.

24. Nunn, "Magellan's Route," 615–63.

25. "Mucho mas cerca que por el cabo de Esperança . . . no muy lexos de Panama." F. López de Gómara, *Historia general de las Indias* (Zaragoza: Miguel de Papila, 1555), fol. 40v, facs. ed. F. Pease (Lima: Comisión Nacional del Quinto Centenario del Descubrimiento de América, 1993).

26. Navarrete, *Colección de los viajes*, 4:219.

27. Pigafetta, *First Voyage*, ed. Stanley, 9.

28. F. Lopes de Castanheda, *História do descobrimento e conquista da India pelos portugueses*, 7 vols. (Lisbon: Rollandiana, 1833), 6:16 (bk. 6, chap. 8).

29. Pigafetta, *Primo viaggio*, ed. A. Da Mosto, 67, and *First Voyage*, ed. Cachey, 28.

30. A. Rodríguez González, "La expedición de Loaysa: una guerra en las Antípodas," in *V Centario del nacimiento de Andrés de Urdaneta: Ciclo de conferencias* (Madrid: Instituo de Historia y Cultura Naval, 2009), 113; Andres de Urdaneta, *El sueño de Cipango*, accessed February 4, 2021, https://artsandculture.google.com/asset/el-sue%C3%B1o-de-cipango-andres-de-urdaneta/2AGNG61SRc3vMQ.

31. Pigafetta, *Primo viaggio*, ed. A. Da Mosto, 67, and *First Voyage*, ed. Cachey, 28, 150–51.

32. Nunn, "Magellan's Route," 630.

33. L.C. Wroth, "The Early Cartography of the Pacific," *Papers of the Bibliographical Society of America* 38 (1944): 207–27.

34. Kozo Yamamura and Tetsuo Kamiki, "Silver Mines and Sung Coins," in *Precious Metals in the Late Medieval and Early Modern Worlds*, ed. J.F. Richards (Durham: University of North Carolina Press, 1983), 329–62.

35. Pigafetta, *Primo viaggio*, ed. A. Da Mosto, 67, and *First Voyage*, ed. Cachey, 28.

36. "Mangiavamo biscoto, non piú biscoto ma polvere di quello con vermi a pugnate, perchè essi havevano mangiato il buono, puzava grandemente de orina de sorzi, et bevevamo hacqua ialla già putrifata per molti giorni, et mangiavamo certe pelle de bove, che erano sopra lántena mangiore, açiò che lántena non rompesse la sarzia durissime per il solle, piogia et vento. le lasciavamo per quatro ho cinque giorni nel mare et poi le meteva uno pocho sopra le braze et così le mangiavamo, et ancora assay volte segature de ase. li sorgi se vendevano mezo ducato lo uno, et se pur ne avessemo potuto havere." Pigafetta, *Primo viaggio*, ed. A. Da Mosto, 65, and *First Voyage*, ed. Cachey, 26.

37. Pigafetta, *First Voyage*, ed. Cachey, 29.

38. P.K. Himmelman, "The Medicinal Body: An Analysis of Medicinal Cannibalism in Europe, 1300–1700," *Dialectical Anthropology* 22 (1997): 183–203.

39. A.W.B. Simpson, *Cannibalism and the Common Law* (Chicago: University of Chicago Press, 1984).

40. "Cressivano le gengive ad alguni sopra li denti cosí de soto como des sovra che per modo alguno non potevano mangiare. . . . per la grazia de Dio, yo non hebi algunna infirmitade." Pigafetta, *Primo viaggio*, ed. A. Da Mosto, 65, and *First Voyage*, ed. Cachey, 26–27.

41. R.F. Rogers and D.A. Ballendorf, "Magellan's Landfall in the Mariana Islands," *Journal of Pacific History* 24 (1989): 193–208.

42. Blázquez y Delgado-Aguilera, *Descripción*, 194.

43. "Y asi vimos muchas velas pequeñas que venian a nos, y andaban tanto que parecia que volasen, y tenían las velas de esteera hechas en triángulo, y andaban por ambas partes que hacían de la popa proa y de la proa popa cuando querían." Navarrete, *Colección de los viajes*, 4:219; Albo, "Derrotero," 13.

44. "Pensavano a li segni che facevano, non fossero altri homini al mondo se non loro." Pigafetta, *Primo viaggio*, ed. A. Da Mosto, 69, and *First Voyage*, ed. Cachey, 31.

45. "Sono como delfini nel saltar a l'acqua de onda in onda." Pigafetta, *Primo viaggio*, ed. A. Da Mosto, 69.

46. Albo, "Derrotero," 13.

47. Pigafetta, *Primo viaggio*, ed. A. Da Mosto, 68, *First Voyage*, ed. Cachey, 29, and *First Voyage*, ed. Stanley, 9.

48. Blázquez y Delgado-Aguilera, *Descripción*, p. 194.

49. Pigafetta, *First Voyage*, ed. Stanley, 31.

50. Pigafetta, *Primo viaggio*, ed. A. Da Mosto, 68, and *First Voyage*, ed. Cachey, 30.

51. Pigafetta, *Primo viaggio*, ed. A. Da Mosto, 68, and *First Voyage*, ed. Cachey, 30.

52. "Quando ferivamo alguni de questi con li veretuni, che li passavano li fianqui da l'una banda al'altra, triravano il veretone mo di qua, mo di là, gardandolo, poi lo tiravano fuora maravigliandose molto, et cossì morivano." Pigafetta, *Primo viaggio*, ed. A. Da Mosto, 68, and *First Voyage*, ed. Cachey, 30.

53. Pigafetta, *Primo viaggio*, ed. A. Da Mosto, 68, and *First Voyage*, ed. Cachey, 29.

54. Pigafetta, *Primo viaggio*, ed. A. Da Mosto, 68, and *First Voyage*, ed. Cachey, 30.

55. Navarrete, *Colección de los viajes*, 4:169.

56. G. V. Scammell, "Indigenous Assistance in the Establishment of Portuguese Power in Asia in the Sixteenth Century," *Modern Asian Studies* 14 (1980): n119; M. Restall, *Seven Myths of the Spanish Conquest* (Oxford: Oxford University Press, 2003), 139–43.

57. Navarrete, *Colección de los viajes*, 4:5, 168.

58. D. Edgerton, *The Shock of the Old: Technology and Global History since 1900* (Oxford: Oxford University Press, 2007).

59. D. Headrick, *Power over Peoples: Technology, Environments, and Western Imperialism, 1400 to the Present* (Princeton, NJ: Princeton University Press, 2010).

60. Navarrete, *Colección de los viajes*, 4:4.

61. Navarrete, *Colección de los viajes*, 167–68.

62. Pigafetta, *Primo viaggio*, ed. A. Da Mosto, 68, and *First Voyage*, ed. Cachey, 29.

63. Pigafetta, *First Voyage*, ed. Stanley, 10.

64. Pigafetta, *Primo viaggio*, ed. A. Da Mosto, 68, and *First Voyage*, ed. Cachey, 29.

65. Albo, "Derrotero," 13.

66. Transylvanus, *Maximiliani Transyluani Cæsaris a secretis epistola*, 22; Pigafetta, *First Voyage*, ed. Stanley, 198.

67. "Et se Ydio, et sa la sua Madre bennedeta non ne dava cosí bon tempo, morivamo tucti de fame in questo mare grandissimo. credo certamente non si farà may piú tal viagio." Pigafetta, *Primo viaggio*, ed. A. Da Mosto, 66, and *First Voyage*, ed. Cachey, 27.

CHAPTER 9. DEATH AS ADVERTISED

1. J. L. Blussé, "Kongkoan and Kongsi: Representations of Chinese Identity and Ethnicity in Early Modern Southeast Asia," in *Shifting Communities and Identity Formation in Early Modern Asia*, ed. J. L Blussé and F. Fernández-Armesto (Leiden: CNWS, 2002), 93–106.

2. "Y fuimos a dar en otra isla pequeña y allí surgimos, . . . y esta isla se llama Suluan." F. Albo, "Derrotero del viaje al Maluco, formado por Francisco Albo, piloto de la nao Trinidad y, posteriormente, de la nao Victoria," 1519, in "Documentos para el quinto centenario de la primera vuelta al mundo: La huella archivada del viaje y sus protagonistas," transcribed by Cristóbal Bernal, 2019–

22, http://sevilla.2019–2022.org/wp-content/uploads/2016/03/8.ICSevilla2019_ Derrotero-de-Francisco-Albo-fi5.pdf, 13.

3. A. Pigafetta, *Il primo viaggio intorno al globo di Antonio Pigafetta, e le sue Regole sull'arte del navigare*, ed. A. Da Mosto, Raccolta Colombiana, pt. 5, vol. 3 (Rome: Ministero della pubblica istruzione, 1894), 70, and *The First Voyage around the World (1519–22): An Account of Magellan's Expedition by Antonio Pigafetta*, ed. T. Cachey (New York: Marsiglio, 1995), 33.

4. Pigafetta, *Primo viaggio*, ed. Da Mosto, 69, and *First Voyage*, ed. Cachey, 152.

5. Albo, "Derrotero," 13.

6. The mistaken but widespread belief that Albo located the Philippines in the Portuguese zone derives from a rare misreading by R. A. Laguarda Triás, "Las longitudes geográficas de la membranza de Magallanes y del primer viaje de circunnavegación," in *A viagem de Fernão de Magalhães e a ques.tão das Molucas: Actas do II Congreso luso-espanhol de História Ultramarina*, ed. A. Teixeira da Mota (Lisbon: Junta de Investigacões Científicas, 1975), 169, who took Albo as referring to the Tordesillas line rather than the Seville meridian, and who therefore assumed that Albo's reading of 106 degrees from Cabo Deseado was a transcription error for 160 degrees.

7. Pigafetta, *Primo viaggio*, ed. Da Mosto, 70, and *First Voyage*, ed. Cachey, 33.

8. "Che niuno si movesse nè dicesse parolla alguna senza sua licentia." Pigafetta, *Primo viaggio*, ed. Da Mosto, 69, and *First Voyage*, ed. Cachey, 31.

9. Pigafetta, *First Voyage*, ed. Cachey, 31–33, and *Primo viaggio*, ed. Da Mosto, 69.

10. A. Pigafetta, *The First Voyage round the World, by Magellan*, ed. H. E. J. Stanley, Lord Stanley of Alderley (London: Hakluyt Society, 1874), 10, *Primo viaggio*, ed. Da Mosto, 70, and *First Voyage*, ed. Cachey, 33–34.

11. Pigafetta, *First Voyage*, ed. Stanley, 11.

12. Pigafetta, *First Voyage*, ed. Cachey, 31, and *Primo viaggio*, ed. Da Mosto, 69.

13. Pigafetta, *Primo viaggio*, ed. Da Mosto, 70, *First Voyage*, ed. Cachey, 33, and *First Voyage*, ed. Stanley, 11.

14. D. V. and H. C. Hart, "'Maka-andog': A Reconstructed Myth from Eastern Samar, Philippines," *Journal of American Folklore* 79 (1966): 84–108.

15. "Nudi con tella de scorsa dárbore intorno le sue vergonie, se non alguni principali, con telle de banbazo lavorate ne i capi, con seda a guchia." Pigafetta, *Primo viaggio*, ed. Da Mosto, 70, and *First Voyage*, ed. Cachey, 34.

16. Pigafetta, *Primo viaggio*, ed. Da Mosto, 70, and *First Voyage*, ed. Cachey, 34.

17. L. Tormo Sanz, "El mundo indígena conocido por Magallanes en las Islas de San Lázaro," in *A viagem de Fernão de Magalhães e a questão das Molucas: Actas do II Congreso luso-espanhol de História Ultramarina*, ed. A. Teixeira da Mota (Lisbon: Junta de Investigacões Científicas, 1975), 379–409; Pigafetta, *First Voyage*, ed. Cachey, 67.

18. J. T. Medina, *Colección de documentos inéditos para la historia de Chile*, 14 vols. (Santiago: Ercilla, 1888–1902), 2:295.

19. Medina, *Colección*, 1:211.

20. F. López de Gómara, *Historia general de las Indias* (Zaragoza: Miguel de Papila, 1555), fol. 43r, facs. ed. F. Pease (Lima: Comisión Nacional del Quinto Centenario del Descubrimiento de América, 1993).

21. "Perchè in questa parte li re sanno piú linguaggii che li altri." Pigafetta, *Primo viaggio*, ed. Da Mosto, 71, and *First Voyage*, ed. Cachey, 35.

22. "Una bara de oro grande et una sporta piena de gengero." Pigafetta, *Primo viaggio*, ed. Da Mosto, 71, and *First Voyage*, ed. Cachey, 35.

23. Pigafetta, *First Voyage*, ed. Cachey, 42, 50. "Lo capitano generale non volse se pigliasse troppo oro, perquè sarebe stato alguno marinaro che haverebe dato tuto lo suo per uno poco de oro, et haverla disconciato lo trafigo per sempre." Pigafetta, *Primo viaggio*, ed. Da Mosto, 77.

24. Gómara, *Historia general*, fol. 43v.

25. "Iubet eum bono animo esse, si modo Christo devoveret protinus salutem et pristinum valetudinem recuperaturum." Maximilianus Transylvanus, *Maximiliani Transyluani Cæsaris a secretis epistola, de admirabili & nouissima Hispanorum in Orientem nauigatione* (Rome: Calvi, 1524), 23; Oviedo's version is virtually identical: G. Fernández de Oviedo, *Historia general y natural de las Indias*, ed. J. Amador de los Ríos, 14 vols. (1851–55; repr., Asunción: Guarania, 1944–45), 2:14.

26. Fernández de Oviedo, *Historia general*; Pigafetta, *First Voyage*, ed. Stanley, 199.

27. Pigafetta, *Primo viaggio*, ed. Da Mosto, 78, and *First Voyage*, ed. Cachey, 56.

28. Pigafetta, *Primo viaggio*, ed. Da Mosto, 78, and *First Voyage*, ed. Cachey, 55.

29. Pigafetta, *Primo viaggio*, ed. Da Mosto, 72, and *First Voyage*, ed. Cachey, 38.

30. Pigafetta, *First Voyage*, ed. Cachey, 44, and *Primo viaggio*, ed. Da Mosto, 73.

31. Peter Martyr, *Cartas sobre el Nuevo Mundo*, fol. 78v, trans. J. Bauzano (Madrid: Polifemo, 1990).

32. Pigafetta, *Primo viaggio*, ed. Da Mosto, 76, and *First Voyage*, ed. Cachey, 49.

33. Pigafetta, *Primo viaggio*, ed. Da Mosto, 72, and *First Voyage*, ed. Cachey, 36.

34. Pigafetta, *First Voyage*, ed. Cachey, 38, 41, and *Primo viaggio*, ed. Da Mosto, 73, 75.

35. "Fece portare uno plato de carne de porco con uno vazo grande pienno de vino. bevevamo ad ogni boconne una tassa de vino; lo vino che li avansava qualche volta, ben che foscono poche, se meteva en uno vazo da per sì ... inanzi che lo re pigliasse la tassa per bere, alzava li mani giunte al çielo e verso de nui, et quando voleva bere, extendeva lo pugnio de la mano sinistra verso di me (prima pensava me volesse dare un pognio) et poi beveva; faceva così yo

verso il re. questi segni fanno tuti l'uno verso de l'altro, quando beveno. con quest cerimonie et altri segni de amistia merendasemo. mangiay nel vennere sancto carne per non potere fare altro." Pigafetta, *Primo viaggio*, ed. Da Mosto, 72, and *First Voyage*, ed. Cachey, 37.

36. Pigafetta, *First Voyage*, ed. Cachey, 37, and *Primo viaggio*, ed. Da Mosto, 72.

37. "Otro día que era viernes de la Cruz, el señor de aquella isla vino ala nao y habló muy bien a Magallanes y a todos e hizo paces con ellos a la costumbre de la tierra, ques es sangrarse 408: del pecho ambos, echada en un vaso la sangre junta, revuelta co vino, bebe cada uno la mitad. Esto paresce que es ceremonia de buena amistad." Tormo Sanz, "Mundo indígena," 407–8; A. Blázquez y Delgado-Aguilera, ed., *Descripción de los reinos, costas, puertos e islas* (Madrid: Torrent, 1921), 198.

38. Blázquez y Delgado-Aguilera, *Descripción*, 36, 45, 67, 70; Pigafetta, *Primo viaggio*, ed. Da Mosto, 71, 76, 83, 85.

39. "Andassemo in terra forse cinquanta huomini, non armati la persona, ma con le altre nostre arme, et meglio vestiti che potessemo. inanzi que arivassemo a la riva con li bateli forenno scaricati sei pezi de bombarde in segnio de pace. saltassmo in terra: le dui re abbrassarono lo capitano genrale et lo messeno in mezo di loro: andassemo in ordinanza fino al locho consacrato, non molto longi de la riva. Inanzi se comensasse la messa il capitano bagnò tuto el corpo de li dui re con hacqua moscada." Pigafetta, *Primo viaggio*, ed. Da Mosto, 72, and *First Voyage*, ed. Cachey, 39.

40. "inanzi se comensasse la messa il capitano la bagniò con aliquante sue dame de hacqua rosa muschiata: molto se delectavano de talle odore." Pigafetta, *Primo viaggio*, ed. Da Mosto, 78, and *First Voyage*, ed. Cachey, 54.

41. "Se oferse a la messa: li re andoronoa bassiare la croce como nuy, ma non oferseno. quando selevava lo corpo de Nostro Signior stavano in genoquioni et adoravanlo con le mae giunte. le nave tirarono tuta la artigliaria in uno quando se levò lo corpo de Christo, dandoge lo segnio de la tera con li schiopeti." Pigafetta, *Primo viaggio*, ed. Da Mosto, 72–73, and *First Voyage*, ed. Cachey, 39.

42. "Che non ce fàcero Christiani per paura, nè per compiacerne, ma volontariamente." Pigafetta, *Primo viaggio*, ed. Da Mosto, 75, and *First Voyage*, ed. Cachey, 46–47.

43. Pigafetta, *First Voyage*, ed. Cachey, 55, and *Primo viaggio*, ed. Da Mosto, 78.

44. "Per sua utilità, perchè, se venissero algune nave de le nostre, saperianno, con questa croce, noi essere stati in questo locho, et non farebenno despiacere a loro nè a le cose et, se pigliasseno alguno de li soi, subito, mostrandoli questo segnialle, le lasserianno andare." Pigafetta, *Primo viaggio*, ed. Da Mosto, 73, and *First Voyage*, ed. Cachey, 40.

45. Pigafetta, *Primo viaggio*, ed. Da Mosto, 73, and *First Voyage*, ed. Cachey, 40–41.

46. Pigafetta, *Primo viaggio*, ed. Da Mosto, 82–83, and *First Voyage*, ed. Cachey, 65–66; Tormo Sanz, "Mundo indígena," 405.

47. Pigafetta, *Primo viaggio*, ed. Da Mosto, 74, and *First Voyage*, ed. Cachey, 42.

48. "Li re tanto mangiorono et beveteno [sic] che dormitero tuto il giorno. alguni, per escusarli, dicero che havevano un pocho de malle." Pigafetta, *Primo viaggio*, ed. Da Mosto, 74, and *First Voyage*, ed. Cachey, 37–38, 41.

49. Pigafetta, *First Voyage*, ed. Cachey, 47, 49; "assay belle et bianque, casi como le nostre, et così grande." Pigafetta, *Primo viaggio*, ed. Da Mosto, 76.

50. Pigafetta, *Primo viaggio*, ed. Da Mosto, 79, and *First Voyage*, ed. Cachey, 58.

51. Peter Martyr, *Cartas*, fol. 78v.

52. Tormo Sanz, "Mundo indígena," 391; Pigafetta, *First Voyage*, ed. Cachey, 104.

53. "Nel mezo dil fero è un buso per il quale urinano . . . Loro dicono che le sue moglie voleno cussì et, se fossero de altra sorte, non uzariano con elle." Pigafetta, *Primo viaggio*, ed. Da Mosto, 79, and *First Voyage*, ed. Cachey, 58.

54. "Sempre sta dentro fin que diventa molle, perchè altramenti non lo pori-ann cavare fuora." Pigafetta, *Primo viaggio*, ed. Da Mosto, 79, and *First Voyage*, ed. Cachey, 58.

55. Pigafetta, *First Voyage*, ed. Cachey, 59. 57, and *Primo viaggio*, ed. Da Mosto, 79.

56. "Questi populi viveno con iustitia, peso et mesura; amano la pace, l'otio etla quiete." Pigafetta, *Primo viaggio*, ed. Da Mosto, 76, and *First Voyage*, ed. Cachey, 50.

57. Pigafetta, *First Voyage*, ed. Cachey, 41, and *Primo viaggio*, ed. Da Mosto, 74.

58. Pigafetta, *Primo viaggio*, ed. Da Mosto, 73–74, and *First Voyage*, ed. Cachey, 41.

59. Pigafetta, *First Voyage*, ed. Cachey, 43, and *Primo viaggio*, ed. Da Mosto, 74.

60. Pigafetta, *First Voyage*, ed. Cachey, 44, and *Primo viaggio*, ed. Da Mosto, 74.

61. Tormo Sanz, "Mundo indígena," 409; "Cabezudo capitán." Blázquez y Delgado-Aguilera, *Descripción*, 198.

62. "Si bene si li fa, ben se à [sic], se male, male, et pegio como ànno a Cali-cut et a Malaca." Pigafetta, *Primo viaggio*, ed. Da Mosto, 75, and *First Voyage*, ed. Cachey, 44–46.

63. "Per lo habito che haveva, li prometeva che li dava la pace perpetua col re de Spagnia: resposero que lo simille prometevano." Pigafetta, *Primo viaggio*, ed. Da Mosto, 75–76, and *First Voyage*, ed. Cachey, 47.

64. Pigafetta, *Primo viaggio*, ed. Da Mosto, 76, and *First Voyage*, ed. Cachey, 49.

65. "Se non hobedivano al re como suo re, li farebe amazare et daia la sua roba al re." Pigafetta, *Primo viaggio*, ed. Da Mosto, 77, and *First Voyage*, ed. Cachey, 51.

66. Pigafetta, *Primo viaggio*, ed. Da Mosto, 78, and *First Voyage*, ed. Cachey, 53.

67. "Cavòla sua spade, inanzi la ymagine de Nostra Donna, et disse al re, quando cossì se iurava, più presto doveriasi morire, que a romper uno simil iuramento . . . per questa ymagine, per la vita de l'imperatore suo signore et per il suo habito." Pigafetta, *Primo viaggio*, ed. Da Mosto, 78, and *First Voyage*, ed. Cachey, 54.

68. Pigafetta, *Primo viaggio*, ed. Da Mosto, 78, and *First Voyage*, ed. Cachey, 54.

69. "Con tanto potere che lo faria lo magior re di quella parte." Pigafetta, *Primo viaggio*, ed. Da Mosto, 77, and *First Voyage*, ed. Cachey, 51.

70. "Al capitán Magallanes cómo le mataron los indios; porque algunos de los que allá quedan, y en esta nao vienen, dicen que fue muerto de otra manera." M. Fernández de Navarrete, *Colección de los viajes y descubrimientos que hicieron por mar los españoles* (Madrid: Imprenta Nacional, 1837),4:286.

71. J. Ibañez Cerdá, "La muerte de Magallanes," in Teixeira, *Viagem*, 411–33.

72. Pigafetta, *First Voyage*, ed. Cachey, 62, and *Primo viaggio*, ed. Da Mosto, 80–81.

73. Navarrete, *Colección de los viajes*, 4:290.

74. Medina, *Colección*, 2:450.

75. Pigafetta, *First Voyage*, ed. Cachey, 52, and *Primo viaggio*, ed. Da Mosto, 77.

76. Cf C. Jostmann, *Magellan oder Die erste Umsegelung der Erde* (Munich: Beck, 2019), 230.

77. Blázquez y Delgado-Aguilera, *Descripción*, 200.

78. Peter Martyr, fol. 78v.

79. Pigafetta, *Primo viaggio*, ed. Da Mosto, 78, and *First Voyage*, ed. Cachey 53.

80. J. L. Phelan, *The People and the King: The Comunero Revolution in Colombia* (Madison: University of Wisconsin Press, 1978), 83.

81. Navarrete, *Colección de los viajes*, 4:294.

82. "Que no rescibiese pena dello, porque con el tiempo aquel rebelde se amansaría y que éllo procuraría porque era casado con su hermana." Blázquez y Delgado-Aguilera, *Descripción*, 193.

83. "Que le parecia que non tratase de aquella jornada, porque demás de que de ella no se seguia provecho, las naves quedaban con tan mal recado, que poca gente las tomaría; y que si todavía quería que se hiciese que no fuese, sino que envias a otro en su lugar." A. de Herrera, *Historia general de los hechos de los castellanos en las islas y tierra-firme de el mar occeano [sic]*, ed. J. Natalicio Gómez, 10 vols. (Asunción: Guarania, 1945)5:19 (Dec. I, bk. 1, chap. 4).

84. Medina, *Colección*, 2:450.

85. Peter Martyr, *Cartas*, fol. 78v.

86. Blázquez y Delgado-Aguilera, *Descripción*, 202.

87. "Emquanto foè vivo não quis q o rey seu amigo lhe acudisse com a sua gente que para isso ahy tinh, dizendo que abastanão os christãos com o favor divino pa vencer toda aquella canalha." Ibañez Cerdá, "Muerte de Magallanes," 426.

88. Fernández de Oviedo, *Historia general*, 2:13; he makes Magellan refer to a battle "pocos días antes" in Juvagana, where dosçientos españoles avían

puesto en fuga dosçientos y tres cientos mil indios" (14) but he also says (13) that Juvagana was uninhabited.

89. T. Mazón, "¿Llegó Magallanes a saber de la conquista de México por Hernán Cortés?," *Historia y Mapas* (blog), April 21, 2019, https://historiaymapas.wordpress.com/2019/04/21/llego-magallanes-a-saber-de-la-conquista-de-mexico-por-hernan-cortes/.

90. Peter Martyr, *Cartas*, fol. 78.

91. "Por su demaiada valentia, i haver querido, sin causa, tentar la Fortuna." Herrera, *Historia general*, 4:20 (Dec. III, bk. 1, chap. 4).

92. Peter Martyr, *Cartas*, fol. 78v.

93. Medina, *Colección*, 2:450; Ibañez Cerdá, "Muerte de Magallanes," 429; Herrera, *Historia general*, 4:20 (Dec. III, bk. 1, chap. 4); Pigafetta, *First Voyage*, ed. Cachey, 62, and *Primo viaggio*, ed. Da Mosto, 81.

94. "Declaración de las personas fallecidas en el viaje al Maluco," from December 20, 1519, to July 29, 1522, in "Documentos para el quinto centenario de la primera vuelta al mundo: La huella archivada del viaje y sus protagonistas," transcribed by Cristóbal Bernal, 2019–22, http://sevilla.2019–2022.org/wp-content/uploads/2016/03/4.ICSevilla2019_Declaracion-de-fallecidos-en-el-viaje-s14.pdf, 5.

95. J. de Barros, *Decadas da Asia,* 24 vols. (Lisbon: Regia Officina, 1777–88),5:649 (Dec. III, bk. 5, chap. 10).

96. Pigafetta, *First Voyage*, ed. Cachey, 63, and *Primo viaggio*, ed. Da Mosto, 81.

97. "Pur piagendo, ne disse che non haveressemo così presto facto vella, che láverianno amazat, et disse che pregava Idio, nei iorni del iuditio, dimandasse l'anima sua a Iohan Carvalo, suo compadre. Subito se partissemo, no so se morto o vivo lui restasse." Pigafetta, *Primo viaggio*, ed. Da Mosto, 81, and *First Voyage*, ed. Cachey, 64.

98. Pigafetta, *First Voyage*, ed., 51, and *Primo viaggio*, ed. Da Mosto, 77.

CHAPTER 10. AFTERMATH AND APOTHEOSIS

1. J. M. Logsdon, *John F. Kennedy and the Race to the Moon* (Basingstoke: Palgrave, 2011).

2. Maximilianus Transylvanus, *Maximiliani Transyluani Cæsaris a secretis epistola, de admirabili & nouissima Hispanorum in Orientem nauigatione* (Rome: Calvi, 1524), 35.

3. G. Fernández de Oviedo, *Historia general y natural de las Indias*, ed. J. Amador de los Ríos, 14 vols. (1851–55; repr., Asunción: Guarania, 1944–45), 2:21.

4. "Hoc inauditum hactenus tentatum nunquam ab initio mundi . . . terram unversam circuivit." Peter Martyr, *Cartas sobre el Nuevo Mundo,* fol. 78v, trans. J. Bauzano (Madrid: Polifemo, 1990), fol. 79v (Dec. V, chap. 7).

5. "No haviendo hasta entonces, entre los Famosos Antiguos, ni en los Modernos, ninguno que se le puede comparar." A. de Herrera, *Historia general de los hechos de los castellanos en las islas i tierra frma del mar occeano* [*sic*], ed. J. Natalicio Gómez, 10 vols. (Asunción: Guarania, 1945), 3:189 (Dec III, bk. 4, chap. 1).

6. "Ninguno altro havere avuto tanto ingenio, ni ardire de saper dare una volta al mondo como ià cazi lui haveva dato." A. Pigafetta, *Il primo viaggio intorno al globo di Antonio Pigafetta, e le sue Regole sull'arte del navigare*, ed. A. Da Mosto, Raccolta Colombiana, pt. 5, vol. 3 (Rome: Ministero della pubblica istruzione, 1894), 80, *The First Voyage around the World (1519–22): An Account of Magellan's Expedition by Antonio Pigafetta*, ed. T. Cachey (New York: Marsiglio, 1995), 62.

7. C. Jostmann, *Magellan oder Die erste Umsegelung der Erde* (Munich: Beck, 2019), 319.

8. "No llegó a las islas de Maluco y Especería, este loor a sólo Magallanes se le debe, y a él se le atribuye este grand viaje y descubrimiento." Fernández de Oviedo, *Historia general*, 2:34.

9. "Prima ego velivolis ambivi cursibus orbem Magellane novo te duce ducta freto." Image of *Victoria* from Abraham Ortelius's map "Maris Pacifici . . ." (Antwerp, 1595). Historic Maps Collection, Princeton Library. https://library. princeton.edu/special-collections/file/map-pacific-ortelius-1589-victoriajpg-0.

10. H. Keazor, "Theodore De Bry's Images for *America*," *Print Quarterly* 15 (1998): 131–49; S. Bernabeu, "Magallanes: Retrato de un hombre," in *A 500 años del hallazgo del Pacífico la presencia novohispana en el Mar del Sur*, ed. C. Yuste López and G. Pinzón Ríos (Mexico City: UNAM, 2016), 22–23, www .historicas.unam.mx/publicaciones/publicadigital/libros.

11. *Circumnavigators; Sir Francis Drake; Oliver van Noort; Ferdinand Magellan; Wilhelm Schouten; Thomas Cavendish; George Spilman*, etching by unknown artist, early seventeenth century, National Portrait Gallery, London, www.npg.org.uk/collections/search/portrait/mw127354/Cicumnavigators- Sir-Francis-Drake-Oliver-van-Noort-Ferdinand-Magellan-Wilhelm-Schouten- Thomas-Cavendish-George-Spilman?LinkID=mp72990&role=sit&rNo=0, also shown on Middle Temple Library site as *Circumnavigators, 16th to 17th Century*, https://fineartamerica.com/featured/circumnavigators-16th-to-17th- century-middle-temple-library.html.

12. "The Strait of Magellan: 250 Years of Maps (1520–1787)," Princeton Library, accessed December 17, 2021, https://library.princeton.edu/visual_ materials/maps/websites/pacific/magellan-strait/magellan-strait-maps.html; Nicolas de Larmessin, "Ferdinand Magellan (1480–1521)," engraving after an original painting at Toledo, Spain, 1695, https://prints.rmg.co.uk/products/fer- dinand-magellan-1480–1521-pu2328.

13. Fernández de Oviedo, *Historia general*, 2:8.

14. S. E. Morison, *The Southern Voyages*, vol. 2 of *The European Discovery of America* (Oxford: Oxford University Press, 1974),381–82, 398.

15. H. Kelsey, *The First Circumnavigators: Unsung Heroes of the Age of Discovery* (New Haven, CT: Yale University Press, 2016), 141.

16. J. E. Chaplin, *Round about the Earth: Circumnavigation from Magellan to Orbit* (New York: Simon and Schuster, 2012), 1–300.

17. Chaplin, *Round about the Earth*, xix.

18. "Qualquiera hombre de razon, aunque no tenga letras, cayia luego en quanto los tales estropeçavan en la llanura del mundo: y alli no es menester mas

declaracion." F. López de Gómara, *Historia general de las Indias* (Zaragoza: Miguel de Papila, 1555), fol. 1v, facs. ed. F. Pease (Lima: Comisión Nacional del Quinto Centenario del Descubrimiento de América, 1993).

19. F. Fernández-Armesto, *Pathfinders: A Global History of Exploration* (New York: Norton, 2006), 283–86.

20. "Denn mit dem seit einem Jahrtausend vergeblich gesuchten Mass des Umfangs unsere Erde gewinnt die ganze Menschheit zum erstenmal ein neues Mass ihrer Kraft, an der Grösse des überwundenen Weltraums wurde ihr erst mit neuer Lust und neuem Mut ihre eigene Grösse bewusst." S. Zweig, *Magellan: Der Mann und seine Tat* (Frankfurt: Fischer, 1938), 280.

21. "Quod summa replebit admiratione legentes, eos quae praecipue qui se putant vagos coelorum cursus habere prae manibus suis." Peter Martyr, *Cartas*, fol. 80r., Dec. V, chap. 7; *De Orbe Novo* (London: Dawson, 1612), 218.

22. F. de Danville, *La géographie des humanistes* (Paris: Paris : Beauchesne, 1940), 92n3, quoted in J. H. Elliott, *Illusión and Disillusionment: Spain and the Indies* (London: University of London, 1992), 7.

23. R. A. Laguarda Triás, "Las longitudes geográficas de la membranza de Magallanes y del primer viaje de circunnavegación," in *A viagem de Fernão de Magalhães e a ques.tão das Molucas: Actas do II Congreso luso-espanhol de História Ultramarina*, ed. A. Teixeira da Mota (Lisbon: Junta de Investigacões Científicas, 1975)" 151–73.

24. R. Padrón, *The Indies of the Setting Sun: How Early Modern Spain Mapped the Far East as the Transpacific West* (Chicago: University of Chicago Press, 2020).

25. "Mostró que toda la grandeza de la tierra—por mayor que se pinte—está sujeta a los pies de un hombre, pues la pudo medir." J. de Acosta, *Historia natural y moral de las Indias*, ed. E. O'Gorman (Mexico City: Fondo de Cultura Económica, 2006), 20.

26. F. Albo, "Derrotero del viaje al Maluco, formado por Francisco Albo, piloto de la nao Trinidad y, posteriormente, de la nao Victoria," 1519, in "Documentos para el quinto centenario de la primera vuelta al mundo: La huella archivada del viaje y sus protagonistas," transcribed by Cristóbal Bernal, 2019–22, http://sevilla.2019-2022.org/wp-content/uploads/2016/03/8.ICSevilla2019_Derrotero-de-Francisco-Albo-f15.pdf, 14.

27. J. T. Medina, *Colección de documentos inéditos para la historia de Chile*, 14 vols. (Santiago: Ercilla, 1888–1902), 2:155.

28. Pigafetta, *First Voyage*, ed. Cachey, 70, and *Primo viaggio*, ed. Da Mosto, 85.

29. Pigafetta, *First Voyage*, ed. Cachey, 78, and *Primo viaggio*, ed. Da Mosto, 89.

30. Pigafetta, *First Voyage*, ed. Cachey, 80, and *Primo viaggio*, ed. Da Mosto, 89.

31. "Per il que rengratiassemo Ydio et per allegrezza descaricassemo tuta la artigliari." Pigafetta, *Primo viaggio*, ed. Da Mosto, 92, and *First Voyage*, ed. Cachey, 83.

32. Albo, "Derrotero," 16–19. The glosses provided by the excellent transcriber, C. Bernal, are misconceived, on the assumption that by "meridian" the pilot meant the Tordesillas line: if that were so, it would be contrary to the usage adopted throughout the rest of the log.

33. Pigafetta, *First Voyage*, ed. Cachey, 103, and *Primo viaggio*, ed. Da Mosto, 100.

34. Medina, *Colección*, 2:287–88.

35. "Esta es la nao Trinidad que queriendo venir a la mar del sur subió hasta 42 grados por allar vientos contrarios, e de alli se volvió a Maluco otra vez por que avía ya 6 meses que andava en la mar, i hazia agua, i le faltava mantenimientos."

36. "Carta de Gonzalo Gómez de Espinosa a Carlos I, narrando las vicisitudes del periplo en solitario de la nao Trinidad por el Pacífico Norte, y su prisión por los portugueses," 1528, Archivo General de Indias, Portal de Archivos Españoles (PARES), http://pares.mcu.es/ParesBusquedas20/catalogo /show/304066?nm; Medina, *Colección*, 2:153–62, and M. Fernández de Navarrete, *Colección de los viajes y descubrimientos que hicieron por mar los españoles* (Madrid: Imprenta Nacional, 1837), 4:378–88.

37. "Porque la gente no tenía pan que comer enflaquesió la más parte della." "Carta de Gonzalo Gómez de Espinosa a Carlos I."

38. Medina, *Colección*, 2:144.

39. Medina, *Colección*, 2:159–60.

40. Medina, *Colección*, 2:180.

41. Medina, *Colección*, 2:167.

42. J. Denucé, *Magellan, la question des Moluques et la première circumnavigation du monde* (Brussells: Hayez, 1911), 349.

43. Pigafetta, *First Voyage*, ed. Cachey, 113–21, and *Primo viaggio*, ed. Da Mosto, 106–11.

44. Albo, "Derrotero," 24–25.

45. "Ma alguni de li altri, più desiderosi del suo honore, che de la propria vita, deliberono, vivi or morti, volere andare in Spagna." Pigafetta, *Primo viaggio*, ed. Da Mosto, 111, and *First Voyage*, ed. Cachey, 122.

46. Albo, "Derrotero," 32.

47. Albo, "Derrotero," 36.

48. Medina, *Colección*, 2:235.

49. L. Díaz-Trechuelo, "La organización del viaje magallánico: Financiación, engaches, acopios y preparativos," in Teixeira da Mota, *Viagem*, 267–314.

50. I am unable to identify the sources of the figures in Jostmann, *Magellan*, 307–9.

51. G. F. Dille, ed., *Spanish and Portuguese Conflict in the Spice Islands: the Loaisa Expedition to the Moluccas* (London: Hakluyt Society, 2021), 19–22.

52. L. Díaz-Trechuelo, "El tratado de Tordesillas y so proyección en el Pacífico," *Revista española del Pacífico* 4 (January-December 1994), www .cervantesvirtual.com/obra-visor/revista-espanola-del-pacifico—14/html /02546916-82b2-11df-acc7-00218 5ce6064_23.htm. For the treaty reproduced in facsimile, see "Tratado de Zaragoza," April 17, 1529, PARES, http:// pares.mcu.es/ParesBusquedas20/catalogo/show/122513.g.

53. Morison, *Southern Voyages,* 320.

54. "Regressaque iam diu est blasphemiis infectans Magaglianum. Non impune talem inobservantiam admisisse navis duces arbitramur." Peter Martyr, *Cartas,* fol. 78r (Dec. V, chap. 7).

55. M. Dennefield, "A History of the Magellanic Clouds and the European Exploration of the Southern Hemisphere," *The Messenger* 181 (2020): 41.

56. "St. Catherine's College: Magellan Prize," Postgraduate Funding website, accessed July 20, 2021, www.postgraduatefunding.com/award-2894.

57. Washington and Jefferson College, "The Magellan Project," accessed December 2020, www.washjeff.edu/student-life/the-magellan-project/.

58. University of South Carolina, Office of the Vice President for Research, "Magellan Ten Scavanger Hunt," accessed December 17, 2020, https://sc.edu /about/offices_and_divisions/research/internal_funding_awards/students /undergraduate/magellanten/scavenger_hunt.php.

59. Marca Chile, "Strait of Magellan Award for Innovation and Exploration with Global Impact," accessed December 17, 2020, https://marcachile.cl /magallanes-500-a%C3%B1os/index_en.php.

60. Circumnavigators Club, "Magellan Award," accessed December 16, 2020, https://circumnavigators.org/magellan-award/.

61. Quoted in Chaplin, *Round about the Earth,* 263.

62. Northrop & Johnson, "Yacht for Sale: Project Magellan," accessed December 16, 2020, www.northropandjohnson.com/yachts-for-sale/projectmagellan-135-jfa.

63. Global Insights, "Startup Success for Magellan Life Science," interview of Dr. Abhiram Dukkipati, September 2, 2019, https://insights.figlobal.com /startups/startup-success-magellan-life-science-interview.

64. See, e.g., "Magellan: Race Record and Form," Racing Post, accessed December 16, 2020, racehorse www.racingpost.com/profile/horse/1433453 /magellan/form; MAGELLAN Consortium, "Return from the Game Jam," accessed December 16, 2020, www.magellanproject.eu/news2/173-gamejam-return.html; Adam Schoon, "Studies on Europe and Asia, 1400–2019," Magellan Centre, accessed December 17, 2020, www.themagellancentre.co.uk/studies /east-india-company#; "Most Innovative Product of the Year: Magellan, Kuzco Lighting," in Madeleine D'Angelo, "LightFair Announces 2020 Innovation Award Winners," *Architect Magazine,* June 3, 2020, www.architect magazine.com/technology/lighting/lightfair-announces-2020-innovation-award-winners_o.

65. Pigafetta, *First Voyage,* ed. Cachey, 62, and *Primo viaggio,* ed. Da Mosto, 80.

66. T. Carlyle, *On Heroes and Hero-Worship,* ed. David R. Sorensen and Brent E. Kinser (New Haven, CT: Yale University Press, 2013), 22.

67. Jostmann, *Magellan.*

68. M. Cattan, "Fernando de Magallanes: La creación del mito del héroe," *Hipogrifo* 6 (2018): 535–53.

69. J.M. Latino Coelho, *Fernão de Magalhães* (Lisbon: Santos and Vieira, 1917), 138, quoted in Cattan, "Fernando de Magallanes"; D. Barros Arana, *Vida i viajes de Fernando de Magallanes* (Santiago: Nacional, 1864), 2.

70. Vizconde de Lagôa, *Fernao de Magalhaes: A sua vida e a sua viagem*, 2 vols. (Lisbon: Serra Nova, 1938); J. M. Queiroz Velloso, *Fernão de Magalhães: A sua vida e a viagem* (Lisbon: Imperio, 1941).

71. F. H. H. Guillemard, *The Life of Magellan and the First Circumnavigation of the Globe* (Londo: Philip, 1891), 19, 257.

72. Morison, *Southern Voyages*, 372; Chaplin, *Round about the Earth*, 13.

73. E. F. Benson, *Ferdinand Magellan* (London: Bodley Head, 1929), vii, 241.

74. "Ein geschworener Feind alles unnöttigen Blutvergiessens, der wahre Antipode all der andern schlächterischen Konquistadoren." Zweig, *Magellan*, 232.

75. "Zum heiligen Menschheitskrieg wider das Unbekannte . . . vom Genius beschwingt sich stärker erweist als alle Elemente der Natur." Zweig, *Magellan*, 280. I am grateful to Professor Manuel Lucena Giraldo for his just observations of Zweig.

76. C. M. Parr, *So Noble a Captain: The Life and Times of Ferdinand Magellan* (New York: Crowe, 1953).

77. Morison, *Southern Voyages*, 313–437. See above, 000, 000.

78. Denucé, *Magellan*, 320.

79. Medina, *Colección*, 1:178–213.

80. T. Joyner, *Magellan* (Camden, ME: International Marine, 1992).

81. Jostmann, *Magellan*, 8.

82. Jostmann, *Magellan*, 251–55. Dr Jostmann provides the most comprehensive available bibliography at "Magellan: A Bibliography," January 2019, https://jostmann.at/wp-content/uploads/2019/01/Magellan_Bibliography.pdf.

83. A. Blázquez y Delgado-Aguilera, ed., *Descripción de los reinos, costas, puertos e islas* (Madrid: Torrent, 1921), 190.

84. C. Gamble, *Timewalkers: The Prehistory of Global Colonization* (Cambridge, MA: Harvard University Press, 1994).

85. F. Fernández-Armesto, "Before the Farmers: Culture and Climate from te Emergence of Homo Sapiens to about Ten Thousand Years Ago," in *The Cambridge World History*, vol. 1, ed. D. Christian (Cambridge: Cambridge University Press, 2015), 313–38.

86. Herodotus, *Histories* 4.11–12.

87. A. N. Oikonomidès et al., eds., *Periplus: or, Circumnavigation of Africa: Greek Text with Facing English Translation, Commentary, Notes and Facsimile of Codex Palatinus Gr. 398* (Chicago: Ares, 1977).

88. F. Fernández-Armesto, *Civilizations: Culture, Ambition, and the Transformation of Nature* (New York: Free Press, 2001), 339–40.

89. G. Orlandi and R. Guglielmetti, eds., *Navigatio Sancti Brendani: Alla scoperta dei segreti meravigliosi del mondo* (Florence: Fondazione Ezio Franceschini, 2014).

90. F. Fernández-Armesto, *Before Columbus* (Philadelphia: University of Pennsylvania Press, 1987), 172–73, 196–97.

91. F. Fernández-Armesto, "Inglaterra y el Atlántico en la baja edad media," in *Canarias e Inglaterra a través de l historia*, ed. A. Bethencourt Massieu et al. (Las Palmas: Cabildo Insular, 1995), 11–28.

92. Fernández-Armesto, *Before Columbus*, 185–92.

93. I. A. Leonard, *Books of the Brave: Being an Account of Books and of Men in the Spanish Conquest and Settlement of the Sixteenth-Century New World* (Berkeley: University of California Press, 1992).

94. F. Fernández-Armesto, "Colón y los libros de caballería," in *Cristóbal Colón*, ed. C. Martínez Shaw and C. A. Parcero Torre (Burgos: Junta de Castilla y León, 2004).

95. Fernández-Armesto, *Pathfinders*, 204–6.

96. G. A. Williams, *Madoc, the Making of a Myth* (London: Eyre Methuen, 1979).

97. On Cook, see T. H. Beaglehole, *The Life of Captain James Cook* (Stanford, CA: Stanford University Press, 1974), 365; on Moreau, see Fernández-Armesto, *Pathfinders*, 284–85; on Lewis, see B. De Voto, ed., *The Journals of Lewis and Clark* (Boston: Houghton, Mifflin, 1953), 92; on Burke and Wills and on Scott, see Fernández-Armesto, *Pathfinders*, 359, 381–84.

Index